Lecture Notes in Computer Science

Lecture Notes in Bioinformatics 14616

The series Lecture Notes in Bioinformatics (LNBI) was established in 2003 as a topical subseries of LNCS devoted to bioinformatics and computational biology.

The series publishes state-of-the-art research results at a high level. As with the LNCS mother series, the mission of the series is to serve the international R & D community by providing an invaluable service, mainly focused on the publication of conference and workshop proceedings and postproceedings.

Celine Scornavacca · Maribel Hernández-Rosales
Editors

Comparative Genomics

21st International Conference, RECOMB-CG 2024
Boston, MA, USA, April 27–28, 2024
Proceedings

 Springer

Editors
Celine Scornavacca 🅳
CNRS, University of Montpellier
Montpellier, France

Maribel Hernández-Rosales 🅳
CINVESTAV-IPN
Irapuato, Mexico

ISSN 0302-9743   ISSN 1611-3349 (electronic)
Lecture Notes in Bioinformatics
ISBN 978-3-031-58071-0   ISBN 978-3-031-58072-7 (eBook)
https://doi.org/10.1007/978-3-031-58072-7

LNCS Sublibrary: SL8 – Bioinformatics

This Springer imprint is published by the registered company Springer Nature Switzerland AG
The registered company address is: Gewerbestrasse 11, 6330 Cham, Switzerland

Paper in this product is recyclable.

Preface

The RECOMB Satellite Conference on Comparative Genomics, established in 2003, serves as a gathering point for leading researchers across diverse fields, including mathematics, computational sciences, and life sciences. This conference is dedicated to advancing research in comparative genomics, emphasizing the importance of computational methodologies and innovative experimental results. The 2024 edition of the conference upheld this tradition of excellence.

This volume comprises the articles presented at the 21st RECOMB Satellite Conference on Comparative Genomics (RECOMB-CG), held on April 27–28, 2024, in Boston, Massachusetts, USA. Co-located with the annual RECOMB conference, RECOMB-CG 2024 provided researchers with a unique platform to contribute to and engage in discussions on the latest theoretical and empirical approaches in the field.

We received a total of 21 submissions from authors representing regions spanning North America, Latin America, Europe, Asia, and Oceania. Each submission underwent rigorous review by 3–5 members of the program committee and invited subreviewers. Based on their evaluations, 13 submissions were selected for presentation at the conference and are included in these proceedings.

RECOMB-CG 2024 extends heartfelt gratitude to our distinguished keynote speakers, Katja Nowick from the Free University of Berlin, and Devaki Bhaya from Stanford University. The conference program also featured poster presentations.

We extend our sincerest appreciation to the program committee members and subreviewers for their exceptional dedication during the review process, providing comprehensive discussions and review reports. Furthermore, we would like to acknowledge the guidance provided by the Steering Committee members. Special recognition goes to the local organizers, Lenore Cowen, Ezgi Ebren, and Jian Ma, for their tireless efforts in coordinating the conference and providing financial support.

We utilized the EasyChair conference system for managing submissions, reviews, and formatting.

Lastly, we express our profound gratitude to all scientists who submitted papers and posters, as well as to those whose enthusiastic participation contributed to making RECOMB-CG 2024 an enriching and memorable experience.

April 2024

Celine Scornavacca
Maribel Hernández-Rosales

Organization

Program Committee Chairs

Maribel Hernández-Rosales	CINVESTAV-IPN, Irapuato, Mexico
Céline Scornavacca	CNRS, University of Montpellier, France

Steering Committee

Marília Braga	Bielefeld University, Germany
Dannie Durand	Carnegie Mellon University, USA
Jens Lagergren	KTH Royal Institute of Technology, Sweden
Aoife McLysaght	Trinity College Dublin, Ireland
Luay Nakhleh	Rice University, USA
David Sankoff	University of Ottawa, Canada

Program Committee

Maribel Hernández-Rosales	CINVESTAV-IPN, Irapuato, Mexico
Celine Scornavacca	CNRS, University of Montpellier, France
Max Alekseyev	George Washington University, USA
Lars Arvestad	Stockholm University, Sweden
Katia Aviña-Padilla	CINVESTAV-IPN, Irapuato, Mexico
Mukul S. Bansal	University of Connecticut, USA
Sèverine Bérard	Université de Montpellier, France
Anne Bergeron	Université du Québec à Montréal, Canada
Paola Bonizzoni	Università di Milano-Bicocca, Italy
Marilia Braga	Bielefeld University, Germany
Broňa Brejová	Comenius University, Bratislava, Slovakia
Cedric Chauve	Simon Fraser University, Canada
Siyu Chen	Princeton University, USA
Miklós Csűrös	University of Montreal, Canada
Daniel Doerr	Heinrich Heine University Düsseldorf, Germany
Mohammed El-Kebir	University of Illinois at Urbana-Champaign, USA
Nadia El-Mabrouk	University of Montreal, Canada
Oliver Eulenstein	Iowa State University, USA
Guillaume Fertin	University of Nantes, France
Martin Frith	University of Tokyo, Japan

Pawel Gorecki	University of Warsaw, Poland
Katharina Jahn	Freie Universität Berlin, Germany
Asif Javed	University of Hong Kong, China
Lingling Jin	University of Saskatchewan, Canada
Jaebum Kim	Konkuk University, South Korea
Manuel Lafond	Université de Sherbrooke, Canada
Kevin Liu	Michigan State University, USA
Istvan Miklos	Renyi Institute, Hungary
Siavash Mirarab	University of California, San Diego, USA
Aïda Ouangraoua	Université de Sherbrooke, Canada
Teresa Przytycka	National Center of Biotechnology Information, USA
Aakrosh Ratan	University of Virginia, USA
Mingfu Shao	Carnegie Mellon University, USA
Sagi Snir	University of Haifa, Israel
Giltae Song	Pusan National University, South Korea
Yanni Sun	City University of Hong Kong, China
Wing-Kin Sung	The Chinese University of Hong Kong, China
Krister Swenson	CNRS, Université de Montpellier, France
Oznur Tastan	Sabanci University, Turkey
Olivier Tremblay-Savard	University of Manitoba, Canada
Tamir Tuller	Tel Aviv University, Israel
Fábio Henrique Viduani Martinez	Federal University of Mato Grosso do Sul, Brazil
Tomas Vinar	Comenius University, Bratislava, Slovakia
Yong Wang	Academy of Mathematics and Systems Science, China
Tandy Warnow	University of Illinois at Urbana-Champaign, USA
Yufeng Wu	University of Connecticut, USA
Xiuwei Zhang	Georgia Institute of Technology, USA
Louxin Zhang	National University of Singapore, Singapore
Fa Zhang	Chinese Academy of Science, China
Jie Zheng	ShanghaiTech University, China

Subreviewers

Eloi Araujo	Askar Gafurov
Leonard Bohnenkämper	Marina Herrera Sarrias
Dehan Cai	Kaari Landry
Guowei Chen	Alitzel Lopez-Sanchez
Mattéo Delabre	Diego P. Rubert
Riccardo Dondi	Jarosław Paszek
Yoav Dvir	Qian Shi

Jessica Stockdale
Siyu Tao
Sriram Vijendran

Sanket Wagle
Wend Yam Donald Davy Ouedraogo
Xiaofei Zang

Contents

Genome Rearrangements

Genome Evolution

Phylogenetic Networks

Statistically Consistent Estimation of Rooted and Unrooted Level-1 Phylogenetic Networks from SNP Data

Tandy Warnow[1]([✉]) [iD], Yasamin Tabatabaee[1] [iD], and Steven N. Evans[2] [iD]

[1] University of Illinois Urbana-Champaign, Urbana, IL 61822, USA
{warnow,syt3}@illinois.edu
[2] University of California at Berkeley, Berkeley, CA 94720, USA
evans@stat.berkeley.edu

Abstract. We address the problem of estimating a rooted phylogenetic network, as well as its unrooted version, from SNPs (i.e., single nucleotide polymorphisms), allowing for multiple crossover events. Thus, each SNP is assumed to have evolved under the infinite sites assumption down some tree inside the phylogenetic network. We prove that level-1 phylogenetic networks can be reconstructed uniquely from any set of SNPs that cover all bipartitions of the rooted trees contained in the network, even when the ancestral state is unknown. To the best of our knowledge, this is the first result to establish that the unrooted topology of a level-1 network is uniquely recoverable from SNPs without known ancestral states. We present a stochastic model for DNA evolution, and we prove that Gusfield's algorithms in JCSS 2005 (one for the case where the ancestral state is known, and the other when it is not known) can be used in polynomial time, statistically consistent pipelines to estimate level-1 phylogenetic networks when all cycles are of length at least five, under the stochastic model we propose, provided that we have access to an oracle for indicating which sites in the DNA alignment are SNPs.

Keywords: phylogenetic network · level-1 · galled tree

1 Introduction

Phylogenetic networks are models of evolutionary relationships that allow for reticulations, such as hybridizations and horizontal gene transfers, and so expand the range of allowed evolutionary events beyond the standard "tree-like" evolutionary model [15]. In the population genetics research community, the assumption is made that SNPs (single nucleotide polymorphisms) can be identified; and then used for the inference of phylogenetic networks using methods designed for SNPs [8–10]. However, to the best of our knowledge, no methods for estimating rooted or unrooted phylogenetic networks have been evaluated and characterized for their statistical properties under any formal stochastic model of SNP evolution. Such methods are the main focus of this study.

C. Scornavacca and M. Hernández-Rosales (Eds.): RECOMB-CG 2024, LNBI 14616, pp. 3–23, 2024.
https://doi.org/10.1007/978-3-031-58072-7_1

SNPs are sites in a DNA alignment that exhibit only two states and are assumed to evolve under the infinite sites assumption (i.e., the characters are homoplasy-free, and so do not evolve with parallel evolution or back mutation). When the evolutionary process is purely treelike, this is the *perfect phylogeny* model [2], and the characters are said to be *compatible*. However, when the evolutionary process involves recombination, then the SNPs are assumed to evolve down some tree contained inside the phylogenetic network. Thus, phylogenetic network estimation from SNPs in biology operates in two steps from the given DNA sequence alignment: first, the biologist assesses each site in the alignment and selects those sites considered likely to be SNPs (i.e., they exhibit only two states and are deemed likely to have evolved without homoplasy) and then the SNPs are used to estimate the phylogenetic network. See [9] for a textbook on the problem of inferring phylogenetic networks from SNPs.

One direction for phylogenetic network estimation has sought to construct networks in which the overlap between cycles in the network (where cycles are defined by the unrooted version of the network) is bounded; the *level-1 phylogenetic network* or *galled tree* is the simplest of these variants, and assumes that no two cycles share any nodes in common (see Fig. 1).

Methods to construct a rooted level-1 phylogenetic network N have been developed, some with provable guarantees. For example, there are polynomial time methods that construct the unique level-1 rooted phylogenetic network under the assumption that no cycle is of length less than five [11–13, 16–18].

In contrast, phylogenetic networks of level-2 or greater level are more challenging to construct; for example, Gambette *et al.* [5] show that two level-2 networks may induce the same set of triplets and soft-wired clusters but be non-isomorphic (see Figs. 11 and 12 in that paper), which means they are not identifiable from their triplets or clusters. Thus, much of the focus in the literature has been on the construction of level-1 phylogenetic networks.

In comparison to the "root-known" case (i.e., when the ancestral state is known for SNPs, or when rooted triplet trees or clades are available), much less is known about recovering the unrooted topology of a level-1 network from SNPs when the ancestral state is unknown, or from bipartitions and quartet trees of the network. This is the more biologically realistic question, compared to the root-known case, and hence results for this question are more relevant to biological phylogenetics. Gusfield [8] presented an $O(mn + n^3)$ time algorithm for constructing unrooted level-1 phylogenetic networks with n leaves that are consistent with m SNPs with unknown ancestral state, but even when restricted to networks where the smallest cycle has length at least five, he did not establish that the returned unrooted network topology was topologically identical to the unrooted topology of the true network. Gambette, Berry, and Paul [4] presented an $O(n^4)$ polynomial time method to construct unrooted level-1 phylogenetic networks from $Q(N)$, the set of all quartet trees in the unrooted network for N. However, we will show that $Q(N)$ contains quartet trees that are not in the set of quartet trees found in at least one tree down which a SNP can evolve (we refer to this set as $Q(N_r)$), so that on the face of it, the method by Gambette,

Berry, and Paul is not directly applicable. Thus, constructing level-1 networks from SNPs without a known ancestral state is incompletely understood.

Our paper makes the following contributions. We prove that for the root-unknown case, there are level-1 phylogenetic networks in which all cycles have length at least 6 on which the algorithm of Gambette, Berry, and Paul [4] can fail if only provided the quartet trees in $Q(N_r)$. We also establish that an algorithm from Gusfield [8] is guaranteed to return the unrooted topology of the level-1 network N if given a set of SNPs covering the bipartitions of the trees contained in the network, and achieves this in $O(mn + n^3)$ time where m is the number of SNPs and n is the number of leaves. We use these results to establish statistical identifiability of the level-1 phylogenetic network from SNPs without known ancestral state, and to present algorithms for estimating the network topology that are statistically consistent.

The rest of the paper is organized as follows. In Sect. 2 we present some preliminary material, including notation and already established theoretical properties about level-1 phylogenetic networks. In Sect. 3 we present new theoretical results for computing unrooted level-1 phylogenetic network topologies from the set $Q(N_r)$. In Sect. 4 we present the stochastic model for DNA site evolution down a level-1 phylogenetic network, and then discuss statistical guarantees for pipelines for estimating both rooted level-1 phylogenetic networks and unrooted level-1 network topologies from SNPs (identified by an oracle) under this model. Finally, we conclude with a discussion of future work in Sect. 5.

2   Preliminary Material

2.1   Phylogenetic Networks

Here we describe the basic graphical models of evolution: phylogenetic trees and phylogenetic networks, as well as terminology associated to each.

Phylogenetic Networks. A phylogenetic network N is a rooted directed graph that extends the phylogenetic tree model to allow for lateral transmission. Thus, as with a phylogenetic tree, N has a root r and leaf set S. Since we allow for lateral transmission, we have extra edges that, when considering the network N as an undirected graph (i.e., by ignoring the directionality on the edges), would form cycles. Furthermore, if γ is an undirected cycle when considering N as an undirected graph, we will also refer to γ as a cycle in N, when it is considered as the directed rooted network N. Here we provide a more detailed definition of a phylogenetic network.

The nodes of the network are in four categories: the root, the leaves, the tree nodes, and the reticulate nodes. As with the phylogenetic tree model, every leaf node has indegree 1 and outdegree 0, the root has indegree 0 and outdegree 2, and all tree nodes have indegree 1 and outdegree 2. However, the reticulate nodes have indegree 2 and outdegree 1.

Every edge in the network is directed, and they are designated as either tree edges or reticulate edges, as follows. If $e = v \rightarrow w$ is a directed edge with w a tree node, then e is a tree edge. Furthermore, we designate at least one of the two incoming edges into a reticulate node as a reticulate edge; however, some papers require both edges to be denoted as reticulate edges (e.g., see [8]).

Finally, a phylogenetic tree is a special case of a phylogenetic network in which all nodes are tree nodes and hence there are no reticulate edges.

Definition 1. *If N is a phylogenetic network, then its* unrooted *version is obtained by ignoring the orientation of the edges and suppressing the root. The cycles in the unrooted version thus define a set of nodes in the rooted version that we refer to as* **cycles in the rooted network**. *A phylogenetic network N is said to be* **level-1** *(or equivalently a* **galled tree**) *if no two cycles in the network share any nodes in common (see Fig. 1).*

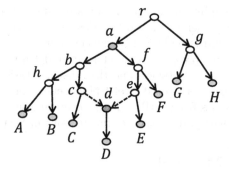

Fig. 1. Level-1 phylogenetic network with one cycle of size six. Node a is the root of the cycle and node d is a reticulate node and is the bottom of the cycle; the edges entering d are both dashed, indicating that they are reticulate edges.

Observation 1. *Suppose N is a level-1 network, and that when N is considered as an undirected graph, it has a cycle γ. Then there is a unique node $a \in \gamma$ that has outdegree two in γ and a unique node $b \in \gamma$ that has indegree two in γ, which we refer to as the root and the bottom of γ, respectively. The cycle consists of exactly two paths that begin at the root and end at the bottom, and that intersect only at the root and bottom.*

Proof. The proof follows easily from noting that the cycle cannot contain a chord, as otherwise there are two cycles in N that share nodes.

Definition 2. *Given a level-1 network N and a cycle γ of length at least three in N with root a and bottom node b, we will say γ is* **one-sided** *if one of the two paths from a to b in γ is a single edge and otherwise we say that γ is* **two-sided**.

According to this definition, the cycle in the network in Fig. 1 is two-sided.

Notation Related to Rooted Trees. When T is a rooted tree, the set $Clades(T)$ is the set of all clades in T, where a clade in T is the set of leaves below some node in T. Given the rooted T, it is easy to construct $Clades(T)$ in polynomial time, and further, T can be uniquely recovered from $Clades(T)$ in linear time [7].

If we treat the rooted tree T as an unrooted tree, we can define its set of bipartitions on the leafset of T (each formed by deleting one edge from T) by $Bip(T)$. Furthermore, we can also define its set of quartet trees using the bipartitions as follows. Let $A|B \in Bip(T)$, and let $\{a_1, a_2\} \subseteq A$ and $\{b_1, b_2\} \subseteq B$. Then we say that $a_1, a_2|b_1, b_2$ is a quartet tree of T, and the set of all such quartet trees is denoted by $Q(T)$. Note that $a_1, a_2|b_1, b_2$ is a quartet tree of T if and only if there is an edge of T that separates a_1, a_2 from b_1, b_2. Equivalently, there are two node-disjoint paths in T where one connects a_1 and a_2 and the other connects b_1 and b_2.

Definition 3. *A phylogenetic network defines a set of trees that it contains. These trees are obtained by deleting, for every reticulate node, exactly one incoming edge. We refer to this set of trees by $\mathcal{T}_N$, where N is the phylogenetic network.*

Note that according to this definition, there are exactly two trees contained in the network in Fig. 1, each obtained by deleting one of the edges entering node d.

Definition 4. *Given phylogenetic network N, $Clades(N) = \bigcup_{T \in \mathcal{T}_N} Clades(T)$ (note that the set of clades is also referred to as the soft-wired clusters of the network [18]). We let $Bip(N_r) = \bigcup_{T \in \mathcal{T}_N} Bip(T)$ and $Q(N_r) = \bigcup_{T \in \mathcal{T}_N} Q(T)$.*

Note that the terms in Definition 4 are based on all the rooted trees contained in N, where N is a rooted network. Thus, the use of the subscript r in $Bip(N_r)$ and $Q(N_r)$ serves to emphasize that these sets depend on N being rooted.

Definition 5. *Let N be a rooted phylogenetic network. We define $Q(N)$ to be the set of all quartet trees $ab|cd$ where N, treated as an undirected graph, contains two node-disjoint paths P_{ab} and P_{cd}, where P_{ab} connects a and b and P_{cd} connects c and d.*

Note that by construction, $Q(N_r) \subseteq Q(N)$.

Definition 6. *Let N be a level-1 phylogenetic network and let v be a node on a cycle γ in N and w a neighbor of v that is not in γ; therefore v is not the root of N. Note that (v, w) is a cut edge for N. Any leaf x in N for which the path to v passes through w is said to* **attach to** γ **at** v. *Each such leaf is also said to* **label** *the node it attaches to, but when we label a node v in a cycle, we arbitrarily pick one leaf that attaches to v for this labeling. Note that every leaf in N attaches to exactly one node in γ, and that the root of γ can be labeled by a leaf using this technique if and only if it is not also the root of N.*

Note that according to this definition, leaves A and B both attach to the cycle in Fig. 1 at node b and so either leaf could be used to label node b. Note also that leaves G and H both attach to the root of the cycle. Finally, only one leaf attaches to the cycle at the other nodes. Note also that every leaf in the network attaches to exactly one node of the cycle.

Theorem 1. *Let N be a rooted level-1 network that contains a rooted two-sided cycle γ with at least two nodes (other than the root and bottom) on each side of γ. Then $Q(N_r) \subset Q(N)$; that is, $Q(N)$ contains $Q(N_r)$ as a proper subset.*

Proof. Let N be such a network, let r denote the root of the cycle, and let x denote the bottom node of the cycle. Now suppose we label the internal nodes of the cycle using leaves of N (see Definition 6) so that on the left-hand side we have the sequence $r, a_1, a_2, \ldots, x$ and on the right-hand side we have the sequence $r, b_1, b_2, \ldots, x$. It is easy to see that there are two node-disjoint paths P_1 and P_2 where P_1 connects a_1 and b_1 and P_2 connects a_2 and b_2 (namely, P_1 goes across the root and P_2 goes across the bottom node). Since these two paths are node-disjoint, by Definition 5, quartet tree $a_1, b_1 | a_2, b_2 \in Q(N)$.

We now show that $a_1, b_1 | a_2, b_2$ is not in $Q(N_r)$. Suppose to the contrary, that there is some rooted tree T in $\mathcal{T}_N$ that has this quartet tree. By Definition 3, T is formed by deleting one of the incoming edges to every reticulate node, and hence, in particular, requires the deletion of one of the incoming edges (say e_1) to the bottom node x of γ. Then $a_1, b_1 | a_2, b_2$ is a quartet tree in $Q(T)$ if and only if there is some edge e_2 in T whose removal from T separates a_1, b_1 from a_2, b_2. Since N is level-1, e_2 must be an edge in γ. Hence, there are two node-disjoint paths in T that connect the relevant pairs and that do not use either e_1 or e_2. Therefore, neither of the two paths can go through the bottom node, which means they must cross each other – and hence share a node in common. This violates the assumption that these four leaves label the cycle nodes as indicated, and so we derive a contradiction. Hence, $a_1, b_1 | a_2, b_2 \notin Q(N_r)$.   □

As an example of a quartet tree in $Q(N)$ that is not in $Q(N_r)$, see Fig. 1, and consider the set of four leaves $\{A, C, E, F\}$. There are exactly two trees in $\mathcal{T}_N$, but they differ only in the placement of leaf D. Both these trees induce $AC|EF$, and so $Q(N_r)$ contains exactly one quartet tree on these four leaves. On the other hand, $Q(N)$ also contains $AF|CE$, since there are two node-disjoint paths, one connecting A and F and the other connecting C and E. Hence $Q(N)$ contains $Q(N_r)$ as a proper subset.

2.2   Established Theory for Constructing Level-1 Networks

Although much is known about constructing level-1 phylogenetic networks from clades (equivalently, from SNPs with known ancestral state), as described in the previous section, much less is known about constructing unrooted level-1 network topologies when we have bipartitions or quartet trees (equivalently, from SNPs without known ancestral states). We are interested in the case where we have enough SNPs to recover the unrooted object. Since these SNPs are given without known ancestral state, each SNP defines a bipartition, and hence also a set of (unrooted) quartet trees. Thus, related to the question of whether we can recover N from enough SNPs, and recalling the definitions of $Q(N_r)$ and $Bip(N_r)$ (see Definition 4), we ask whether we can determine the unrooted network topology of a level-1 network N given $Bip(N_r)$ or $Q(N_r)$. Since this relates to the SNPs, we make the following definitions:

Definition 7. *Let N be a level-1 phylogenetic network. We will say that a set of SNPs **covers** $Bip(N_r)$ if for every bipartition in $Bip(N_r)$, there is a SNP that produces that bipartition. Note that if a set of SNPs covers $Bip(N_r)$ then it also covers (in the same sense) $Q(N_r)$.*

Two prior algorithms are of interest, both described earlier. One is the algorithm presented by Gusfield in [8] for constructing unrooted level-1 phylogenetic network topologies from SNPs without known ancestral state, which we refer to henceforth as GUSFIELD-CONSTRUCT-UNROOTED. The other is by Gambette, Berry, and Paul [4], which we refer to as the GBP algorithm. Recall that the GBP algorithm, if given $Q(N)$, is guaranteed to reconstruct the unrooted topology of N in $O(n^4)$ time, where N is level-1 and no cycle has length smaller than five. However, it is by no means obvious that the GBP algorithm could be used to construct the unrooted topology of network N from SNPs when we do not know the ancestral state, as the set $Q(N_r)$ of quartet trees derivable from such SNPs might not be identical to $Q(N)$.

3   New Theory: When the Ancestral State is Unknown

3.1   Overview

The ability to correctly identify the unrooted topology of a level-1 phylogenetic network from bipartitions or quartet trees is the subject of this section. We will provide examples of level-1 networks N on which the GBP algorithm applied to $Q(N_r)$ will fail to return the unrooted topology for N (Theorem 2), even though all the cycles in the network are of size at least six. However, we present CUPNS, a new algorithm that we guarantee recovers the unrooted topology of N when given $Q(N_r)$ (Theorem 3), provided all cycles are of length at least five. This implies in turn that $Q(N_r)$ and $Bip(N_r)$ are each sufficient to define unrooted level-1 phylogenetic network topologies. Hence, any algorithm that produces an unrooted level-1 phylogenetic network topology consistent with a set of SNPs is guaranteed to return the unrooted topology of the true network N when the SNPs cover $Bip(N_r)$ (see Definition 7) and all cycles in N have at least five nodes. One of the nice consequences of this analysis is that it establishes that Gusfield's $O(mn + n^3)$ algorithm for constructing unrooted level-1 phylogenetic networks with n leaves from m SNPs without known ancestral state, but allowing for multiple crossovers [8], is guaranteed to return the true unrooted network topology when given a set of SNPs that covers $Q(N_r)$ (see Definition 7).

3.2   The GBP Algorithm Fails on $Q(N_r)$

We begin with a high-level description of the GBP algorithm (see [4] for full details), given an input set Q of unrooted quartet trees. The GBP algorithm has two steps. In Step 1, an unrooted tree, referred to as the "SN-tree", is computed from Q. The SN-tree is the maximally resolved tree that has the property that all quartet trees in the SN-tree are in Q and no resolved quartet tree in the SN-tree conflicts with a quartet tree in Q.

In Step 2, the polytomies in the SN-tree are expanded into cycles, as we now describe. If v is a polytomy with q neighbors, then GBP replaces v by a cycle of length q. Given a polytomy v and a neighbor w, consider the split of the SN-tree into two parts formed by deleting the edge (v, w). Any leaf in the part that contains w can be used to label the node in the cycle corresponding to w. One such leaf is picked for every neighbor of v, and the quartet trees on these selected leaves are then used to construct a cycle that replaces the polytomy v. This second step applies the node ordering algorithm from [4] to determine how to replace each polytomy with a cycle.

The node ordering algorithm makes a graph whose nodes are pairs of leaves, each labeling a different node in the cycle. There is an edge between (a, b) and (b, c) if and only if there is no leaf f labeling a node in the cycle such that $ac|fb$ is a quartet tree in Q. After constructing this graph, if it forms a cycle, then the polytomy is replaced by the cycle, and otherwise, the algorithm rejects the input. The node ordering algorithm uses the result that if a, b, c are three leaves that label the nodes in a cycle and if $Q = Q(N)$, then the nodes they label are contiguous in the cycle and appear in the order a, b, c if and only if there is no leaf f labeling a node in the cycle for which quartet tree $ac|bf$ is in Q (see proof of Lemma 5 in [4]).

This algorithm explicitly assumes that the input set Q is identical to $Q(N)$, in order to be guaranteed to return the unrooted topology for N. In Theorem 2, we will prove that GBP can fail if the quartet tree set is limited to $Q(N_r)$.

We continue with a lemma that we will find useful.

Lemma 1. *Let N be a rooted level-1 phylogenetic network in which all cycles have at least four nodes. Let Z be any set of four leaves that attach to the same cycle at different nodes. Then $Q(N_r)$ has exactly two quartet trees on Z if and only if at least one of these leaves labels the bottom node in γ.*

Proof. First note that if we have four leaves that attach to different nodes in the same cycle γ and none of them is the bottom node, then there is only one quartet tree in $Q(N_r)$ on these leaves since they are not impacted by the choice of which edge entering the bottom node is used. Furthermore, for any set of four leaves that attach to different nodes in the cycle, if one of them attaches to the bottom node x, then there are two quartet trees in $Q(N_r)$ for that set of four leaves, as we now show. Suppose that the four leaves are x, a, b, c and appear in that order as you move clockwise from the bottom node. There are two edges incoming to the bottom node, which we will refer to as e_L and e_R, with e_L on the left (and so in between the bottom node and a) and e_R on the right (and so in between the bottom node and c). To obtain a quartet tree on a, b, c, x, we must delete one of e_L or e_R as well as one other edge, thus separating two of the set from the other two. It is easy to see that if we delete e_L we can create $ab|cx$ but no other quartet tree, while if we delete e_R we can create $ax|bc$ but no other quartet tree. Thus, we can create two quartet trees for any set of four leaves labeling different nodes in the same cycle, where one of the leaves labels the bottom node. Thus, any four leaves involving the bottom node always has two quartet trees in $Q(N_r)$, but every set of four leaves that does not include the bottom node has only one quartet tree in $Q(N_r)$.   □

Theorem 2. *There are level-1 networks N in which all cycles are of length at least six, on which the GBP algorithm applied to $Q(N_r)$ will fail to correctly construct the unrooted topology of N.*

Proof. Let N be a level-1 network rooted at r_0 with a cycle γ having root $r \neq r_0$ and bottom node x, where one side of γ has exactly four nodes in between r and x. We will show that for any such network N, the GBP algorithm will fail to correctly construct the unrooted topology of N.

For each node in γ, pick one leaf to label the node (see Definition 6), so that the path is labeled r, a, f, b, c, x. Note that a, b, c are not consecutive. The node ordering algorithm in GBP would infer a, b, c appear consecutively, without any intervening node, if and only if there is no node z such that $ac|bz$ is in the input set of quartet trees, which is $Q(N_r)$. We will show that there is no quartet tree of the form $ac|bz$ in $Q(N_r)$, so that the node ordering algorithm will find that a, b, c appear consecutively, and this would mean the algorithm makes a mistake.

Our proof is by contradiction, and so assumes that there is some leaf z so that quartet tree $ac|bz$ is in $Q(N_r)$. The analysis is a case by case analysis, based on what node in γ can be labeled by z. Note that because N is level-1, every leaf will label exactly one node in γ, and so we can examine which node in γ the node z labels. There are three cases to consider: z labels the bottom node, z labels node f, and z labels some other node (not f and not the bottom node). Before we consider each case in turn, recall that if u, v, u', v' are labels of four nodes in a cycle γ, then quartet tree $uv|u'v' \in Q(N_r)$ if and only if we can find two edges to delete from the cycle, one of which enters the bottom node in the cycle, so that after deleting these two edges we have u, v separated from u', v'.

- Case (1): z labels the bottom node (i.e., $z = x$). If we delete the edge incident with x that is on the same side as a, b, c, we obtain quartet tree $xa|bc$, and if we delete the other edge we obtain $ab|cx$. Thus, in neither case do we obtain $ac|bx$.
- Case (2): z labels f (i.e., $z = f$). Clearly, $af|bc \in Q(N_r)$, but since none of these nodes is the bottom node, it follows that this is the unique quartet tree in $Q(N_r)$ by Lemma 1.
- Case (3): z labels some node other than the bottom node and f. It is easy to see that $az|bc$ is the unique quartet tree on these four leaves in $Q(N_r)$.

Hence, there is no node z for which $ac|bz \in Q(N_r)$, so the node ordering algorithm used in GBP will infer that the nodes abc appear consecutively, which is a mistake.   □

3.3 Constructing Unrooted Phylogenetic Networks from SNPs

Throughout this section we assume N is a level-1 network in which all cycles have length at least five. We present the CONSTRUCTING UNROOTED PHYLOGENETIC NETWORKS FROM SNPs algorithm (short form: CUPNS), that takes as input an arbitrary set of SNPs without ancestral state information, and we prove that if the SNPs cover $Bip(N_r)$, then CUPNS returns the unrooted topology of N.

At a high level, CUPNS can be described as having three phases. In the first phase, we produce a set of quartet trees corresponding to the SNPs. In the second phase, we apply the first step of GBP to that set of quartet trees, to obtain an SN-tree. In the third phase, each polytomy of the SN-tree is expanded into a cycle, thus producing a phylogenetic network that is level-1. The algorithm can fail during the third phase if it is not possible to perform the expansion of the polytomies into cycles in such a way that is consistent with the assumptions of the algorithm: the network is level-1, every cycle is of size at least 5, and the SNPs cover $Bip(N_r)$ (see Definition 7).

We now describe CUPNS in greater detail.

- Phase 1. We use the SNPs to produce a set of quartet trees (all the quartet trees defined by the SNPs), which we denote by Q_{SNP}.
- Phase 2. We run the first step of GBP on Q_{SNP}, to obtain the SN-tree T.
- Phase 3. For each polytomy v in T:
  - For every neighbor w of v, we select a single leaf to label v (where that leaf can be reached from w without going through v), and we denote the set of leaves labeling its neighbors by $L(v)$.
  - If $L(v)$ has fewer than five nodes, we return FAIL. Otherwise, we examine all the quartet trees on $L(v)$ to find a leaf x such that every quartet involving x has at least two quartet trees in Q_{SNP}. If there is no such leaf or more than one such leaf, we return FAIL, indicating that no level-1 phylogenetic network is consistent with the input. Otherwise, we let x be the unique leaf that has at least two quartet trees in Q_{SNP}.
  - We remove x from $L(v)$. We construct a perfect phylogeny for the remaining leaves in $L(v)$, if it exists, using [7]; if no perfect phylogeny exists, we return FAIL. If the perfect phylogeny exists, it will be a path with two cherries (pairs of leaves that are siblings). We use the quartet trees in Q_{SNP} to determine how to add b and close the cycle.

It is worth noting that CUPNS has features from the GBP algorithm and Gusfield's algorithm for the root-unknown case: constructing the SN-tree is from GBP, finding the unique bottom node is our contribution (and follows from Lemma 1), and then replacing the polytomy by a cycle (after finding the bottom node) follows arguments in Gusfield [8]. We continue with a lemma that we will need for Theorem 3.

Lemma 2. *Let N be a rooted level-1 phylogenetic network in which all cycles are of length at least five. Then the SN-tree computed using $Q(N_r)$ is identical to the SN-tree computed using $Q(N)$.*

Proof. Recall that the SN-tree computed in Step 1 of GBP, given quartet set Q as input, is the maximally resolved unrooted tree T such that (1) $Q(T) \subseteq Q$ and (2) no quartet tree in $Q(T)$ conflicts with any quartet tree in Q. Consider then the tree obtained by taking the unrooted topology for N and collapsing every cycle into a polytomy (i.e., contracting all the edges in the cycle). Clearly, this tree satisfies the two properties above (and is maximally resolved subject to satisfying (1) and (2) when $Q = Q(N)$). Here we show that this tree also satisfies the two required properties and is maximally resolved subject to this when $Q = Q(N_r)$.

It is easy to see that the tree satisfies (1) and (2). If some refinement T^* also satisfies (1) and (2), then there is a polytomy v in T, some pair of neighbors of v (which we will call v_1, v_2) so that adding a new node v' and making it adjacent to v, v_1, and v_2 satisfies (1) and (2). Equivalently, that means finding two leaves a, b that attach to different nodes (v_1 and v_2) in the cycle that was collapsed to this polytomy, so as adding an edge that separates a, b from the rest of the leaves labeling the neighbors of the polytomy does not violate the two conditions above.

Note that the bottom node of the cycle is labeled by some leaf, x. Recall that, by Lemma 1, any quartet involving x and three other nodes in the cycle has two quartet trees in $Q(N_r)$. If $x \in \{a, b\}$, then picking any two other leaves labeling different nodes in the cycle than those labeled by a, b produces a set of four leaves for which there are two quartet trees. That means that this refinement, T^*, produces a quartet tree that conflicts with a quartet tree in $Q(N_r)$. If $x \notin \{a, b\}$, then we pick x and any other leaf y labeling some other node in the cycle, to get the set of four leaves a, b, x, y. This set of four leaves also has two quartet trees in $Q(N_r)$, meaning that the refinement T^* produces a quartet tree that conflicts with a quartet tree in $Q(N_r)$. Thus, no matter how a, b is picked, the tree T^* we obtain violates property (2). Hence, no refinement of this tree will satisfy the two required properties, and thus the SN-trees defined using $Q(N)$ and defined using $Q(N_r)$ will be the same.   □

We now continue with our major theorems.

Theorem 3. *Let N be a rooted level-1 phylogenetic network in which all cycles are of length at least five. If we apply CUPNS to a set of SNPs that covers $Q(N_r)$, then we obtain the unrooted topology for N, and the bottom node for each cycle is also identified.*

Proof. By Lemma 2, the SN-tree computed using $Q(N_r)$ is identical to the SN-tree computed using $Q(N)$. We now show that the second step where we replace the polytomies in the SN-tree by cycles returns the unrooted topology of N, when the SNPs cover $Q(N_r)$. Let v be a polytomy and $L(v)$ be the set of leaves labeling the neighbors of v. Since every cycle in N has at least five nodes, for all sets of four leaves labeling different nodes in the cycle, where none of them label the bottom node, there is exactly one quartet tree in $Q(N_r)$ (Lemma 1). Furthermore, for any set of four leaves labeling different nodes in the cycle, if one of them labels the bottom node, there are two quartet trees in $Q(N_r)$ (Lemma 1). Hence, the algorithm correctly identifies the leaf x labeling the bottom node of each cycle.

The algorithm computes the set $L(v) \setminus \{x\}$ of leaves that label the remaining nodes of the cycle. It is easy to see (but see [8] for additional discussion and algorithmic details) that this set has a perfect phylogeny (i.e., a tree on which all the SNPs evolve without homoplasy), and since binary character perfect phylogenies are unique [7], the perfect phylogeny is the caterpillar tree with two cherries formed by deleting the edges incident with the bottom node from the cycle. As described in [8], there is also a unique way to add back in the leaf labeling the bottom node to create the cycle that is consistent with the quartet trees in $Q(N_r)$, and it can be computed efficiently. Hence, the algorithm produces the correct cycle for each polytomy in the SN-tree, establishing correctness.   □

Hence, we obtain the following theorem and corollary.

Theorem 4. *Let N be a level-1 rooted phylogenetic network with all cycles of length at least five. Then, the set $Q(N_r)$, as well as the set $Bip(N_r)$, are each sufficient to define the unrooted topology of N.*

Proof. Given the rooted level-1 phylogenetic network N in which all cycles are of size at least five and a set of SNPs that covers $Q(N_r)$, Theorem 3 establishes that CUPNS will return the unrooted version of N. Thus, the set $Q(N_r)$ is sufficient to define the unrooted topology of N. Since $Q(N_r)$ can be constructed from $Bip(N_r)$, it follows that $Bip(N_r)$ is also sufficient to define the unrooted topology of N.   □

Corollary 1. *Let N be a level-1 rooted phylogenetic network with all cycles of length at least five. GUSFIELD-CONSTRUCT-UNROOTED will return the unrooted topology of N if the SNPs cover $Bip(N_r)$. Furthermore, GUSFIELD-CONSTRUCT-UNROOTED runs in $O(mn + n^3)$ time, where N has n leaves and there are m SNPs.*

Proof. In [8], Gusfield showed that GUSFIELD-CONSTRUCT-UNROOTED runs in $O(mn + n^3)$ time, where there are m SNPs and n taxa. He also proved that the unrooted level-1 phylogenetic network N' that is constructed by GUSFIELD-CONSTRUCT-UNROOTED will be consistent with the input SNPs, provided that these SNPs can evolve down trees within a rooted version of N' without homoplasy. In Theorem 4, we showed that there is a unique unrooted level-1 network consistent with all the bipartitions of the network. Since GUSFIELD-CONSTRUCT-UNROOTED produces an unrooted network N' that is consistent with the input SNPs, it follows that when the input SNPs cover $Bip(N_r)$, N' will be topologically identical to the unrooted version of N.   □

Thus, we have discussed two methods for constructing unrooted level-1 phylogenetic networks from SNPs: CUPNS (the method we introduced) and GUSFIELD-CONSTRUCT-UNROOTED. These methods have very different runtimes: GUSFIELD-CONSTRUCT-UNROOTED runs in $O(mn+n^3)$ time where there are m SNPs and n leaves, while CUPNS has a step that requires it compute all the quartet trees, and so its runtime is higher than GUSFIELD-CONSTRUCT-UNROOTED. Finally, we also described a method where we compute all the quartet trees implied by the SNPs and give this set to GBP; however, we proved this will fail to be correct on some inputs (Theorem 2). We summarize this discussion for constructing unrooted level-1 phylogenetic networks from SNPs in Table 1.

Table 1. Properties of algorithms for constructing unrooted level-1 networks from SNPs. The "required conditions" are that the network be level-1 and its unrooted topology cannot have any cycles of size less than five, and that the SNPs cover $Bip(N_r)$. The runtimes for GBP-SNPs and CUPNS assume that it takes $O(mn^4)$ time to compute all the quartet trees from the input set of m SNPs, and that this dominates the total runtime.

Method	Comments if required conditions hold	Runtime
Gusfield–Construct-Unrooted	Guaranteed correct	$O(mn + n^3)$
CUPNS	Guaranteed correct	$O(mn^4)$
GBP-SNPs	Will fail on some networks	$O(mn^4)$

4   Model of SNP Evolution and Network Estimation

4.1   Preliminary Discussion

In this section, we describe the stochastic model of nucleotide evolution down a level-1 phylogenetic network, and then show how we can estimate the model phylogenetic network from the alignment that is produced. We will refer to the leaves of the phylogenetic network as "species", but in some contexts (e.g., population genetics) they may be individuals rather than species.

A character c that has states assigned to the leaves of the tree T is said to evolve without homoplasy on T if it is possible to assign character states to the internal nodes so that all the nodes with the same state are path connected (i.e., the path in T between two nodes with the same state only uses nodes with that same state). Otherwise, the character is said to evolve homoplastically. This concept is extended to phylogenetic networks by saying that the character c evolves without homoplasy down N if there is at least one tree $T \in \mathcal{T}_N$ on which c evolves without homoplasy. If c also exhibits only two states, then c is called a SNP for N. Therefore, for a SNP c, there is a single edge in T on which c

changes, and hence also a clade of T that contains the derived state. We will refer to that clade as $clade(c)$.

In a biological analysis, the input is a multiple sequence alignment and the biologist must select from the sites of the alignment those sites that they consider to have evolved under the infinite-sites assumption (so that the characters exhibit two states and are homoplasy-free). Here we model this selection process by saying that the biologist has access to an "Oracle", which tells them which sites evolve under the infinite-sites assumption as well as the ancestral state for the site. Thus, the SNP data we will examine will have two states – the ancestral state, represented by 0, and the derived state, represented by 1.

The character evolution model we will describe is 4-state, and allows for homoplasy. Each character evolves down the given level-1 phylogenetic network N (see Sect. 2.1) by "picking" one tree in $\mathcal{T}_N$ down which to evolve, and then evolving down this tree under a substitution model. Hence, to define the character evolution model, we need to define how each character "picks" the tree down which to evolve, and also the substitution model. Here we describe the numeric parameters of the level-1 phylogenetic network N and the character evolution. We assume that the characters evolve identically and independently ($i.i.d.$), so that it suffices to model the evolution of a single character. The model thus defines the probability, for each character, which tree within $\mathcal{T}$ it "selects", and then how the character evolves (changing state) down that tree. We note that this model does not enforce that the characters are two-state and homoplasy-free.

4.2   Single Site Evolution Model

Recall that the sites evolve down a level-1 phylogenetic network, which is a rooted network with edges denoted as either tree edges or reticulate edges. The network numeric parameters and the constraints on these parameters are as follows:

- The state at the root is random and selected under the uniform distribution.
- Every edge e entering a reticulate node has a transmission probability $\kappa(e)$ with $0 < \kappa(e) < 1$ and the transmission probabilities of the edges entering a reticulate node must sum to 1.
- Every tree edge e has a substitution probability $p(e)$ with $0 < p(e) < 1$ for whether a character changes state on that edge; for the reticulate edges, the substitution probability is always 0.
- If a character changes state on edge e, it changes to one of the other nucleotides with equal probability (i.e., this is similar to the Jukes-Cantor [14] model).

We now provide some comments on how these parameters influence the evolution of a single character down the network.

- Because the transmission probabilities of the edges entering a reticulate node must sum to 1, for each character, each reticulate node inherits from exactly one of its parents. Thus, each character evolves down a tree contained in $\mathcal{T}_N$.

- Because characters evolve *i.i.d.*, different characters can be inherited from different parents.
- Because character state changes on reticulate edges are not allowed (they represent lateral gene transfer events), when a transmission of character c occurs on a reticulate edge $v \rightarrow w$, the state at w for character c is the same as at v, and so $c(w) = c(v)$.

Overall, therefore, this character evolution model can be seen as enforcing that every character evolves down a single tree in $\mathcal{T}_N$, with a probability that depends on the parameter values for $\kappa(e)$ for all edges e entering reticulation nodes; and that once the tree is selected, the substitution process is specified by the substitution probabilities on the edges. Moreover, every tree in $\mathcal{T}_N$ has a strictly positive probability of being selected.

4.3   Estimating the Rooted Level-1 Phylogenetic Network

We are interested in estimating a rooted level-1 phylogenetic network from SNPs, and so we assume we know the ancestral state for each SNP. The input to the algorithm is an $n \times k$ DNA alignment where the n rows correspond to the leaves of the network and the k columns correspond to sites. From this alignment, the Oracle identifies m of them to be SNPs. Thus, each SNP exhibits two states and evolved without homoplasy down some tree contained in the level-1 model phylogenetic network.

Definition 8. *Let N be a level-1 phylogenetic network. A set of rooted SNPs is said to cover $Clades(N)$ if and only if for every clade in $Clades(N)$ there is at least one rooted SNP defining the same bipartition (with the nodes in the clade having the derived state).*

We now describe the use of Dan Gusfield's algorithm for the estimation problem.

- Stage 1: Given the k DNA characters (equivalently, a DNA alignment of k sites), the Oracle determines which of them are SNPs (i.e., evolved without homoplasy down some tree contained in the network and exhibit only two states) and identifies the ancestral state for each such SNP (thus producing rooted SNPs).
- Stage 2: Apply GUSFIELD-CONSTRUCT ROOTED PHYLOGENETIC NETWORK FROM ROOTED SNPs to the rooted SNPs.

We call this two-step approach GUSFIELD - STATISTICAL ESTIMATION OF ROOTED PHYLOGENETIC NETWORK FROM ROOTED SNPs (or more simply GUSFIELD-ESTIMATE-ROOTED), to emphasize that this is a statistical estimation approach. Note that this technique uses the input alignment in a two-stage process, where the first stage extracts the rooted SNPs, and these are then used to construct a level-1 phylogenetic network in the second stage.

4.4   Estimating the Unrooted Level-1 Phylogenetic Network

Here we wish to estimate the unrooted topology of the level-1 phylogenetic network, and so we do not need to know the ancestral state for each SNP. For the case where we do not have the ancestral state, GBP does not have theoretical guarantees for estimating the network from SNPs, but by Theorem 3 and Corollary 1, both CUPNS and GUSFIELD-CONSTRUCT-ROOTED are guaranteed to return N if the SNPs cover $Bip(N_r)$ (and hence cover $Q(N_r)$) and the network does not have cycles with length less than five. This is summarized in Table 1.

Therefore, we now describe the use of GUSFIELD-CONSTRUCT-UNROOTED for *statistical estimation* of the unrooted phylogenetic network from a DNA multiple sequence alignment.

Statistical Estimation of Unrooted Phylogenetic Network from Unrooted SNPs

- Step 1: Given the k DNA characters, the Oracle determines which of them are SNPs (i.e., evolved without homoplasy down some tree contained in the network and exhibit only two states).
- Step 2: Apply GUSFIELD-CONSTRUCT UNROOTED PHYLOGENETIC NETWORK FROM UNROOTED SNPs to the SNPs.

We call this two-stage approach the GUSFIELD-STATISTICAL ESTIMATION OF UNROOTED PHYLOGENETIC NETWORK FROM UNROOTED SNPs, or more simply as GUSFIELD-ESTIMATE-UNROOTED, to emphasize that this is a statistical estimation approach that operates on a given multiple sequence alignment.

4.5   Statistical Consistency

Theorem 5. *Let N be a level-1 phylogenetic network with n leaves where all cycles have length at least five. Assume that character evolution down N is under the model from Sect. 4, and that the Oracle specifies those characters that are SNPs and indicates their ancestral state. Then, as the number k of characters increases, GUSFIELD-ESTIMATE-ROOTED will return the rooted network topology for N with probability converging to 1 (i.e., it is a statistically consistent estimator of the rooted network topology). Furthermore, the total runtime is $O(kn + n^3)$, where k is the alignment length in the input.*

Proof. The proof of correctness for GUSFIELD-CONSTRUCT-ROOTED if the SNPs cover all the clades of the level-1 network is given in [8]. Hence, all we need to establish is that as the number k of SNPs increases, with probability converging to 1, the SNPs in the input will cover $Clades(N)$ (see Definition 8).

Let $A \in Clades(N)$ and let S denote the set of leaves in N. Hence, $A \in Clades(T)$ for some $T \in \mathcal{T}_N$. Every edge in T is either a tree edge or a reticulate edge. Since $0 < \kappa(e) < 1$ for all reticulate edges, the probability of evolving down T is strictly positive.

Now consider a character c that evolves down T. Let $v = lca_T(A)$. By construction, v will have outdegree two, and hence is a tree node. Let w be the parent of v in T. The edge $e = w \to v$ is a tree edge, and so $p(e) > 0$. Since $0 < p(e') < 1$ for all tree edges $e' \in E(T)$, the probability that a character c evolves down T and changes on e but on no other edge in T is strictly positive. Hence, the probability of A appearing as $clade(c)$ for some homoplasy-free binary character c is strictly positive.

For consistency, since the Oracle correctly identifies the characters that are SNPs in the input alignment sites and tells us the ancestral state for these characters, we can construct a (possibly proper) subset of $Clades(N)$ from the observed characters. Since every clade in $Clades(N)$ has strictly positive probability of appearing as $clade(c)$ for a binary non-homoplastic character c, as the number of characters increases, with probability going to 1, the set of clades for these characters converges to $Clades(N)$. The result follows.

Runtime: The $O(mn + n^3)$ runtime for GUSFIELD-CONSTRUCT-ROOTED to construct a network from m SNPs on n leaves is provided in [8]. The input alignment has size $O(nk)$ and must be processed; after the Oracle selects the SNPs, we are left with m SNPs. Since $m \leq k$, the result follows.   □

Theorem 6. *Let N be a level-1 phylogenetic network with n leaves, and assume all cycles have at least five nodes. Assume that character evolution down N is under the model from Sect. 4. If the Oracle specifies the SNPs but not the ancestral state, then as the number k of characters increases, GUSFIELD-ESTIMATE-UNROOTED will return the unrooted topology of N with probability converging to 1 (i.e. it is a statistically consistent estimator of the unrooted network topology). Furthermore, the total runtime is $O(kn + n^3)$, where k is the alignment length in the input.*

Proof. The correctness and runtime of GUSFIELD-CONSTRUCT-UNROOTED is given in Corollary 1 (see also [8]). The rest of the proof of consistency and runtime analysis follows as for the rooted case.   □

4.6   Sample Complexity

Theorem 7. *Let N be a level-1 phylogenetic network without cycles of length less than five. Assume that the Oracle identifies the SNPs and defines the ancestral state for every SNP. For a given $\epsilon > 0$, GUSFIELD-ESTIMATE-ROOTED returns N with probability at least $1 - \epsilon$ when the number k of characters satisfies*

$$k > \frac{\log(\epsilon/4n)}{\log(1 - \pi)} \tag{1}$$

for $\pi = p_1 \times (1 - p_2)^{2n-3} \times \min\{\kappa_1, 1 - \kappa_2\}$, where $p_1, p_2, \kappa_1, \kappa_2$ are constants such that $0 < p_1 \leq p(e) \leq p_2 < 1$ for all edges e in N that are not reticulation edges and $0 < \kappa_1 \leq \kappa(e) \leq \kappa_2 < 1$ for all edges e entering reticulate nodes in N. The same guarantee holds for GUSFIELD-ESTIMATE-UNROOTED.

Proof. We begin with the argument for GUSFIELD-ESTIMATE-ROOTED. Recall that here we assume that the Oracle correctly identifies the SNPs and also identifies the root state. We already know from [8] that GUSFIELD-CONSTRUCT-ROOTED will return the correct rooted network N if the input set of SNPs cover $Clades(N)$. From this, it follows that GUSFIELD-ESTIMATED-ROOTED will also return the correct rooted network N if the alignment has SNPs that cover $Clades(N)$. Therefore, we only need to compute the minimum number k of characters (sites) that guarantees that the SNPs cover $Clades(N)$ with probability at least $1 - \epsilon$.

We begin with some observations that will be useful in this proof. Each edge in the network that is not in any cycle defines a unique bipartition, and hence clade (since we know the root state). Each edge in the network that is in a cycle, but is not incident to the bottom node of the cycle, defines two bipartitions (and hence two clades). Specifically, if γ is a cycle with bottom node x, then we let X denote the clade below the bottom node, and we refer to this as the bottom clade. Then any edge e in γ, with node v at the bottom end of e, denotes two clades: one containing X and one disjoint from X. We refer to these two as the *full clade* and *partial clade*, respectively, associated with edge e and hence also with node v. Furthermore, any clade c that is not the clade below the bottom node in a cycle or (in the case of a one-sided cycle) the clade below the root of the cycle is associated with a unique edge (and hence a unique node at the bottom of the edge). The clade that is below the bottom node of a cycle is associated with all three edges incident with the bottom node of the cycle.

The probability of the event that a clade A appears for the network as input from a given character is of the general form $\sum_T P(\mathcal{E}(T)) \times P(\mathcal{E}(A,T) \mid \mathcal{E}(T))$, where $\mathcal{E}(T)$ is the event that the character evolves down the tree T, $\mathcal{E}(A,T)$ is the event that the character changes state on the edge in T associated with the clade A, and the sum is over trees T belonging to some suitable subset of $\mathcal{T}_N$. The first probability in the sum is a product over the transmission probabilities $\kappa(e)$ for edges e that enter nodes in T that were reticulate nodes in N. If the sum of these products is over all of $\mathcal{T}_N$ then it is, of course, 1. The second (conditional) probability is

$$p(e) \times \prod_{e' \in E(T), e' \neq e} (1 - p(e')) \geq p_1 \times (1 - p_2)^{2n-3},$$

where e is the edge in T associated with the clade A.

We analyze the following cases: (1) clades associated with edges that are not in any cycle, (2) the partial clade associated with an edge in a cycle, and (3) the full clade associated with an edge in a cycle.

- **Case 1: The clade A is associated with an edge e_A that is not in any cycle.** Since e_A is not in any cycle, such an e_A is an edge in every tree contained in N. Hence, when there is a substitution on edge e_A and no other edge in the network, clade A appears in the input, no matter which tree the character evolves down. It follows from the discussion above that the probability we seek is $p(e_A) \times \prod_{e' \in E(T), e' \neq e_A} (1 - p(e')) \geq p_1 \times (1 - p_2)^{2n-3}$.

- **Case 2: The clade A is a partial clade associated with an edge in a cycle.** Let X be the bottom clade (i.e., the clade associated with the bottom node) and let A and $A \cup X$ be the partial clade and full clades associated with the node v_A and the edge e_A. Let e_r be the edge entering the bottom node that is on the same side of the cycle as v_A; note that $e_r \neq e_A$, since v_A is not the bottom node. The probability of transmission on edge e_r is $\kappa(e_r)$. When a character only changes state on the edge e_A and does not transmit on the reticulation edge e_r, clade A appears in the data; and when the character transmits on e_r then clade $A \cup X$ appears in the input. Now we must restrict our sum to trees T that do not transmit on edge e_r. The sum of the corresponding set of products is then $1 - \kappa(e_r)$. The conditional probability that clade A appears in the input given that we evolve down T is $p(e_A) \times \prod_{e' \in E(T), e' \neq e_A}(1 - p(e'))$. Thus, the probability we seek is at least $(1 - \kappa_2) \times p_1 \times (1 - p_2)^{2n-3}$.
- **Case 3: The clade A is a full clade associated with an edge in a cycle.** Let T be a tree in the network down which a character evolves that yields clade $A \cup X$. Hence T must include the edge e_r. The probability that clade $A \cup X$ appears in the input given that we evolve down T is $p(e_A) \times \prod_{e' \in E(T), e' \neq e_A}(1 - p(e'))$. Note that we must transmit on edge e_r. Thus, the probability we seek is at least $\kappa_1 \times p_1 \times (1 - p_2)^{2n-3}$.

Write $\pi = p_1 \times (1 - p_2)^{2n-3} \times \min\{\kappa_1, 1 - \kappa_2\}$ for the lower bound we have obtained for the probability that an arbitrary given clade appears for a given character. It follows that the probability that an arbitrary given clade appears for at least one of k characters is at least $1 - (1 - \pi)^k$. By Boole's inequality (sometimes known as the union bound), an upper bound on the probability that some clade is missing from the list of clades for k characters is $c(N)(1 - \pi)^k \leq 4n(1 - \pi)^k$ where $c(N)$ is the number of clades for the network N and the upper bound $c(N) \leq 4n$ is a consequence of Lemma 8.8.1 in [9]. A lower bound on the probability that all clades are present in the list of clades for k characters is thus $1 - 4n(1 - \pi)^k$. This last quantity is greater than $1 - \epsilon$ if and only if k is greater than the quantity on the right-hand side in the statement of the theorem.

We now address the unrooted case, where the Oracle does not indicate the ancestral state. Note that Theorem 6 establishes that GUSFIELD-ESTIMATE-UNROOTED will return the correct unrooted network N if the input alignment has SNPs that cover $Bip(N_r)$; this is the same set of SNPs that would cover the clades in N. Hence, the argument provided for the rooted SNP case can be modified for use in the unrooted SNP case.   □

5   Summary and Discussion

In this paper, we have provided statistically consistent polynomial time methods for estimating both rooted level-1 phylogenetic networks and their unrooted network topologies from SNPs. When the ancestral state is known for all the SNPs, multiple recombinations are allowed, and all cycles have at least five nodes, we can construct the unique level-1 phylogenetic network on which the SNPs evolved in $O(kn+n^3)$ time, where there are k sites (only some of which are SNPs) and n species, provided that the SNPs cover the clades in the network. For the more biologically relevant case where the ancestral state is unknown, we provide statistically consistent polynomial time methods for estimating the unrooted topology of the level-1 network, provided all cycles are of length at least five, and in the same $O(kn + n^3)$ running time.

The key insight that led to these methods and proofs of statistical consistency is that the unrooted topology of the level-1 phylogenetic network N is identifiable from the sets $Bip(N_r)$ and $Q(N_r)$, the bipartitions and (unrooted) quartet trees of the rooted trees contained in N (our Theorem 4), provided that all cycles are of length at least five. This is a different result than the well-known theorem that the set $Q(N)$ of all quartet trees of the unrooted network topology (which may contain additional quartet trees beyond those in $Q(N_r)$) defines an unrooted level-1 phylogenetic network topology. The importance of using $Q(N_r)$ instead of $Q(N)$ is that the set of quartet trees that can be inferred from data are those in $Q(N_r)$, rather than those in $Q(N)$. This is key to developing statistically consistent pipelines for phylogenetic network estimation from quartet trees (equivalently from bipartitions).

A direct consequence of Theorem 4 is that any method that returns a level-1 network consistent with the input set of SNPs (or bipartitions or quartets) is statistically consistent for estimating the unrooted level-1 network topologies for which all cycles are of length at least five. Thus, in particular, we establish in Theorem 5 that the pipeline that detects the SNPs and then runs GUSFIELD-CONSTRUCT-UNROOTED (i.e., Gusfield's algorithm [8] for the root-unknown case) on the SNPs is not only very fast (specifically, $O(kn + n^3)$ time, where there are k sites) but also statistically consistent, provided that all cycles in N are of length at least five.

This study suggests several directions for future work. For example, we would like to develop estimation methods that are not based on the infinite sites assumption. Related to this, two studies that have examined the related question of identifiability of level-1 phylogenetic networks from sites that evolve down trees within the network (but without assuming the infinite sites assumption) [1,6] or from the set of trees contained in the network [19]. Another study of interest, though not for level-1 phylogenetic network estimation, is [3]; this study proves that the class of tree-child phylogenetic networks are identifiable under a recombination-mutation model of evolution. These identifiability results are valuable but do not directly lead to estimation methods that are statistically consistent, and these papers do not provide such methods. Future work should explore whether these identifiability results can be used to develop estimation methods that are polynomial time and have statistical consistency guarantees.

Acknowledgments. This study was supported in part by the Grainger Foundation to TW. The authors also thank Dan Gusfield for valuable discussions and the anonymous reviewers for helpful feedback.

Disclosure of Interests. The authors have no competing interests to declare that are relevant to the content of this article.

References

1. Allman, E.S., Baños, H., Rhodes, J.A.: Identifiability of species network topologies from genomic sequences using the logdet distance. J. Math. Biol. **84**(5), 35 (2022)
2. Felsenstein, J.: Numerical methods for inferring evolutionary trees. Q. Rev. Biol. **57**(4), 379–404 (1982)
3. Francis, A., Moulton, V.: Identifiability of tree-child phylogenetic networks under a probabilistic recombination-mutation model of evolution. J. Theor. Biol. **446**, 160–167 (2018)
4. Gambette, P., Berry, V., Paul, C.: Quartets and unrooted phylogenetic networks. J. Bioinform. Comput. Biol. **10**(04), 1250004 (2012)
5. Gambette, P., Huber, K.T.: On encodings of phylogenetic networks of bounded level. J. Math. Biol. **65**(1), 157–180 (2012)
6. Gross, E., van Iersel, L., Janssen, R., Jones, M., Long, C., Murakami, Y.: Distinguishing level-1 phylogenetic networks on the basis of data generated by Markov processes. J. Math. Biol. **83**, 1–24 (2021)
7. Gusfield, D.: Efficient algorithms for inferring evolutionary trees. Networks **21**(1), 19–28 (1991)
8. Gusfield, D.: Optimal, efficient reconstruction of root-unknown phylogenetic networks with constrained and structured recombination. J. Comput. Syst. Sci. **70**(3), 381–398 (2005)
9. Gusfield, D.: ReCombinatorics: The Algorithmics of Ancestral Recombination Graphs and Explicit Phylogenetic Networks. MIT Press, Cambridge (2014)
10. Gusfield, D., Eddhu, S., Langley, C.: Efficient reconstruction of phylogenetic networks with constrained recombination. In: Computational Systems Bioinformatics. CSB2003. Proceedings of the 2003 IEEE Bioinformatics Conference. CSB2003, pp. 363–374. IEEE (2003)
11. Habib, M., To, T.H.: Constructing a minimum phylogenetic network from a dense triplet set. J. Bioinform. Comput. Biol. **10**(05), 1250013 (2012)
12. Huber, K.T., Van Iersel, L., Kelk, S., Suchecki, R.: A practical algorithm for reconstructing level-1 phylogenetic networks. IEEE/ACM Trans. Comput. Biol. Bioinf. **8**(3), 635–649 (2010)
13. Jansson, J., Nguyen, N.B., Sung, W.K.: Algorithms for combining rooted triplets into a galled phylogenetic network. SIAM J. Comput. **35**(5), 1098–1121 (2006)
14. Jukes, T.H., Cantor, C.R.: Evolution of protein molecules. Mammalian Protein Metabolism **3**, 21–132 (1969)
15. Morrison, D.A.: An Introduction to Phylogenetic Networks. RJR Productions (2011)
16. Poormohammadi, H., Eslahchi, C., Tusserkani, R.: TripNet: a method for constructing rooted phylogenetic networks from rooted triplets. PLoS ONE **9**(9), e106531 (2014)
17. Van Iersel, L., Keijsper, J., Kelk, S., Stougie, L., Hagen, F., Boekhout, T.: Constructing level-2 phylogenetic networks from triplets. IEEE/ACM Trans. Comput. Biol. Bioinf. **6**(4), 667–681 (2009)
18. Van Iersel, L., Kelk, S., Rupp, R., Huson, D.: Phylogenetic networks do not need to be complex: using fewer reticulations to represent conflicting clusters. Bioinformatics **26**(12), i124–i131 (2010)
19. Xu, J., Ané, C.: Identifiability of local and global features of phylogenetic networks from average distances. J. Math. Biol. **86**(1), 12 (2023)

Galled Perfect Transfer Networks

Alitzel López Sánchez[✉] and Manuel Lafond

Department of Computer Science, Université de Sherbrooke,
2500 boul de l'Université, Quebec J1K2R1, Canada
{alitzel.lopez.sanchez,manuel.lafond}@usherbrooke.ca

Abstract. Predicting horizontal gene transfers often requires comparative sequence data, but recent work has shown that character-based approaches could also be useful for this task. Notably, *perfect transfer networks* (PTN) explain the character diversity of a set of taxa for traits that are gained once, rarely lost, but that can be transferred laterally. Characterizing the structure of such characters is an important step towards understanding more complex characters. Although efficient algorithms can infer such networks from character data, they can sometimes predict overly complicated transfer histories.

With the goal of recovering the simplest possible scenarios in this model, we introduce *galled perfect transfer networks*, which are PTNs that are galled trees. Such networks are useful for characters that are incompatible in terms of tree-like evolution, but that do fit in an almost-tree scenario. We provide polynomial-time algorithms for two problems: deciding whether one can add transfer edges to a tree to transform it into a galled PTN, and deciding whether a set of characters are galled-compatible, that is, they can be explained by some galled PTN.

Keywords: Galled trees · Horizontal Gene Transfer · Algorithms

1 Introduction

Trees have served as a conventional representation of evolution for centuries in biology. However, contemporary evidence has found frequent exchanges of genetic material between co-existing species, indicating that evolution should rather be expressed as a "web of life". Horizontal Gene Transfer (HGT) is an important force of innovation between and within in all the domains of life [40]. They are known to occur routinely between procaryotes [26,41] but also happen between different domains. For example, the thermotogale bacteria, which thrive in extreme environments, are believed to have acquired several genes from archea [15,33]. HGTs also affect eukaryotes [24], with examples including the acquisition of fructophily from bacteria by yeasts [16] and transfers from parasitic plants to their hosts [42].

Owing to their central role in evolution, several algorithmic approaches have been developed to identify HGTs [37]. *Parametric methods* seek DNA regions that exhibit a signature that differs from the rest of the genome [28], whereas *phylogenetic methods* rely on the comparison of reconstructed species and gene trees,

C. Scornavacca and M. Hernández-Rosales (Eds.): RECOMB-CG 2024, LNBI 14616, pp. 24–43, 2024.
https://doi.org/10.1007/978-3-031-58072-7_2

often using reconciliation [4,30]. Some approaches also use sequence divergence patterns to infer timing discrepancies that correspond to transfers [14,23,27,39].

The vast majority of these methods rely on sequence comparisons. However, sequence-based methods are known to struggle when highly divergent sequences are involved, especially in the presence of ancient transfer events [6]. An alternative is to predict HGTs with *characters*, which are morphological or molecular traits that a taxa may possess or not. Character-based methods have been successfully applied to recover recombination or hybridation events [18]. A fundamental example of character-based data is gene expression, where the trait is whether or not a gene is expressed in a condition of interest [9,36,38], which can sometimes exhibit better phylogenetic signals than similarity measures [1].

These approaches aim to explain the diversity of a set of taxa S that each possess a subset of characters from a set C. Ideally, there should be a phylogeny in which, for each character $C \in C$, the taxa that possess C form a clade. If such a tree exists, it is called a *perfect phylogeny* [3,5,12,22]. In this setting, perfect phylogenies assume that each character has a unique origin (no-homoplasy), and is always inherited vertically once acquired (no-losses). Of course, these are strong assumptions that rarely apply to real biological datasets. However, understanding this theoretical model has led to multiple extensions with practical applications. Examples include the reconstruction of evolution from Short Interspersed Nuclear Elements (SINE) using partial characters [34]; haplotyping [3]; or the inference of cancer phylogenies, which were modeled as an extension of perfect phylogenies in [11], and broadened to Dollo parsimonies in [10]. Therefore, gaining a deeper understanding of such restricted models can often serve as a stepping stone to reconstruct more complex evolutionary scenarios.

When a perfect phylogeny does not exist for a set of characters, one may instead consider network-like structures to explain this diversity. To this end, Nakhleh et al. introduced *Perfect Phylogenetic Networks* (PPNs) in [31,32]. These networks allow multi-state characters and require that, for each character C, the network contains a tree in which nodes in the same state are connected. This is a powerful model that is, unfortunately, difficult to work with, since even deciding whether a known network explains a set of characters is NP-hard.

On the other hand, PPNs were introduced as trees with additional transfer edges. This implies knowledge of the vertical versus horizontal transmissions, and that these networks belong to the class of *tree-based networks* [13,35] (in fact, they are *LGT networks*, see next section). In [29], we introduced *perfect transfer networks* (PTNs), a specialization of PPNs where characters satisfy the same assumptions as perfect phylogenies: they are binary, have a unique origin, and are assumed to always be inherited vertically once acquired. An important difference of PTNs is that one cannot choose which subtree of the network can be used to explain a character, as the tree of vertical inheritance is fixed. Because of this, as we showed in [29], even PPNs that forbid losses are different from PTNs. In the latter, a character may or may not be transferred horizontally to other species. This can be useful for transferrable characters that are difficult to revert, such as material acquired horizontally from mitochondria or chloroplasts resulting from endosymbiotic events [2,43], or in the case of metabolites when HGT plays a role in the generation of new metabolic pathways in bacteria [17]. From an algorithmic standpoint, an advantage of the more restricted PTN model is that the problems of deciding whether a network explains a set of characters, or whether a tree can be augmented with transfers to do so, become polynomial-time solvable[1].

Thus, PTNs are a promising model for HGT inference from characters. However, PTNs have demonstrated a tendency to introduce an excessive number of transfer events. In [29], we provide an algorithm that shows that *any tree* can be made to explain *any set of characters* by adding transfers, although the resulting networks may be overly complicated and bloated with HGT edges. This raises the question of explaining a set of characters with the *simplest* possible network. One way would be to build a network or augment a tree with a minimum number of transfers, which is an unsolved complexity problem. Another direction is to impose structural conditions on the desired network.

Our Contribution. In this work, we focus on the second direction. We explore the evolutionary structure that many consider as the simplest beyond trees, namely *galled trees*. These consist of networks in which all underlying cycles are independent, and were first used in the context of hybridization and recombination [19]. When they fit the data, galled trees are desirable because of their parsimonious nature and ease of interpretation. They are also a popular graph-theoretical structure that serves as a first-step towards the development of more structurally complex networks [8,20]. In our case, characters that can be explained by a galled tree can be thought of "not quite tree-like, but almost".

We present *galled perfect transfer networks* (galled PTNs) which are galled tree-based networks with unidirectional edges that explain a set of characters. We then provide polynomial-time algorithms for two problems. In the *galled-completion* problem, we ask whether it is possible to complete a given tree with transfers to obtain a galled PTN. We also study the *galled-compatibility* problem in which, given a set of characters, we must decide whether a galled PTN can explain them. During the process, we provide several structural characterizations of characters that can be displayed on a given tree or a network. Due to space constraints, the detailed proofs are deferred to a full version of the paper.

Related Work. Aside from PPNs, other models have been proposed to explain characters via networks. In *recombination networks* [18], characters are explained by an *ancestral recombination graph* (ARG) in which hybridation nodes represent recombination events through crossovers. A fundamental is that such hybrids do not consider a donor/recipient relationship whereas in HGTs, it can be important to distinguish between the parental and lateral acquisition. As we showed in [29, Figure 3], PTNs and recombination graphs explain different sets of characters, even on galled trees with a single transfer/hybridation event, the main reason being that crossover events are different from transfer events. Nonetheless, it

---

[1] Note that we are not aware of complexity results for binary characters on PPNs without the no-loss condition.

is worth mentioning that in [19], the author shows how to reconstruct a galled ancestral recombination graph from a set of m characters and n taxa, if possible, in time $O(nm+n^3)$. In a similar vein, in [21,25] the authors study the question of reconstructing a network that displays a set of characters in the *softwired* sense, meaning that for each character, some tree contained in the network contains it as a clade (characters are called *clusters* therein). In particular, one can reconstruct in polynomial time a level-1 network that explains a set of characters, if one exists. As level-1 networks are closely related to galled trees, it is likely that the latter can be used in our setting. But, as also argued in [29, Figure 2], softwired characters can explain different sets of characters than PTNs. The main reason is that distinguishing between horizontal and vertical edges does not allow choosing which tree of the network should be used to explain a character.

2   Preliminaries

A *phylogenetic network*, or simply a *network* for short, is a directed acyclic graph $N = (V, E)$ with one node $\rho(N)$ of in-degree zero, called the *root*. A node of in-degree one and outdegree zero is a *leaf*, and $L(N)$ denotes the set of leaves of N. A node of in-degree and out-degree 1 is a *subdivision node*, which we allow. For $W \subseteq V$, $N - W$ is the directed graph obtained after removing W and incident edges. An *underlying cycle* of N is a set of nodes and edges that form a cycle when ignoring the edge directions. A network N is a *galled tree* if no two distinct underlying cycles of N contain a common node (see Fig. 1(c)).

A *tree* T is a network with no underlying cycle. We write $u \preceq_T v$ if v is on the path from $\rho(T)$ to u, in which case v is an ancestor of u and u a descendant of v. Note that v is an ancestor and descendant of itself. Two nodes u,v are incomparable in T if none descends from the other, i.e. if neither $u \preceq v$ nor $v \preceq u$. The parent of a node v in T is $p_T(v)$. For $v \in V(T)$, we use $T(v)$ for the subtree of T rooted at v, that is, $T(v)$ contains v and all of its descendants. We may write $L_T(v)$ as a shorthand for $L(T(v))$, or just $L(v)$ if T is understood. The set $L(v)$ is called a *clade* of T.

An *LGT network* (where *LGT* comes from *Lateral, or Horizontal, Gene Trasnfers*) [7] is a network $N = (V, E_S \cup E_T)$, where $\{E_S, E_T\}$ is a specified partition of the edge-set of N, such that the subgraph $\mathcal{T}_N := (V, E_S)$ is a tree with the same set of nodes as N. The tree $\mathcal{T}_N$ is called the *support tree* of N. The edges in E_S are called *support edges* and the edges in E_T are called *transfer edges*. A node that is the endpoint of a transfer edge is called a *transfer node*. We assume that for each transfer edge $(u, v) \in E_T$, the nodes u and v are incomparable in $\mathcal{T}_N$. We also assume that transfer nodes have exactly one out-neighbor in $\mathcal{T}_N$. The tree obtained from $\mathcal{T}_N$ by suppressing its subdivision nodes is called the *base tree* of G^2. For $v \in V(N) \setminus \{\rho(N)\}$, we use $p_N(v)$ to denote the unique in-neighbor of v in the support tree of N. For simplicity, an LGT network that is also a galled tree will be called a *galled LGT network*.

---

[2] Suppressing a subdivision node u with parent p and child v consists of removing u and adding an edge from p to v.

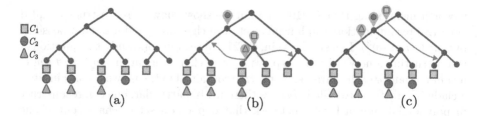

Fig. 1. (a) A tree and characters C_1, C_2, C_3. (b) A PTN with T as base tree that is *not* a galled tree (two underlying cycles contain the left child of the root). (c) A galled-completion of T. Note that in (b) and (c) for every character C, there exists a node (in gray) that transmits to all its vertical descendants, and reaches every leaf in C.

2.1 Perfect Transfer Networks

Let S be a set of taxa. A *character* C is a subset of S, which represents the set of taxa that possess the common trait. We usually denote a set of characters by $\mathcal{C}$. To formalize PTNs, given an LGT network N, a $\mathcal{C}$-*labeling* of N is a function $l : V(N) \rightarrow 2^{\mathcal{C}}$ that maps each node to the subset of characters it possesses.

Definition 1. *Let S be a set of taxa, let $\mathcal{C} \subseteq 2^S$ be a set of characters, and let $N = (V, E_S \cup E_T)$ be an LGT network with leafset S. We say that a $\mathcal{C}$-labeling l of N explains $\mathcal{C}$ if the following conditions hold:*

1. *for each leaf x, $l(x) = \{C \in \mathcal{C} : x \in C\}$ (leaves are labeled by their characters);*
2. *for each support edge $(u, v) \in E_S$, $C \in l(u)$ implies that $C \in l(v)$ (never lost once acquired);*
3. *for each $C \in \mathcal{C}$, there exists a unique node $v \in V$ that reaches every node w satisfying $C \in l(w)$ (single origin).*

Furthermore, we call N a perfect transfer network (PTN) for $\mathcal{C}$ if there exists a $\mathcal{C}$-labeling of N that explains $\mathcal{C}$.

Figure 1 shows two PTNs for the same set of characters. It does not exhibit the full $\mathcal{C}$-labeling, but a possible origin of each character is annotated on the internal nodes. In [29], it is shown that any set of characters can be explained by some PTN. In fact, any tree can become a PTN by adding enough transfer edges. Our goal is to constrain the PTNs to avoid overcomplicated solutions.

Let T be a tree on leafset S, and let $\mathcal{C}$ be a set of characters. We say that T is *galled-completable* for $\mathcal{C}$ if there exists a galled LGT network N that explains S and whose base tree is T. We call such a galled PTN a *galled-completion* of T. An example of this problem is shown on Fig. 1. For a set of characters $\mathcal{C}$ on taxa S, we say that $\mathcal{C}$ is *galled-compatible* if there exists a galled tree on leafset S that explains $\mathcal{C}$. Our problems of interest are the following:

- The GALLED COMPLETION problem: given a tree T on leafset S and a character set $\mathcal{C}$, is T galled-completable for $\mathcal{C}$?
- The GALLED COMPATIBILITY problem. Given a set of taxa S and a character set $\mathcal{C}$, is $\mathcal{C}$ galled-compatible?

2.2 Properties of Galled PTNs

We begin by stating properties of galled PTNs that will be useful throughout. Let N be a network that explains a set of characters $\mathcal{C}$. Observe that if a leaf l does not possess a character $C \in \mathcal{C}$, then by the "never lost once acquired" condition, the parent of l in T_N cannot possess C either (otherwise, the parent would be required to transfer C to v). In fact, no ancestor of l in T_N can possess C. This lets us deduce a subset F_C of nodes that forbid C, as follows:

$$F_C(N) = \{v \in V(N) : \exists l \in L_{T_N}(v) \text{such that } l \notin C\}.$$

It follows that the origin of C must be a node of $N - F_C$. By the "single origin" condition, it is also necessary that this origin reaches every leaf in C. As shown in [29], the existence of such a node is also sufficient to explain C. Since it is easier to deal with, we will heavily use the following characterization[3].

Lemma 1 ([29]). *Let N be an LGT network and let $\mathcal{C}$ be a set of characters. Then N is a perfect transfer network for $\mathcal{C}$ if and only if for every character $C \in \mathcal{C}$, the network $N - F_C(N)$ contains a node v that reaches every leaf in C.*

Proof (sketch). If N is a PTN for $\mathcal{C}$, then for a character $C \in \mathcal{C}$, as mentioned the nodes in $F_C(N)$ cannot possess C. This means that the origin of C, that is, a node that satisfies the third condition of PTNs, must be in $N - F_C(N)$. Conversely, if some node of $N - F_C(N)$ reaches every leaf, then it can serve as the origin of the character. □

We also describe two useful generic properties of galled LGT networks. The first one says that a node v cannot have two descending transfer nodes that each go outside of the v subtree (otherwise, v would be part of two distinct cycles).

Lemma 2. *Let $N = (V, E_S \cup E_T)$ be a galled LGT network and let $v \in V(N)$ be a node with two distinct descendants x and y that are transfer nodes (with $v \in \{x, y\}$ being possible). Then one of the transfer edges incident to x or y has both endpoints in the subtree $T_N(v)$.*

Proof (sketch). If v has two descendants x, y with transfer neighbors that are external to the v subtree of T_N, then the two outgoing transfers inevitably create two distinct cycles that both contain v, contradicting the galled property. □

The next property states that if some x is able to reach a node y that does not descend from it, then x must of course achieve this through a transfer edge.

Lemma 3. *Let $N = (V, E_S \cup E_T)$ be a galled LGT network. Let $x, y \in V$ be two nodes that are not comparable in T_N and such that x reaches y in N. Then there exists an edge $(x', y') \in E_T$ such that $x' \preceq_{T_N} x$ and $y \preceq_{T_N} y'$.*

---

[3] Note that we adapted this characterization, since in the original definition of PTNs, the taxa were treated as sets of characters instead of the other way around.

Proof (sketch). Consider the first transfer edge (x', y') on a path from x to y in N. Note that x' must descend from x in T_N. If y is not a descendant of y', then from y' the path needs to borrow another transfer edge that goes out of the y' subtree of T_N. Thus, the y' subtree has two transfers with external endpoints, contradicting Lemma 2. □

3   The Galled Completion Problem

In this section, let T be a tree on taxa S and let C be a set of characters. We assume that T has no subdivision node. We will first describe the necessary conditions for T to be galled-completable. The key factor lies in the ancestor relationship that exists between any set of *first-appearance (FA)* nodes for distinct characters.

Definition 2. *Let T be a tree on leafset S and let $C \subseteq S$ be a character. A node $v \in V(T)$ is a* first-appearance (FA) *node for C if $L_T(v) \subseteq C$ and v is either the root, or its parent u satisfies $L_T(u) \nsubseteq C$. The set of FA nodes for C in T is denoted as $\alpha_T(C)$.*

In other words, v is a FA for C if it roots a maximal subtree of taxa that contain C. An example of this definition is shown in Fig. 2.

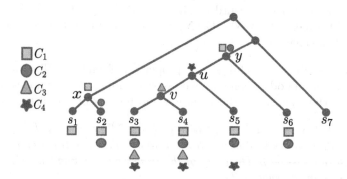

Fig. 2. A tree T on species $S = \{s_1, s_2, s_3, s_4, s_5, s_6, s_7\}$ with character set $C = \{C_1, C_2, C_3, C_4\}$. Every colored shape indicates to which character a specific taxa belongs to. The corresponding sets of FAs for every character are as follows: $\alpha_T(C_1) = \{x, y\}, \alpha_T(C_2) = \{y, s_2\}, \alpha_T(C_3) = \{v\}, \alpha_T(C_4) = \{u\}$. (Color figure online)

For a character $C \in C$, we will say that a node v is an *origin* for C if v reaches every leaf in C in $N - F_C(N)$. By Lemma 1, our goal is to ensure that every character has an origin. We first show that in any galled-completion of a tree T, an origin of a character C must descend from a FA node (or it could be a transfer node added just above it), and it must "give" its character to all other FAs by transferring just above them.

Lemma 4. *Let N be a galled-completion of a tree T that explains C. Let $C \in C$ be a character and let w be an origin for C in N. Then there is $\alpha_i \in \alpha_T(C)$ such that both of the following hold:*

- either $w \preceq \alpha_i$ or w is a transfer node whose child in T_N is α_i.
- for every $\alpha_j \in \alpha_T(C) \setminus \{\alpha_i\}$ there is a transfer edge (u, v) such that $u \preceq w$ and $v = p_N(\alpha_j)$.

Proof (sketch). By definition, in T the ancestors of FAs are in $F_C(N)$ since they have descendants not in C. Thus, an origin w in N must be in a FA subtree, or it could be a transfer node added above when adding transfer edges from T to N. If there are multiple FAs, w must be able to reach the other ones, and Lemma 3 lets us establish that transfer edges are needed to achieve this.   □

We next argue that the galled requirement places an important limitation on FAs, as there can be at most two per character. This means that a character C must either be a clade of T, or it could be split in two clades.

Lemma 5. *Let T be a galled-completable tree for a character set C. Then $\alpha_T(C) \leq 2$ for any character $C \in C$.*

Proof (sketch). By Lemma 4, an origin for C must be in one of the FA subtrees (or just above). If there are three FAs, that origin must also reach the other two FAs, but this requires two descending transfers that contradict Lemma 2.   □

We next show that FAs of distinct characters have limited ancestry relationships. The proof relies on a case analysis and the previous properties.

Lemma 6. *Let T be a galled-completable tree for C. Suppose that there exists two distinct characters A and B with $\alpha_T(A) = \{a_1, a_2\}$ and $\alpha_T(B) = \{b_1, b_2\}$. Then the two following statements hold:*

1. *If $b_1 \prec a_1$ and b_2 is not comparable to a_1, then $a_2 = b_2$.*
2. *If $a_1 = b_1$, then either $b_2 \prec a_2$ or $a_2 \prec b_2$.*

Proof (sketch). For the first statement, if $b_1 \prec a_1$ and b_2 is outside of the a_1 subtree, then b_1 needs to use the same transfer edge as a_1 to exchange material (because of the galled property). This is only possible if they send the character to the same clade, that is, if $a_2 = b_2$.

For the second statement, if $a_1 = b_1$, then again the two characters must use the same transfer edge to exchange material. In fact, one can argue that $a_1 = b_1$ must be the receiving end of the transfer, and that a_2, b_2 must be comparable to be able to use the same node to send the character.   □

Note that the first statement of the proof of Lemma 6 requires transfer edges to be unidirectional. If we allowed bidirectional transfer edges, then we can devise examples in which the lemma does not hold (not shown due to space constraints). Allowing such edges already leads to more complex structures.

We can begin describing our algorithmic strategy. The first step is to locate and count the FAs for each character to verify the condition established by Lemma 5. A subtree which is rooted at a FA for a specific character C is in fact a maximal clade for C. An intuitive way of joining these clades is to add transfer edges between the different subtrees to fulfill the connectivity requirement in Lemma 1. However, this may add superfluous transfer edges, as only the *minimal* ones are required.

To make this precise, let $v \in V(T)$. We say that another node $w \in V(T)$ is an *FA neighbor* of v if there exists a character $C \in \mathcal{C}$ such that $\alpha_T(C) = \{v, w\}$. We also say that w is a *minimal* FA neighbor of v if $w \preceq_T w'$ for every FA neighbor of v. A pair of FA nodes $\{v, w\}$ is called *simple* if w is the unique FA neighbor of v and v is the unique FA neighbor of w.

Definition 3. *Let T be a tree on leafset S with character set $\mathcal{C}$. Let T' be the tree obtained by subdividing every edge of T once. Then the* redundancy-free network *for T is the LGT-network $N = (V, E_S \cup E_T)$ with T' as support tree obtained as follows:*

- *for each $v \in V(T)$ with at least two FA neighbors, and for every minimal FA neighbor w of v, add the transfer edge $(p_{T'}(w), p_{T'}(v))$;*
- *for each pair $\{v, w\}$ of simple FA nodes, add one of the transfer edges $(p_{T'}(v), p_{T'}(w))$ or $(p_{T'}(w), p_{T'}(v))$ arbitrarily (but not both).*

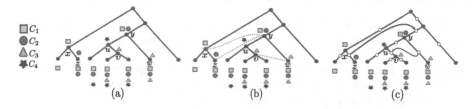

Fig. 3. (a) A given tree T on four characters. (b) The FA neighbors between every FAs for each character are indicated by the dashed lines. (c) The redundancy free network of T. Note that C_3 has simple FAs, so the direction for the transfer edge is arbitrary. On the other hand, for C_1 and C_2 note that z is the minimal FA neighbor of y so the direction of the transfer edge is not arbitrary.

The idea of this network is that if a node y has multiple FA neighbors, as in Fig. 3, then there are many characters with y as a FA. By adding a transfer edge whose sender is the parent of the minimal FA neighbor z of y, again as in the figure, all those characters can have their origin on the z side and transfer to y. In the figure, we add a transfer from the parent of z to that of y for character C_2. At the same time, C_1 can use that same transfer edge because it also wants y and its descendants to inherit the character. For characters whose FAs correspond to a simple pair, no such transfer re-use is needed and the direction is arbitrary.

Before proving our main characterization in terms of redundancy-free networks, we need an intermediary result.

Lemma 7. *Let T be a tree on leafset S and character set $\mathcal{C}$. If T is galled-completable, then in the redundancy-free network $N = (V_T, E_T)$ of T, every node is incident to at most one transfer edge.*

Proof (sketch). Suppose that some node u' of N is incident to two distinct transfer edges, with v' and w' as the other endpoints (where u', v', w' are subdivision nodes added when transforming T to N). Since N only adds transfers between minimal FA neighbors, this means that v' and w' are incomparable nodes of $\mathcal{T}_N$ (a special case arises when $v' = w'$, see the full proof for details). Roughly speaking, this means that two characters need to exchange material between the clades of u' and v', and also between the clades of u' and w'. This is only possible in a galled-completion if the u' subtree contains at least two transfers, one exchanging with v' and the other with w'. This creates two transfers below u' with external endpoints, contradicting the galled property by Lemma 2.   □

We finally arrive to our characterization of galled-completable trees.

Lemma 8. *A tree T on leafset $\mathcal{S}$ with character set $\mathcal{C}$ is galled-completable if and only if its redundancy-free network $N = (V, E_S \cup E_T)$ is a galled tree.*

Proof (sketch). In the forward direction, if T can be completed into some LGT network N', we can argue that for every transfer edge (u, v) added to N, there is a corresponding transfer edge in N' whose endpoints are either u and v, or their support tree descendants. Roughly speaking, this means that the cycles of N have corresponding cycles in N', and since N' is a galled tree, so is N. Conversely, if N is galled, it explains $\mathcal{C}$ by construction. That is, for every character split into two clades in T, we either added an edge above the subtrees to connect them, or there is an edge that was added in the descendants due to some minimal FA neighbor, which allow meeting the connectivity requirements of Lemma 1.   □

The previous lemma implies a polynomial time verification algorithm, detailed in Algorithm 1. We build the redundancy-free network and verify that it is a galled tree. We can calculate the set of FAs for each character and use this information to assign the FA neighbors to each node in T. Then, in a postorder traversal pick every node v that has FA neighbors. Note that if those neighbors are not all comparable, then there exists no completion, since two outgoing transfers are necessary to explain them. When v has multiple FA neighbors, we find the minimal one (c_{min} in the algorithm) and add the corresponding transfer edge. If c_{min} is the only FA neighbor of v, we "mark" v. If c_{min} is also marked, then $\{v, c_{min}\}$ is a simple pair, and if not, then either c_{min} will be marked later, or it has multiple FA neighbors and will create its own transfer.

Theorem 1. *Algorithm 1 correctly solves the* GALLED TREE COMPLETION *problem in time $O(|V(T)||\mathcal{C}|)$.*

Let us remark that the complexity of the algorithm is dominated by the computation of the set of FA nodes. Assuming a traversal of T for each $C \in \mathcal{C}$, this takes time $O(|V(T)||\mathcal{C}|)$. The rest of the algorithm only adds a time of $O(|V(T)| + |\mathcal{C}|)$. We leave the problem of computing FAs in linear time, if at all possible, for a future discussion.

```
1  function FindGalledCompletion(T,𝒮,𝒞)
2      //T is a tree on taxa set 𝒮, 𝒞 is the character set.
3      Let T' be the tree that results from subdividing every edge in T, let N = T'
4      // FA(v) is a set that will keep track of FA neighbors present in v.
5      for C ∈ 𝒞 do
6          Compute α_T(C), the set of FAs of C in T
7          if |α_T(C)| > 2 then return "not galled-completable"
8          if α_T(C) = {u,v} then add u to FA(v) and add v to FA(u)
9      for v in postorder(T) do
10         Let c_min be an arbitrary element of FA(v)
11         for u in FA(v) do
12             if u is not comparable to c_min then
13                 return "not galled-completable"
14             if u ≺ c_min then set c_min = u
15         if |FA(v)| ≥ 2 then
16             Add (p_N(c_min), p_N(v)) to N
17         else
18             if c_min is marked as "visited" then
19                 Add (p_N(v), p_N(c_min)) to N.
20             else
21                 Mark v as "visited"
22     if N is a galled tree then return N
23     else return "not galled-completable"
```

Algorithm 1: Check if a given tree T is galled-completable.

4   The Galled Compatibility Problem

Let us recall the galled compatibility problem: we are given a set of characters $\mathcal{C}$ on a set of taxa $\mathcal{S}$ and must decide whether a galled tree N explains $\mathcal{C}$. Note that if such an N exists, then its base tree is galled-completable. Therefore, $\mathcal{C}$ is galled-compatible if and only if there exists a tree T that is galled-completable for $\mathcal{C}$. Instead of aiming to construct N directly, our strategy is to build such a T using the characterizations from the previous section.

Let T be a tree and $v \in V(T)$. Let $C \in \mathcal{C}$ be a character and suppose that C has two FAs x_1, x_2 in T. In this case, we say that C is *split into $L_T(x_1)$ and $L_T(x_2)$* (noting that the union of these two leafsets must be C since there are only two FAs). A character $C \in \mathcal{C}$ is *maximal* if there is no $C' \in \mathcal{C}$ such that $C \subset C'$. Two characters $A, B \in \mathcal{C}$ are *compatible* if there exists a tree T in which A and B are clades, and *incompatible* otherwise. It is well-known that A, B are compatible if and only if one of $A \cap B = \emptyset$ or one of A or B is a subset of the other. This means that if A, B are incompatible, then the sets $A \cap B, A \setminus B, B \setminus A$ are all non-empty. Recall that for any set $\mathcal{C}$ of pairwise-compatible characters, there is a tree T whose set of clades is exactly $\mathcal{C}$, plus the clade of the root and the leaves (see e.g. [18]).

The detailed algorithm is somewhat involved, but the main idea can be described in a few steps that are illustrated in Fig. 4:

$A = \{a, b, c, d, e, f, g, h\}$ $X = \{d, e, f, g, h\}$ $P = \{f, g\}$
$B = \{f, g, h, m\}$ $Y = \{c, d, e, f, g, h\}$ $Q = \{i, j\}$
$C = \{a, b, c, d, e, l\}$ $R = \{j, k\}$

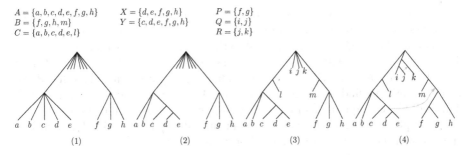

(1)   (2)   (3)   (4)

Fig. 4. An example instance with characters $\mathcal{C} = \{A, B, C, X, Y, P, Q, R\}$ on taxa $\mathcal{S} = \{a, b, c, d, e, f, g, h, i, j, k, l, m\}$ used to illustrate the main steps 1–4 of the algorithm. Step (1) splits A into $A \setminus B, A \cap B$. Step (2) integrates the clades forced by X, Y, which intersect the two clades of A. Step (3) integrates B and C as clades. Step (4) solves for P recursively, and then for Q, R recursively, then the transfer required by A, X, and Y is added.

(0) if there is a maximal character C that is compatible with all the others, we may assume that T splits the root with child clades C and $\mathcal{S} \setminus C$ (not shown in figure, see Lemma 9), and that each remaining character is contained in one of these clades. We can solve each subset of characters recursively;

(1) otherwise, every maximal character A has some incompatibility with some B, as in Fig. 4. We can show that T must either split A into clades $\{A \setminus B, A \cap B\}$, or split B into clades $\{B \setminus A, B \cap A\}$. We try the first option, and if it leads to a dead-end we try the other option. That is, for each of these two possibilities, we initiate a tree with only the two clades and proceed to the next steps. For the rest of the description, we assume that we start a tree with $A \setminus B, A \cap B$ as in Fig. 4.1. This means that in any completion, there will be a transfer edge between the two clades to provide an origin for A;

(2) each character that intersects both clades must use that same transfer edge in a completion. Such characters enforce new clades that refine the previous ones. For example in Fig. 4.2, X enforces the clade $\{d, e\}$, which is the intersection $X \cap \{a, b, c, d, e\}$, and similarly Y enforces $\{c, d, e\}$. These enforced new clades must be ordered by inclusion;

(3) other characters can enforce further clades, namely those that contain the minimal clades enforced so far. For example, the character C from Fig. 4 contains $\{d, e\}$, and so the clade C is enforced. Likewise, B is forced since it contains $\{f, g, h\}$. These are added in Fig. 4.3;

(4) it turns out that if $\mathcal{C}$ is galled-compatible, then any unforced character so far represents a set of leaves that have the same parent v in T. See P, Q, R in Fig. 4.3. We can recurse into the leaf children of each v and replace the leaves by the resulting network. In Fig. 4.4, the leaf set $\{i, j, k\}$ is replaced by a network that explains Q, R, and $\{f, g, h\}$ by a network that explains P.

An important subtlety arises in Step (1). If we need to recurse on both possible splits $\{A \setminus B, A \cap B\}$ and $\{B \setminus A, B \cap A\}$, the complexity could become exponential. Our algorithm is designed so that we never have to recurse on both. That is, we actually make a series of checks before trying $\{A \setminus B, A \cap B\}$, and we only recurse after all the checks pass. These checks are designed so that if the recursion fails to find a solution, then $\{B \setminus A, B \cap A\}$ would fail too anyways. This will become apparent in the details below, to which we now proceed.

As explained in step 0 above, we can first show that maximal compatible characters are easy to deal with, see Fig. 5.

Lemma 9. *Suppose that $\mathcal{C}$ contains a maximal character C that is compatible with every other character. Let $\mathcal{C}_1 = \{A \in \mathcal{C} : A \subset C\}$ and $\mathcal{C}_2 = \{A \in \mathcal{C} : A \cap C \neq \emptyset\}$. Then $\mathcal{C}$ is galled-compatible if and only if $\mathcal{C}_1$ and $\mathcal{C}_2$ are both galled-compatible.*

Moreover, given galled PTNs N_1, N_2 that explain $\mathcal{C}_1, \mathcal{C}_2$, respectively, one can obtain in time $O(|\mathcal{C}|)$ a galled PTN N that explains $\mathcal{C}$.

Proof (sketch). If C as stated exists, we can start constructing a tree whose root has two children, one with C as a clade, and the other with everything not in C. This automatically explains C. We can then recursively solve for $\mathcal{C}_1$ on the C side, and for $\mathcal{C}_2$ on the non-C side, and replace the child clades with the corresponding networks. Because C is compatible, every character will be on one side, this solves for every character.

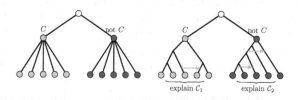

Fig. 5. Illustration of Lemma 9. If C is maximal and compatible, we split the problem into two subproblems on $\mathcal{C}_1$, which contains only subsets of C, and $\mathcal{C}_2$, which do not intersect with C, and put the resulting networks under a common root.

The next step is to handle maximal incompatible characters. We first show a fundamental property on pairs of incompatible characters: one of them must be a clade and the other must be split.

Lemma 10. *Let T be a tree that is galled-completable for $\mathcal{C}$, and let $A, B \in \mathcal{C}$ be a pair of incompatible characters. Then one of the following holds:*

- *A is split into $A \setminus B, A \cap B$ in T, and B is a clade of T;*
- *B is split into $B \setminus A, A \cap B$ in T, and A is a clade of T.*

Proof (sketch). By incompatibility, A and B cannot both be clades of T. If only one of them is, say A is a clade, then we are done since the only way to have two FAs for B is to put $B \cap A$ inside of A, and $B \setminus A$ elsewhere. So assume that both A and B are split, having two FAs a_1, a_2 and b_1, b_2 each. Because A and B intersect, with some effort it can be shown that these FAs must be related by ancestry (i.e., $a_1 \prec b_1$ or $b_1 \prec a_1$, and also $a_2 \prec b_2$ or $b_2 \prec a_2$). The proof shows that each possible case leads to requiring two distinct transfers to explain A and B, which create intersecting cycles.   □

Note that if we allow bidirectional transfers, there are examples in which the above lemma is not always true.

Our algorithm will find maximal incompatible A and B and try splitting A, or B. When trying the split $A_1 = A \cap B$, $A_2 = B \setminus A$, the characters that intersect with both A_1 and A_2 must also be split. Furthermore, one of the FAs of those must be all equal to either that of A_1 or A_2, and the other FAs must form a chain under that of A_1 or A_2. This can be formalized as follows.

Let A be a character and let $\{A_1, A_2\}$ be a partition of A. Let $\mathcal{X} \subseteq \mathcal{C}$ be the set of characters that intersect both A_1 and A_2 (note that $A \in \mathcal{X}$). We say that $\mathcal{X}$ forms an (A_1, A_2)-*chain* if the elements of $\mathcal{X}$ can be ordered as $\mathcal{X} = \{X_1, \dots, X_l\}$ such that $X_l = A$, and both of the following holds:

- $(X_1 \cap A_1) \subset (X_2 \cap A_1) \subset \dots \subset (X_l \cap A_1) = A_1$; and
- for every $X_i \in \mathcal{X}$, $X_i \setminus A_1 = A_2$.

We call $X_1 \cap A_1$ the *bottom of the chain*, and we call A_2 the *stable side of the chain*. In Fig. 4.3, $A_1 = \{a, b, c, d, e\}$, $A_2 = \{f, g, h\}$ and $\mathcal{X} = \{X, Y, A\}$. One can see that $\mathcal{X}$ forms an (A_1, A_2)-chain with bottom $\{d, e\}$ and stable side $\{f, g, h\}$.

Lemma 11. *Let A be a maximal character of $\mathcal{C}$. Suppose that T is a galled-completable tree for $\mathcal{C}$ in which A is split into the clades A_1 and A_2. Let $\mathcal{X} \subseteq \mathcal{C}$ be the subset of characters that intersect with both A_1 and A_2.*

Then, after possibly exchanging the subscripts of A_1 and A_2, $\mathcal{X}$ is an (A_1, A_2)-chain. Moreover, for every $X \in \mathcal{X}$, the clades $X \cap A_1$ and $X \cap A_2$ are in T.

Proof (sketch). Assuming that A is split into A_1, A_2, the maximality of A implies that any $X \in \mathcal{X}$ that intersects with both must also be split. In fact, to avoid creating intersecting cycles, every such X must be explained using the same transfer arc that explains A_1 and A_2. This can only be achieved if, in T, the FAs of the X's are ordered by ancestry on one side, and all lead to the same clade on the other side. Moreover, because A is maximal, every such X must be a subset of A. This results in an (A_1, A_2)-chain.   □

Lemma 11 shows that if we choose to split A into A_1 and A_2, then we know how to split the characters that intersect with both A_1 and A_2. We can make a similar deduction for the other characters that contain the bottom or stable side of the (A_1, A_2)-chain.

Lemma 12. *Let A be a maximal character of $\mathcal{C}$. Suppose that T is a galled-completable tree for $\mathcal{C}$ in which A is split into the clades A_1 and A_2. Let $\mathcal{X} \subseteq \mathcal{C}$ be the subset of characters that intersect with both A_1 and A_2 and suppose that $\mathcal{X}$ is an (A_1, A_2)-chain. Let $X_1 \cap A_1$ be the bottom of the chain and let A_2 be the stable side of the chain.*

If $C \in \mathcal{C} \setminus \mathcal{X}$ contains $X_1 \cap A_1$ or contains A_2, then C is a clade of T.

Proof (sketch). Any C as described intersects with exactly one of $X_1 \cap A_1$ or A_2, but not both (otherwise, it would be in $\mathcal{X}$). In this case, one can show that C and X_1 must be incompatible. Since X_1 is assumed to be split, we know by Lemma 10 that C cannot also be split, and thus it must be a clade.   □

For an example, see the characters B and C in Fig. 4. So far, we have handled characters "forced" by a split of A into A_1 and A_2. As it turns out, the other characters can be handled in a recursive manner. To put this precisely, let A be a maximal character of $\mathcal{C}$ and let $\{A_1, A_2\}$ be a partition of A into two non-empty sets. Let $\mathcal{X}$ be the characters that intersect with A_1 and A_2 and suppose that $\mathcal{X}$ forms and (A_1, A_2)-chain. We say that a character $C \in \mathcal{C}$ is *forced by* $\{A_1, A_2\}$ if either: $C \in \mathcal{X}$; C contains the bottom of the $\mathcal{X}$ chain; or C contains the stable side of the $\mathcal{X}$ chain. Furthermore, a clade $Y \subseteq \mathcal{S}$ is *forced by* $\{A_1, A_2\}$ if either: $Y = X \cap A_1$ or $Y = X \cap A_2$ for some $X \in \mathcal{X}$; or $Y = C$ for some character $C \in \mathcal{C}$ that contains the bottom or the stable side of the $\mathcal{X}$ chain.

The next lemma is crucial: it shows that non-forced characters can be grouped and dealt with recursively according to the tree that contains the forced clades.

Lemma 13. *Let A be a maximal character of $\mathcal{C}$ and suppose that there is a galled-completable tree T^* for $\mathcal{C}$ in which A is split into A_1 and A_2. Let T be the tree whose set of clades is precisely the clades forced by $\{A_1, A_2\}$ (plus the root clade and the leaves).*

If $C \in \mathcal{C}$ is a character not forced by $\{A_1, A_2\}$, then all the taxa in C have the same parent in T.

Proof (sketch). Suppose that T^* is galled-completable for $\mathcal{C}$. By the previous lemmata, all forced clades must be in T^*, and thus T^* is a "refinement" of T (that is, T^* contains the same clades, plus possibly more). We know that a transfer edge is required between the bottom of the chain and the stable side. If there is a non-forced character C with taxa having distinct parents, then no matter how T is refined into T^*, we will need an additional transfer arc to explain C, which will create a cycle that intersects with the one we created near the bottom of the chain. See Fig. 6 and the caption for an illustration.   □

By combining the elements gathered so far, we finally arrive at an algorithmically useful characterization of galled-compatible characters.

Lemma 14. *Let A be a maximal character of $\mathcal{C}$ and let $\{A_1, A_2\}$ be a partition of A. Then there is a tree that is completable for $\mathcal{C}$ and that contains the clades A_1 and A_2 if and only if all the following conditions hold:*

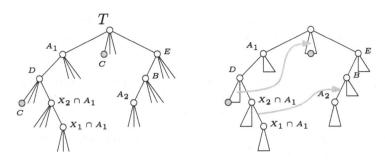

Fig. 6. An illustration of Lemma 13. Left: the tree T that contains the forced clades, with $X_1 \cap A_1$ the bottom of the chain and A_2 the stable side. We assume that some characters B, D, E enforced other clades. If we assume that some character C has elements with distinct parents, the situation on the right is unavoidable. A transfer will be needed to link the bottom and stable side, and another transfer to explain C. These two transfers create intersecting underlying cycles.

1. Let $\mathcal{X}$ be the characters that intersect with A_1 and A_2. Then $\mathcal{X}$ forms an (A_1, A_2)-chain;
2. There exists a tree T on leafset S whose set of clades is precisely the set of clades forced by $\{A_1, A_2\}$ (plus the root clade and leaves).
3. Let $\mathcal{C}_F$ be the set of characters forced by $\{A_1, A_2\}$. Then for any $C \in \mathcal{C} \setminus \mathcal{C}_F$, all the taxa of C have the same parent in T.
4. For any subset $\mathcal{C}' \subseteq \mathcal{C} \setminus \mathcal{C}_F$ such that all taxa that belong to some character of $\mathcal{C}'$ have the same parent in T, $\mathcal{C}'$ is galled-compatible.

Proof (sketch). In the ($\Rightarrow$) direction, we know by all the previous lemmata that all the stated conditions must be satisfied by a galled-completable tree for $\mathcal{C}$. In the ($\Leftarrow$) direction, suppose that all the conditions hold. Let T be a tree that contains all the clades forced by $\{A_1, A_2\}$. Then we can explain all the forced characters by adding a transfer arc just above the bottom of the $\mathcal{X}$ chain, going just above the stable side. As for the other characters, for a node v of T, let $\mathcal{C}_v \subseteq \mathcal{C}$ be the non-forced characters having taxa whose parent is v. We can find a galled network N_v that explains $\mathcal{C}_v$ and "attach" N_v as a child of v. Doing this for every v does not create intersecting cycles and explains all the remaining characters. $\square$

Our strategy is then to find a maximal character A and some B it is incompatible with. We try to split A into $A \setminus B$ and $A \cap B$ and if all the conditions of Lemma 14 pass, we have succeeded. Otherwise, it is B that we split into $B \setminus A$ and $B \cap A$. We want to check the conditions on this split, but since Lemma 14 only apply to maximal characters, then B must be maximal as well. We thus need the following.

Lemma 15. *Suppose that $\mathcal{C}$ has no maximal character that is compatible with all the others. Then there exists a pair of incompatible characters $\{A, B\}$ of $\mathcal{C}$ such that A and B are both maximal.*

Proof (sketch). Let A be maximal. Then A has some incompatibility. Letting B be the character of maximum cardinality such that A, B are incompatible, one can argue that B is also maximal.   □

This summarizes the elements required for our algorithm. Due to space constraints, the full pseudo-code and analysis is relegated to the Appendix. However, its main steps can be summarized as follows:

1. If there is a maximal character compatible with every other, proceed recursively using Lemma 9.
2. Otherwise, find maximal incompatible characters A, B, guaranteed by Lemma 15.
   Initiate a tree whose clades are $A \cap B$ and $A \setminus B$.
3. Check whether Conditions 1-2-3 of Lemma 14 hold with respect to that partition. If so, consider the tree T that contains all forced clades. For each $v \in V(T)$, list the set of characters $\mathcal{C}_v$ that only contain taxa whose parent is v. Solve for every such $\mathcal{C}_v$ recursively. If they all succeed, then we return that $\mathcal{C}$ is galled-compatible. If one of them fails, we may return that no solution is possible (because if one $\mathcal{C}_v$ is not galled-compatible, then $\mathcal{C}$ is not either).
4. If one of Condition 1-2-3 of Lemma 14 does not hold, then repeat the previous step but with a tree that contains clades $B \cap A, B \setminus A$. If this also fails, $\mathcal{C}$ is not galled-compatible.

Importantly, note that in this algorithm, we do not enter a recursion into $\mathcal{C}_v$ subsets for both partition attempts $\{A \cap B, A \setminus B\}$ and $\{B \cap A, B \setminus A\}$, as otherwise the complexity could be exponential. Since recursions are performed in at most one attempt of applying Step 3, we can argue that this takes polynomial time.

Theorem 2. *The* GALLED COMPATIBILITY *problem can be solved in time* $O(n|\mathcal{C}|^3)$.

5   Conclusion

In this work, we expanded the foundations on perfect phylogenies to galled trees. While compatible character in trees enjoy an elegant mathematical structure, it appears that the difficulty of characterizing compatibility ramps up very quickly as we move to non-tree structure. Nonetheless, this work can serve as a stepping-stone towards the understanding of character evolution on more complex structures. A small step in this direction would be to allow bidirectional transfers in galled trees. This change is not as trivial as it seems, since several of our key results do not hold when such edges are present. It will also be interesting to integrate broader classes of networks, for example level-k networks, and to solve the open problem of finding a PTN with a minimum number of transfer edges. Another direction is to remove the "no-loss" assumption for binary characters. The hardness results on PPNs in [31] apply to characters with arbitrarily many states, and thus the complexity of the completion and reconstruction problems is, to our knowledge, open on binary characters.

In practice, it remains to evaluate the potential of galled PTNs on real datasets. In fact, this work was motivated by our (unpublished) empirical results using unrestricted PTNs, which suggest that without imposing restrictions on the desired network, the previous algorithms highly overestimate the number of predicted transfers. It will be interesting to evaluate whether many incompatible characters fit on a galled tree. Another future direction is to loosen the no-homoplasy and/or no-loss conditions of the model. Allowing both transfers and losses to occur, perhaps with different weights, is a promising direction since efficient reconciliation algorithms could be adapted for this purpose.

References

1. Alexander, P.A., He, Y., Chen, Y., Orban, J., Bryan, P.N.: The design and characterization of two proteins with 88% sequence identity but different structure and function. Proc. Natl. Acad. Sci. **104**(29), 11963–11968 (2007)
2. Anselmetti, Y., El-Mabrouk, N., Lafond, M., Ouangraoua, A.: Gene tree and species tree reconciliation with endosymbiotic gene transfer. Bioinformatics **37**(Supplement_1), i120–i132 (2021)
3. Bafna, V., Gusfield, D., Lancia, G., Yooseph, S.: Haplotyping as perfect phylogeny: a direct approach. J. Comput. Biol. **10**(3–4), 323–340 (2003)
4. Bansal, M.S., Alm, E.J., Kellis, M.: Efficient algorithms for the reconciliation problem with gene duplication, horizontal transfer and loss. Bioinformatics **28**(12), i283–i291 (2012). https://doi.org/10.1093/bioinformatics/bts225, http://dx.doi.org/10.1093/bioinformatics/bts225
5. Bodlaender, H.L., Fellows, M.R., Warnow, T.J.: Two strikes against perfect phylogeny. In: Kuich, W. (ed.) ICALP 1992. LNCS, vol. 623, pp. 273–283. Springer, Heidelberg (1992). https://doi.org/10.1007/3-540-55719-9_80
6. Boto, L.: Horizontal gene transfer in evolution: facts and challenges. Proc. Roy. Soc. B: Biol. Sci. **277**(1683), 819–827 (2010)
7. Cardona, G., Pons, J.C., Rosselló, F.: A reconstruction problem for a class of phylogenetic networks with lateral gene transfers. Algorithms Mol. Biol. **10**(1) (Dec 2015). https://doi.org/10.1186/s13015-015-0059-z, http://dx.doi.org/10.1186/s13015-015-0059-z
8. Cardona, G., Zhang, L.: Counting and enumerating tree-child networks and their subclasses. J. Comput. Syst. Sci. **114**, 84-104 (2020). https://doi.org/10.1016/j.jcss.2020.06.001, http://dx.doi.org/10.1016/j.jcss.2020.06.001
9. De Jong, G.: Phenotypic plasticity as a product of selection in a variable environment. Am. Nat. **145**(4), 493–512 (1995)
10. El-Kebir, M.: SPhyR: tumor phylogeny estimation from single-cell sequencing data under loss and error. Bioinformatics **34**(17), i671–i679 (2018)
11. El-Kebir, M., Satas, G., Oesper, L., Raphael, B.J.: Inferring the mutational history of a tumor using multi-state perfect phylogeny mixtures. Cell Syst. **3**(1), 43–53 (2016)
12. Fernández-Baca, D.: The perfect phylogeny problem. In: Cheng, X.Z., Du, D.Z. (eds.) Steiner Trees in Industry. Combinatorial Optimization, vol. 11, pp. 203–234. Springer, Boston (2001). https://doi.org/10.1007/978-1-4613-0255-1_6

13. Fischer, M., Galla, M., Herbst, L., Long, Y., Wicke, K.: Classes of tree-based networks. Vis. Comput. Ind. Biomed. Art **3**(1) (2020). https://doi.org/10.1186/s42492-020-00043-z, http://dx.doi.org/10.1186/s42492-020-00043-z

14. Geiß, M., Anders, J., Stadler, P.F., Wieseke, N., Hellmuth, M.: Reconstructing gene trees from fitch's xenology relation. J. Math. Biol. **77**(5), 1459–1491 (2018). https://doi.org/10.1007/s00285-018-1260-8, http://dx.doi.org/10.1007/s00285-018-1260-8

15. Gogarten, J.P., Townsend, J.P.: Horizontal gene transfer, genome innovation and evolution. Nat. Rev. Microbiol. **3**(9), 679–687 (2005). https://doi.org/10.1038/nrmicro1204, http://dx.doi.org/10.1038/nrmicro1204

16. Gonçalves, C., et al.: Evidence for loss and reacquisition of alcoholic fermentation in a fructophilic yeast lineage. eLife **7** (2018). https://doi.org/10.7554/elife.33034, http://dx.doi.org/10.7554/eLife.33034

17. Goyal, A.: Horizontal gene transfer drives the evolution of dependencies in bacteria. iScience **25**(5), 104312 (2022). https://doi.org/10.1016/j.isci.2022.104312

18. Gusfield, D.: ReCombinatorics: The Algorithmics of Ancestral Recombination Graphs and Explicit Phylogenetic Networks. MIT Press, Cambridge (2014)

19. Gusfield, D., Eddhu, S., Langley, C.: Optimal, efficient reconstruction of phylogenetic networks with constrained recombination. J. Bioinform. Comput. Biol. **2**(01), 173–213 (2004)

20. Huson, D.H., Rupp, R., Berry, V., Gambette, P., Paul, C.: Computing galled networks from real data. Bioinformatics **25**(12), i85–i93 (2009). https://doi.org/10.1093/bioinformatics/btp217, http://dx.doi.org/10.1093/bioinformatics/btp217

21. van Iersel, L., Kelk, S., Rupp, R., Huson, D.: Phylogenetic networks do not need to be complex: using fewer reticulations to represent conflicting clusters. Bioinformatics **26**(12), i124–i131 (Jun 2010). https://doi.org/10.1093/bioinformatics/btq202, http://dx.doi.org/10.1093/bioinformatics/btq202

22. Iersel, L.V., Jones, M., Kelk, S.: A third strike against perfect phylogeny. Syst. Biol. **68**(5), 814–827 (2019)

23. Jones, M., Lafond, M., Scornavacca, C.: Consistency of orthology and paralogy constraints in the presence of gene transfers (2017). https://doi.org/10.48550/ARXIV.1705.01240, https://arxiv.org/abs/1705.01240

24. Keeling, P.J., Palmer, J.D.: Horizontal gene transfer in eukaryotic evolution. Nat. Rev. Genet. **9**(8), 605–618 (2008). https://doi.org/10.1038/nrg2386, http://dx.doi.org/10.1038/nrg2386

25. Kelk, S., Scornavacca, C., van Iersel, L.: On the elusiveness of clusters. IEEE/ACM Trans. Comput. Biol. Bioinf. **9**(2), 517–534 (2012). https://doi.org/10.1109/TCBB.2011.128

26. Koonin, E.V., Makarova, K.S., Aravind, L.: Horizontal gene transfer in prokaryotes: quantification and classification. Annu. Rev. Microbiol. **55**(1), 709–742 (2001)

27. Lafond, M., Hellmuth, M.: Reconstruction of time-consistent species trees. Algorithms Mol. Biol. **15**(1) (2020). https://doi.org/10.1186/s13015-020-00175-0, http://dx.doi.org/10.1186/s13015-020-00175-0

28. Lawrence, J.G., Ochman, H.: Reconciling the many faces of lateral gene transfer. Trends Microbiol. **10**(1), 1–4 (2002). https://doi.org/10.1016/S0966-842X(01)02282-X, https://www.sciencedirect.com/science/article/pii/S0966842X0102282X

29. López Sánchez, A., Lafond, M.: Predicting horizontal gene transfers with perfect transfer networks. Algorithms Mol. Biol. **19**(1), 6 (2024)

30. Menet, H., Daubin, V., Tannier, E.: Phylogenetic reconciliation. PLoS Comput. Biol. **18**(11), e1010621 (2022)

31. Nakhleh, L.: Phylogenetic networks. Ph.D. thesis, The University of Texas at Austin (2004)

32. Nakhleh, L., Ringe, D., Warnow, T.: Perfect phylogenetic networks: a new methodology for reconstructing the evolutionary history of natural languages. Language **81**(2), 382–420 (2005). http://www.jstor.org/stable/4489897

33. Nesbo, C.L., l'Haridon, S., Stetter, K.O., Doolittle, W.F.: Phylogenetic analyses of two "archaeal" genes in thermotoga maritima reveal multiple transfers between archaea and bacteria. Mol. Biol. Evol. **18**(3), 362–375 (2001)

34. Pe'er, I., Pupko, T., Shamir, R., Sharan, R.: Incomplete directed perfect phylogeny. SIAM J. Comput. **33**(3), 590–607 (2004)

35. Pons, J.C., Semple, C., Steel, M.: Tree-based networks: characterisations, metrics, and support trees. J. Math. Biol. **78**(4), 899–918 (2018). https://doi.org/10.1007/s00285-018-1296-9

36. Pontes, B., Giráldez, R., Aguilar-Ruiz, J.S.: Configurable pattern-based evolutionary biclustering of gene expression data. Algorithms Mol. Biol. **8**(1), 1–22 (2013)

37. Ravenhall, M., Škunca, N., Lassalle, F., Dessimoz, C.: Inferring horizontal gene transfer. PLOS Comput. Biol. **11**(5), e1004095 (2015). https://doi.org/10.1371/journal.pcbi.1004095, http://dx.doi.org/10.1371/journal.pcbi.1004095

38. Rawat, A., Seifert, G.J., Deng, Y.: Novel implementation of conditional coregulation by graph theory to derive co-expressed genes from microarray data. BMC Bioinform. **9**, 1–9 (2008)

39. Schaller, D., Lafond, M., Stadler, P.F., Wieseke, N., Hellmuth, M.: Indirect identification of horizontal gene transfer. J. Math. Biol. **83**(1) (2021). https://doi.org/10.1007/s00285-021-01631-0, http://dx.doi.org/10.1007/s00285-021-01631-0

40. Soucy, S.M., Huang, J., Gogarten, J.P.: Horizontal gene transfer: building the web of life. Nat. Rev. Genet. **16**(8), 472–482 (2015). https://doi.org/10.1038/nrg3962, http://dx.doi.org/10.1038/nrg3962

41. Thomas, C.M., Nielsen, K.M.: Mechanisms of, and barriers to, horizontal gene transfer between bacteria. Nat. Rev. Microbiol. **3**(9), 711–721 (2005)

42. Wickell, D.A., Li, F.: On the evolutionary significance of horizontal gene transfers in plants. New Phytol. **225**(1), 113–117 (2019). https://doi.org/10.1111/nph.16022, http://dx.doi.org/10.1111/nph.16022

43. Zachar, I., Boza, G.: Endosymbiosis before eukaryotes: mitochondrial establishment in protoeukaryotes. Cell. Mol. Life Sci. **77**(18), 3503–3523 (2020). https://doi.org/10.1007/s00018-020-03462-6

Homology and Phylogenetic Reconstruction

Inferring Transcript Phylogenies
from Transcript Ortholog Clusters

Wend Yam D. D. Ouedraogo and Aida Ouangraoua[✉]

Université de Sherbrooke, Sherbrooke, QC J1K2R1, Canada
{wend.yam.donald.davy.ouedraogo,aida.ouangraoua}@usherbrooke.ca

Abstract. Alternative Splicing (AS) is a mechanism in eukaryotic gene expression by which different combinations of introns are spliced to produce distinct transcript isoforms from a gene. Recent studies have highlighted that the transcript isoforms of human genes are often conserved in orthologous genes from various species. The conserved transcripts are referred to as transcript orthologs, and the identification of transcript ortholog groups provides valuable insights for studying their functions. Exploring the evolutionary histories of transcripts enhances our understanding of their proteins functions and their origins. It also allows us to better understand the role of alternative splicing in transcript evolution.

In a previous work, we addressed the problem of inferring orthology and paralogy relations at the transcript level. In this work, we focus on the reconstruction of transcript evolutionary histories. We present a progressive supertree construction algorithm that relies on a dynamic programming approach to infer a transcript phylogeny based on precomputed clusters of orthologous transcripts. A phylogeny is constructed iteratively by performing pairwise supertree construction at each internal node of a guide tree defined for the set of transcript clusters.

We applied our algorithm to transcripts from simulated gene families, as well as to two case studies involving the transcripts of real gene families-specifically, the TAF6 and PAX6 gene families from the Ensembl-Compara database. The results align with those of previous studies aimed at reconstructing transcript phylogenies, while improving the computing time. The results also show that accurate transcript phylogenies can be obtained by first inferring accurately the pairwise homology relationships among transcripts and then using the latter to compute a phylogeny that agrees with the homology relationships. The results obtained for the simulated and real gene families are available at https://github.com/UdeS-CoBIUS/TranscriptPhylogenies. The Supplementary material can be found at https://zenodo.org/records/10798958.

Keywords: Transcriptome · Orthology and paralogy · Tree reconciliation · Transcript evolution

1 Introduction

Reconstructing the evolutionary history of transcripts provides insights into how transcripts diverge or remain conserved across genes and species, thus

C. Scornavacca and M. Hernández-Rosales (Eds.): RECOMB-CG 2024, LNBI 14616, pp. 47–68, 2024.
https://doi.org/10.1007/978-3-031-58072-7_3

contributing to our understanding of the evolution of biological complexity [1]. Understanding of transcript evolution offers valuable insights into the functional diversity of genes, particularly in the context of alternative splicing [4,11]. Exploring transcript evolution is also instrumental in predicting the functions of uncharacterized genomic elements and identifying potential disease markers, shedding light on the molecular basis of pathologies like cancers and genetic disorders. As transcript evolution is intricately linked with gene evolution, it also provides valuable information for inferring evolutionary history relationships between genes and species.

At the gene level, three main methodological approaches can be used to infer gene phylogenies and different types of pairwise homology relationships between genes [21]. The first, referred to as "Tree-to-Homology", involves first inferring the gene phylogeny based on the comparison of nucleotide or protein sequences. Then, the gene phylogeny is reconciled with a species phylogeny to infer the types of pairwise homology relationships in the gene tree. The second approach, called "Homology-to-Tree", first infers the types of pairwise homology relationships and then reconstructs a gene phylogeny that aligns with the pairwise homology relations. The third approach involves the joint reconstruction of both by directly inferring a reconciled gene phylogeny.

Similarly, at the transcript level, three categories of methods can be defined: 1) *Tree-to-Homology*: infers transcript homology relations based on the reconciliation of a transcript phylogeny with a gene phylogeny; 2) *Homology-to-Tree*: infers the transcript phylogeny from precomputed pairwise transcript homology relations; 3) *Joint-Tree-Homology*: infers both concurrently. Following the Tree-to-Homology approach, Christinat et al., in a pioneering work [6,7], and later Ait-Hamlat et al. [2], proposed reconstructing transcript phylogenies using parsimony-based tree search methods under various models of transcript evolution. The main limitation of this approach is the prohibitive computing time required to find an optimal tree topology in a large search space.

In this paper, we consider the Homology-to-Tree approach for transcript phylogeny reconstruction. In a previous work [16], we introduced a model defining orthology and paralogy relations between transcripts based on the reconciliation model between transcript trees and gene trees. We also presented an algorithm to infer groups of orthologous transcripts in a gene family. Building on this work, we present a progressive algorithm to infer transcript phylogenies based on the inferred ortholog groups. The algorithm relies on a guide tree defined for the set of ortholog groups using the Neighbor-Joining method [20]. It iteratively performs pairwise supertree construction in the internal nodes of the guide tree using a dynamic programming procedure to infer, at each iteration, a transcript tree having a minimum cost reconciliation with the gene tree.

The manuscript is organized as follows. Section 2 provides the definitions and notations required for the remainder of the paper. Section 3 describes our algorithm for the inference of transcript phylogenies based on orthologous groups of transcripts. Section 4 presents the results of the application of the methods to transcripts of simulated gene families generated using SimSpliceEvol [13], as well as to the transcripts of the TAF6 and PAX6 gene families from the Ensembl-Compara database [25].

2 Preliminaries

Trees. All considered trees are rooted and binary unless stated otherwise, with an edge-labeling that associates each edge to its length. Given a tree P, $v(P)$ denotes the set of nodes of P, $e(P)$ the set of edges of P, $l(P)$ its leafset, and $r(P)$ its root node. Given two nodes x and y of P, $\tau_P(x, y)$ and $\mu_P(x, y)$ denote respectively the number of edges on the path between x and y in P and the sum of edge lengths on the path between x and y in P. $\epsilon_P(x, y)$ represents the set of all edges on the path between x and y in P. The complete subtree of P rooted in x is denoted by $P[x]$ and $l(x)$ denotes its leafset.

Given two nodes x and y of a tree, x is an ancestor of y if y is a node of $P[x]$. $inter(x, y)$ denotes the number of nodes located on the path between x and y, excluding x and y. If x is an internal node, x_l and x_r denote its two children. The bipartition of $l(x)$ induced by x is denoted by $(l(x_l), l(x_r))$ or simply by (x_l, x_r). For a subset L' of $l(P)$, $lca_P(L')$ represents the lowest common ancestor (LCA) in P of L'. This is defined as the ancestor that is common to all the nodes in L' and is located furthest from the root. $P_{|L'}$ is the tree with leafset L' obtained from the subtree $P[lca_P(L')]$ by removing all leaves that are not in L', and then all internal nodes of degree 2, except the root. Given a tree P' such that $l(P') \subseteq l(P)$, we say that P displays P' if and only if $P_{|l(P')}$ is isomorphic to P' while preserving the same leaf-labeling.

Tree Versus Live Tree For a Set of Nodes. A tree for a set Σ is a tree P such that $l(P) = \Sigma$. Given a set of trees $\{P_i, 1 \le i \le k\}$ for subsets $\Sigma_i, 1 \le i \le k$ that form a partition of a leafset Σ, a supertree of $\{P_i, 1 \le i \le k\}$ is a tree for Σ that displays all trees $P_i, 1 \le i \le k$.

A live tree for a set Σ is a tree P such that $l(P) \subseteq \Sigma \subseteq v(P)$. In a live tree, elements of Σ can correspond to both leaves or internal nodes in the tree. The terminology is borrowed from [22]. Given a subset V' of $v(P)$, $P[V']$ denotes the minor tree of P obtained by contracting all edges (x, y) of P such that $y \notin V'$ to keep only nodes in the set V'. Note that a minor tree of a binary tree can be non-binary. In particular, if P is a live tree for Σ, then $P' = P[\Sigma]$ is minor tree of P such that $v(P') = \Sigma$.

Transcript, Gene and Species Trees. $\mathbb{S}$, $\mathbb{G}$, and $\mathbb{T}$ denote, respectively, a set of species, a set of homologous genes from a gene family, and a set of homologous transcripts. Considering a gene $g \in \mathbb{G}$ and a transcript $t \in \mathbb{T}$, two functions $s(g)$ and $g(t)$ are defined to establish a relationship between the elements from the three sets. Specifically, $s : \mathbb{G} \to \mathbb{S}$ maps each gene to its corresponding species, and $g : \mathbb{T} \to \mathbb{G}$ maps each transcript to its corresponding gene ensuring that $\{s(\mathbf{g}) : \mathbf{g} \in \mathbb{G}\} = \mathbb{S}$ and $\{g(\mathbf{t}) : \mathbf{t} \in \mathbb{T}\} = \mathbb{G}$.

S represents a species tree for $\mathbb{S}$. G stands for a gene tree for $\mathbb{G}$, with internal nodes representing speciation and gene duplication events leading to

$\mathbb{G}$. T denotes a transcript tree for $\mathbb{T}$, with internal nodes depicting speciation, gene duplication, and transcript creation events that have led to $\mathbb{T}$. The mapping function s is extended from $v(G)$ to $v(S)$, and the mapping function g from $v(T)$ to $v(G)$. This extension is defined as follows: for any node t in $v(T)$, $g(\mathsf{t}) = lca_G(g(\mathsf{t}') : \mathsf{t}' \in l(T[\mathsf{t}]))$ and for any node g in $v(G)$, $s(\mathsf{g}) = lca_S(s(\mathsf{g}') : \mathsf{g}' \in l(G[\mathsf{g}]))$.

LCA-Reconciliation in Gene and Transcript Evolution. rec_G denotes the function of the LCA-reconciliation of G and S defined as $rec_G : v(G) \setminus \mathbb{G} \rightarrow \{Spe, Dup\}$ that labels any internal node g of G as a speciation (Spe) if $s(\mathsf{g}) \neq s(\mathsf{g}_l)$ and $s(\mathsf{g}) \neq s(\mathsf{g}_r)$, and as a duplication (Dup) otherwise. The LCA-reconciliation cost of G and S is the number of gene duplications and losses underlined by rec_G.

Similarly, the LCA-reconciliation of T and G is a function $rec_T : v(T) \setminus \mathbb{T} \rightarrow \{Spec, Dup, Cre\}$ that labels any internal node t of T as a creation (Cre) if $g(\mathsf{t}) = g(\mathsf{t}_l)$ or $g(\mathsf{t}) = g(\mathsf{t}_r)$, otherwise as a duplication (Dup) if $rec_G(g(\mathsf{t})) = Dup$, and as a speciation (Spe) if $rec_G(g(\mathsf{t})) = Spe$. The LCA-reconciliation cost of T and G is the number of transcript creations and losses underlined by rec_T. Figure 1a shows an illustration of the LCA-reconciliation in gene and transcript evolution.

rec_G provides a reconciliation between G and S that minimizes the number of gene duplications and losses [5]. rec_T also provides a reconciliation between T and G that minimizes the number of transcript creations and losses [14].

In the next paragraph, we recall the definition of orthology and paralogy relationships between transcripts, presented in [16].

Orthology and Paralogy at Gene and Transcript Level. Two distinct genes g_1 and g_2 of $\mathbb{G}$ are orthologs if their LCA in G is a speciation, i.e. $rec_G(lca_G(\{\mathsf{g}_1, \mathsf{g}_2\})) = Spe$; recent paralogs if $rec_G(lca_G(\{\mathsf{g}_1, \mathsf{g}_2\})) = Dup$ and $s(lca_G(\{\mathsf{g}_1, \mathsf{g}_2\})) = s(\mathsf{g}_1)$; and ancient paralogs otherwise. Similarly, two distinct transcripts t_1 and t_2 of $\mathbb{T}$ are ortho-orthologs if their LCA in T is a speciation, i.e. $rec_T(lca_T(\{\mathsf{t}_1, \mathsf{t}_2\})) = Spe$; para-orthologs if $rec_T(lca_T(\{\mathsf{t}_1, \mathsf{t}_2\})) = Dup$; recent paralogs if $rec_T(lca_T(\{\mathsf{t}_1, \mathsf{t}_2\})) = Cre$ and $g(lca_T(\{\mathsf{t}_1, \mathsf{t}_2\}) = g(\mathsf{t}_1)$; and ancient paralogs otherwise.

In Fig. 1a, examples of the different types of homology relationships at gene and transcript levels are provided. Specifically, a_{32} and b_{21} are ortho-orthologs, while a_{11} and b_{21} are para-orthologs, b_{21} and b_{22} are recent paralogs, and a_{31} and a_{32} are ancient paralogs.

Isoorthology Relation Between Transcripts. Following the framework presented in [16], given a creation node x in the LCA-reconciliation of T with G, one of the two edges descending to its children is called a conservation edge and corresponds to the original transcript conserved, whereas the other edge is called a divergent edge and corresponds to a newly created divergent transcript. Figure 1a also illustrates divergence edges, depicted as dashed lines in the transcript tree, while the remaining edges are conserved edges. Distinguishing conservation edges and divergence edges in the transcript tree T enables the definition of a specific type of orthology relation between transcripts.

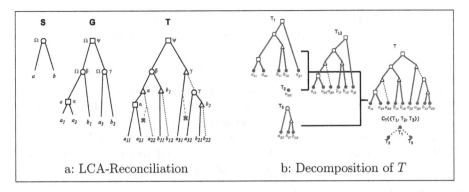

a: LCA-Reconciliation   b: Decomposition of T

Fig. 1. (a) a species tree S on $\mathbb{S} = \{a, b\}$, a gene tree G on $\mathbb{G} = \{a_1, a_2, a_3, b_1, b_2\}$, and a transcript tree T on $\mathbb{T} = \{a_{11}, a_{21}, a_{21}, a_{31}, a_{32}, b_{11}, b_{12}, b_{21}, b_{22}\}$ such that for any species $x \in \mathbb{S}$, gene $x_i \in \mathbb{G}$, and transcript $x_{ij} \in \mathbb{T}$, $s(x_i) = x$ and $g(x_{ij}) = x_i$. Round nodes represent speciations, square nodes gene duplications, triangle nodes transcript creations, and fictive dashed lines that end with a cross symbol are the location of losses in rec_G and rec_T. These reconciliations use green and blue labels for each internal node, respectively. Divergence edges after creation nodes are represented as dotted lines. The sets $\{a_{11}, a_{21}, b_{11}, a_{31}\}$, $\{a_{22}\}$, $\{b_{12}\}$, $\{a_{32}, b_{21}\}$, and $\{b_{22}\}$ represent the 5 isoortholog groups of $\mathbb{T}$. The LCA-reconciliation cost of G and S equals 2 (2 duplications), and the LCA-reconciliation cost of T and G is 6 (4 duplications + 2 losses). (b) Decomposition of the transcript tree T into 3 ortholog trees T_1, T_2, and T_3, which are respectively illustrated in red, blue, and green. The tree T can be reconstructed by first joining T_1 and T_2 the tree T_{12}, and then joining T_{12} and T_3 to obtain T. The connectivity tree $C_T(T_1, T_2, T_3)$ is also depicted.

Two orthologous transcripts t_1 and t_2 of $\mathbb{T}$ (i.e ortho- or para-orthologs) are isoorthologs if there are no divergence edges in $e_T(t_1, t_2)$. The isoorthology relation is transitive, such that, if there exists a third transcript t_3 in $\mathbb{T}$, such that t_2 and t_3 are isoorthologs, then t_1 and t_3 are also isoorthologs. The transitivity property of the isoorthology relation allows to partition the set of transcripts $\mathbb{T}$ into ortholog groups.

Transcript Ortholog Groups. An ortholog group $\mathbb{O}$ of $\mathbb{T}$ is a subset of $\mathbb{T}$ such that any two distinct transcripts t_1 and t_2 belonging to $\mathbb{O}$ are isoorthologs, recent paralogs, or there exist two transcripts t_1' and t_2' in $\mathbb{O}$ such that $t_1 = t_1'$ or t_1 and t_1' are recent paralogs, $t_2 = t_2'$ or t_2 and t_2' are recent paralogs, and t_1' and t_2' are isoorthologs. $O(\mathbb{T})$ denotes the partition of $\mathbb{T}$ into a set of maximum inclusive-wise ortholog groups. For example, in Fig. 1b, $\{a_{11}, a_{21}, b_{11}, b_{12}, a_{31}\}$, $\{a_{22}\}$, and $\{a_{32}, b_{21}, b_{22}\}$ are the maximum inclusive-wise ortholog groups of $\mathbb{T}$.

In the sequel, the maximum inclusive-wise ortholog groups are computed using the heuristic algorithm presented in [16] unless stated otherwise. The algorithm uses a progressive Reciprocal Best Hits approach to infer clusters of transcripts that have retained their function from the lowest common ancestor, i.e. isoorthologous and recent paralogous transcripts.

Definition 1 (Decomposition of the transcript tree into ortholog trees). *The set $O(\mathbb{T}) = \{\mathbb{O}_i, 1 \leq i \leq k\}$ of ortholog groups of $\mathbb{T}$ defines a decomposition of the transcript tree T into a set of ortholog trees $\{T_i, 1 \leq i \leq k\}$ such that for each $1 \leq i \leq k$, $T_i = T_{|\mathbb{O}_i}$.*

For example, in Fig. 1b, a decomposition of the transcript tree T of Fig. 1a into 3 ortholog trees T_1, T_2, T_3 is depicted. In each ortholog tree, the set of leaves represents a maximum inclusive-wise ortholog group.

From Definition 1, it is easy to see that the transcript tree T is a supertree for the set of ortholog trees $\{T_i, 1 \leq i \leq k\}$. The definition of T from the set $\{T_i, 1 \leq i \leq k\}$ can be further refined as follows.

Definition 2. *Given a set of trees $\{P_i, 1 \leq i \leq k\}$ for subsets $\{L_i, 1 \leq i \leq k\}$ that form a partition of a leafset L, a subtree-preserving supertree (SP supertree) of $\{P_i, 1 \leq i \leq k\}$ is a supertree P such that the trees $\{P_i, 1 \leq i \leq k\}$ can be obtained by cutting $k - 1$ edges of the tree P, and then removing all internal nodes of degree 2, except the roots of the resulting trees.*

Lemma 1 (From the transcript tree to the set of ortholog trees and conversely). *(1) The transcript tree T is an SP supertree of its set of ortholog trees $\{T_i, 1 \leq i \leq k\}$, namely $\{T_i, 1 \leq i \leq k\}$ can be obtained by cutting the $k-1$ divergence edges that separate distinct ortholog groups in T, (2) Conversely, the tree T can be reconstructed from the set $\{T_i, 1 \leq i \leq k\}$ by using an iterative process of $k - 1$ steps, each consisting of joining two trees through a subtree grafting operation that attaches one tree to an edge of the other tree.*

Proof. (1) Since two transcripts are isoorthologs if there are no divergence edges on the path between them in T, and recent paralogous transcripts always belong to the same ortholog groups, then the edges to cut in T to obtain the decomposition $T_i, 1 \leq i \leq k$ are the divergence edges that separate distinct ortholog groups in T. (2) follows from (1). $\square$

For example, in Fig. 1b, the process to reconstruct the tree T based on the 3 ortholog trees composing T is illustrated. That T is an SP supertree of its set of ortholog trees allows us to define the connectivity tree of ortholog trees in T.

Definition 3 (Connectivity tree of ortholog trees). *Given a transcript tree T and its set of ortholog trees $T_i, 1 \leq i \leq k$, the connectivity tree $C_T(\{T_i, 1 \leq i \leq k\})$ is the tree obtained from T by contracting all edges except the divergence edges that separate distinct ortholog trees in T. The resulting rooted tree has k nodes, each corresponding to one of the k ortholog trees.*

For example, in Fig. 1b, the connectivity tree $C_T(\{T_1, T_2, T_3\})$ is depicted. Note that the connectivity tree of ortholog trees can be a non-binary tree. In particular, it satisfies the following property.

Property 1. Given a transcript tree T and its set of ortholog trees $T_i, 1 \leq i \leq k$, the connectivity tree $C_T(\{T_i, 1 \leq i \leq k\})$ is live tree for the set of ortholog trees, each ortholog tree being reduced to a single node.

3   Inferring Transcript Phylogenies

In this section, we present a progressive algorithm to construct a transcript phylogeny based on ortholog trees defined for the ortholog groups.

Definition 4 (Ortholog trees reconstruction). *Given $O(\mathbb{T}) = \{\mathbb{O}_i, 1 \leq i \leq k\}$, the set $\Gamma = \{\Gamma_i, 1 \leq i \leq k\}$ denotes the set of ortholog trees such that, for each $1 \leq i \leq k$, Γ_i is a tree for $\mathbb{O}_i$ obtained as follows: (1) consider the set of genes $\mathbb{G}_i = \{g(t) : t \in \mathbb{O}_i\}$; (2) consider the gene tree $G_i = G_{|\mathbb{G}_i}$; (3) for each leaf x in G_i, consider the set of transcripts $\mathbb{T}_x = \{t \in \mathbb{O}_i : g(t) = x\}$ and replace the leaf x in G_i by a transcript tree T_x for $\mathbb{T}_x$ constructed using the Neighbour Joining (NJ) algorithm on the pairwise similarity matrix of $\mathbb{T}_x$.*

The process used to reconstruct ortholog trees from ortholog groups ensures that the pairwise orthology and recent paralogy relations between transcripts within an ortholog group $\mathbb{O}_i$ are consistent with the pairwise homology relations defined by the LCA-reconciliation of the corresponding tree Γ_i with the gene tree G. Thus, each ortholog tree Γ_i reflects the evolutionary history of the set of transcripts composing the ortholog group $\mathbb{O}_i$.

Lemma 1 states that if $\Gamma_i, 1 \leq i \leq k$ is the set of true ortholog trees composing the transcript tree T for $\mathbb{T}$, which we aim to compute, then T must be an SP supertree of $\Gamma_i, 1 \leq i \leq k$. Thus, we seek an SP supertree T of $\Gamma_i, 1 \leq i \leq k$ that can be obtained by using an iterative process that joins two trees at each step until T is obtained. Nethertheless, at each iteration, we must avoid grouping ortholog trees that are not directly connected in the SP supertree T by one divergence edge. To do so, we can consider a phylogenetic distance between ortholog trees as an additional constraint, and then consider the minimum evolution principle [12,19] employed to reconstruct the connectivity tree $C_T(\{\Gamma_i, 1 \leq i \leq k\})$ based on a distance matrix. It remains to define the distance measure between ortholog trees.

The pairwise similarity scores between transcripts of two ortholog groups can be used to compute a distance measure between the corresponding ortholog trees. This distance measure can then be viewed as a proxy for estimating the number of divergence edges and, consequently, the phylogenetic distance separating two ortholog trees. The distance is based on the transcript similarity measure between two transcripts a and b, denoted $tsm(a, b)$, and introduced in [16].

Definition 5 (Phylogenetic distance between ortholog trees). *Given the set $\Gamma = \{\Gamma_i, 1 \leq i \leq k\}$ of ortholog trees, three distance measures between two trees Γ_i and Γ_j are considered:*

- $dist_{min}(\Gamma_i, \Gamma_j) = dist_{avg}(\Gamma_i, \Gamma_j) = dist_{max}(\Gamma_i, \Gamma_j) = 0$ if $i = j$;
  Otherwise:
- $dist_{min}(\Gamma_i, \Gamma_j) = min(\{1 - tsm(t_i, t_j) : (t_i, t_j) \in \Gamma_i \times \Gamma_j\})$;
- $dist_{avg}(\Gamma_i, \Gamma_j) = mean(\{1 - tsm(t_i, t_j) : (t_i, t_j) \in \Gamma_i \times \Gamma_j\})$;
- $dist_{max}(\Gamma_i, \Gamma_j) = max(\{1 - tsm(t_i, t_j) : (t_i, t_j) \in \Gamma_i \times \Gamma_j\})$;

The Minimum Evolution principle [12,19] provides solutions for reconstructing phylogenetic trees when given a matrix of pairwise distances between taxa. Given a distance matrix D for a set $\Gamma = \Gamma_i, 1 \leq i \leq k$, the Minimum Evolution Problem seeks a tree P for Γ with non-negative edge weights that has the minimum sum of edge weights. Additionally, the sum of the weights of the edges belonging to the path between any two leaves Γ_i and Γ_j in P should be greater than or equal to $D(\Gamma_i, \Gamma_j)$.

The Balanced Minimum Evolution (BME) problem [9,17] is a restriction of the Minimum Evolution Problem with a particular branch length estimation model. Namely the BME problem seeks a tree P for Γ that minimizes the function $\sum_{1 \leq i < j \leq k} \frac{D(\Gamma_i, \Gamma_j)}{2^{\tau_P(\Gamma_i, \Gamma_j)}}$. For this problem, the optimal tree is guaranteed to be statistically consistent and characterized by non-negative edge weights if the distance matrix D satisfies the triangle inequality [8].

Since we seek a connectivity tree (Definition 3) that is live tree for the set of orthologs trees, we have extended the BME problem to live phylogenies as follows [22]:

Live Balanced Minimum Evolution (Live BME) Problem
Input. a set $\Gamma = \{\Gamma_i, 1 \leq i \leq k\}$, a symmetric distance matrix for the set Γ.
Output. A live tree P for Γ that minimizes: $\sum_{1 \leq i < j \leq k} \frac{D(\Gamma_i, \Gamma_j)}{2^{\tau_P(\Gamma_i, \Gamma_j)}}$.

In a live tree P for the set of ortholog trees $\Gamma = \Gamma_i, 1 \leq i \leq k$, the internal nodes in $v(P) \setminus \Gamma$ correspond to ancestral ortholog groups whose copies have all been lost in present-day genes. P should then be compatible with the connectivity tree of the set Γ inferred from the true transcript tree T, namely $C_T(\Gamma)$ should correspond to the minor tree $P[\Gamma]$.

We are now ready to introduce the problems considered to reconstruct a transcript tree T given a set of ortholog trees $\Gamma = \{\Gamma_i, 1 \leq i \leq k\}$. We seek an SP supertree T of Γ such that the connectivity tree $C_T(\Gamma)$ is a minor tree of a solution P to the LIVE BME problem given a symmetric distance matrix D for the set Γ. In addition, among all such SP supertrees, T should be of minimum LCA-reconciliation cost with the gene tree G. We present the *Minimum Evolution-Reconciliation SuperTree Problem* (MINEVOLREC) and the labelled version of the problem called *Labelled Minimum Evolution-Reconciliation SuperTree Problem* (L-MINEVOLREC).

MinEvolRec Problem
Input. a set of transcripts $\mathbb{T}$, a consistent set of ortholog trees $\Gamma = \{\Gamma_i, 1 \leq i \leq k\}$ for ortholog groups of $\mathbb{T}$, a symmetric distance matrix D for the set Γ, and a gene tree G for $\mathbb{G}$.
Output. Among all SP supertrees T of Γ such that $C_T(\Gamma)$ is a minor tree $P[\Gamma]$ of a solution P to the LIVE BME problem given D, one of minimum LCA-reconciliation cost with respect to G.

The ortholog trees reflect the evolutionary history of the set of transcripts composing the ortholog groups. The LCA-reconciliation of these ortholog trees with the gene tree defines the relations between any two transcripts of the same ortholog tree as isoorthologs or recent paralogs. Therefore, the labeling of the nodes of the input ortholog trees by the LCA-reconciliation can be considered an additional constraint to the problem.

Definition 6 (Label-compatibility of a supertree). *Given a supertree T of a set of ortholog trees $\Gamma = \{\Gamma_i, 1 \leq i \leq k\}$, T is label-compatible with Γ if for any internal node x' of any tree $\Gamma_i \in \Gamma$ and x of T such that $x = lca_G[l(x')]$, $rec_{\Gamma_i}(x') = rec_T(x)$.*

L-MinEvolRec Problem

Input. a set of transcripts $\mathbb{T}$, a consistent set of ortholog trees $\Gamma = \{\Gamma_i, 1 \leq i \leq k\}$ for ortholog groups of $\mathbb{T}$, a symmetric distance matrix D for the set Γ, and a gene tree G for $\mathbb{G}$.

Output. Among all SP supertrees T of Γ, label-compatible with Γ, and such that $C_T(\Gamma)$ is a minor tree $P[\Gamma]$ of a solution P to the LIVE BME problem given D, one of minimum LCA-reconciliation cost with respect to G.

Lemma 2 (MinEvolRec and L-MinEvolRec complexity). *The* MIN-EVOLREC *problem and the* L-MINEVOLREC *are both NP-hard.*

Proof. By reduction from the LIVE BME problem which is NP-hard [3].   □

3.1   Algorithm for the MINEVOLREC Problem

In this section, we describe a progressive constructive heuristic for the MINEVOL-REC problem. Lemma 1 suggests a heuristic to compute T given $\Gamma = \{\Gamma_i, 1 \leq i \leq k\}$ by using an iterative process of $k - 1$ steps, each consisting of joining two trees through a subtree grafting operation, until a single transcript tree is obtained. Algorithm 1 summarizes this process, but without the details of each step in the algorithm.

In order to reconstruct the transcript tree T from the set of ortholog trees $\Gamma = \{\Gamma_i, 1 \leq i \leq k\}$, it remains to determine (1) the order in which the trees should be joined such that $C_T(\Gamma)$ is a minor tree of a solution to the Live BME problem given D, i.e. which two trees should be selected at each step to be assembled, and (2) which subtree grafting operation should be performed at each step such that the LCA-reconciliation is minimized, i.e. which tree should be attached onto the other and on which edge.

Regarding condition (1), the Neighbor-Joining (NJ) algorithm [20] is a well-known constructive greedy heuristic for the BME problem [9]. It allows for unequal rates of evolution and guarantees the computation of the minimum evolution tree if the pairwise distance matrix is additive. In [3], it was shown that the NJ algorithm is also a heuristic for the LIVE BME, which computes the minimum evolution live tree with zero-length edges when an additive matrix is given. Here, we construct a tree for the set Γ of ortholog trees using the NJ algorithm given the distance matrix D for Γ. This tree serves as a guide to determine the order of tree joining in Algorithm 1 for assembling the ortholog trees into a transcript tree.

Algorithm 1: Construction of the transcript tree

Data: $\Gamma = \{\Gamma_i, 1 \le i \le k\}$, the set of ortholog trees
      D, a symmetric distance matrix for the set Γ
Result: T, a transcript tree for $\mathbb{T}$
while $|\Gamma| > 1$ **do**
   (1) Select two trees T_1, T_2 in Γ based on D;
   (2) Assemble T_1, T_2 into one tree T_{12} by attaching one tree to an edge of the other tree;
   (3) Remove T_1, T_2 from Γ and add T_{12};
end
$T \leftarrow$ The final tree in Γ;
return T;

Regarding condition (2) which regards the process to assemble a pair of trees at each step of Algorithm 1, the minimum reconciliation cost principle is used. Given two input subtrees T_1 and T_2, we used a dynamic programming algorithm to compute a minimum reconciliation cost SP supertree of T_1 and T_2. In order to describe this algorithm, we first formulate the LCA-reconciliation cost of a transcript tree T and a gene tree G as a sum of local reconciliation costs defined at internal nodes of T.

The following Lemma extends a result from [15] (Lemma 1 in the reference) on gene tree - species tree reconciliation.

Lemma 3 (LCA-reconciliation cost formulated as a sum of local costs).
The LCA-reconciliation cost, denoted $cost(T)$ of T, equals a sum of local LCA-reconciliation costs:

$$cost(T) = \sum_{x \in v(T) \setminus l(T)} cost(x)$$

such that for each internal node x of T, $cost(x) = cost((x_l, x_r))$ denoted the local LCA-reconciliation cost at x defined as follows:

$$cost(x) = \begin{cases} 1, & \text{if } g(x) = g(x_l) = g(x_r) \text{ (1)} \\ 2 + inter(g(x), g(x_l)), & \text{if } g(x)! = g(x_l) \text{ and } g(x) = g(x_r) \text{ (2)} \\ 2 + inter(g(x), g(x_r)), & \text{if } g(x)! = g(x_r) \text{ and } g(x) = g(x_l) \text{ (3)} \\ inter(g(x), g(x_l)) + \\ inter(g(x), g(x_r)) & \text{if } g(x)! = g(x_r) \text{ and } g(x)! = g(x_l) \text{ (4)} \end{cases}$$

Proof. In the first, second, and third cases, the node x is a creation node adding 1 to the LCA-reconciliation cost. In the second and third cases, 1 loss should also be added on one branch right after the creation. In the fourth case, the node x is not a creation node. The remaining contributions are losses on one or both of the branches between the node x and its children. Note that $inter(s, t) = 0$ if $s = t$. □

Lemma 3 provides the means to recursively compute a minimum LCA-reconciliation cost SP supertree of two trees T_1 and T_2. The SP supertree can be progressively constructed by exploring, for each of its nodes x from the root to the leaves, all valid bipartitions of $l(x)$. The valid bipartitions of $l(x)$ are described hereafter and directly follow from the definition of an SP supertree.

Property 2 (Valid bipartitions at the root of an SP supertree of two trees). Let T_1 and T_2 be two subtrees for two disjoint subsets of $\mathbb{T}$. Suppose T is an SP supertree of T_1 and T_2, which can be obtained by grafting T_1 onto an edge of T_2, without loss of generality. We have $l(T) = l(T_1) \cup l(T_2)$, and the root of T subdivides its set of leaves, $l(T)$, into a valid bipartition (L_l, L_r) such that:

1. $L_l = l(T_1)$ and $L_r = l(T_2)$; or symmetrically, $L_l = l(T_2)$ and $L_r = l(T_1)$; or
2. $L_l = l(T_1) \cup l(T_{2l})$ and $L_r = l(T_{2r})$; or symmetrically, $L_l = l(T_{2r})$ and $L_r = l(T_1) \cup l(T_{2l})$; or
3. $L_l = l(T_1) \cup l(T_{2r})$ and $L_r = l(T_{2l})$; or symmetrically, $L_l = l(T_{2l})$ and $L_r = l(T_1) \cup l(T_{2r})$;

Given two trees, T_1 and T_2, $B(T_1, T_2)$ denotes the set of all valid bipartitions of $l(T) = l(T_1) \cup l(T_2)$ at the root of an SP supertree of T_1 and T_2, which can be obtained by grafting T_1 onto an edge of T_2. S1-Figure in suplementary material illustrates the set $B(T_1, T_2)$.

Lemma 4. *Given two trees T_1, T_2 for two disjoint subsets of $\mathbb{T}$, $|B(T_1, T_2)| = 3$.*

Proof. Trivial from Property 2:

$$B(T_1, T_2) = \{(l(T_1), l(T_2)), (l(T_{2l}) \cup l(T_1), l(T_{2r})), (l(T_{2l}), l(T_{2r}) \cup l(T_1))\}$$

$\square$

The main recurrence formula of our dynamic programming algorithm to compute a minimum reconciliation cost SP supertree of two trees T_1 and T_2 is provided below.

Lemma 5. *Let T_1 and T_2 be two trees for two disjoint subsets of $\mathbb{T}$.*

$$MinEvolRec(T_1, T_2) = \min\{pMinEvolRec(T_1, T_2), pMinEvolRec(T_2, T_1)\}$$

such that:

(1) $pMinEvolRec(T_1, T_2) = 0$ *if* $l(T_1) = \emptyset$ *and* $l(T_2) = \emptyset$ *(Stop condition)*
(2) $pMinEvolRec(T_1, T_2) = cost(T_1)$ *if* $l(T_2) = \emptyset$ *(Stop condition)*
(3) $pMinEvolRec(T_1, T_2) = cost(T_2)$ *if* $l(T_1) = \emptyset$ *(Stop condition)*
(4) Otherwise, $MinEvolRec(T_1, T_2) = \displaystyle\min_{(L_l, L_r) \in B(\{T_1, T_2\})} \left\{ \begin{array}{l} cost(L_l, L_r) + \\ pMinEvolRec(T_{1|l_l}, T_{2|l_l}) + \\ pMinEvolRec(T_{1|l_r}, T_{2|l_r}) \end{array} \right\}$

The proof for Lemma 5 can be found in S1-Proof in Supplementary Material. The time complexity of the heuristic Algorithm 1 to compute $MinEvolRec(\mathbb{T}, \Gamma, D, G)$ is then the following.

Theorem 1 (Complexity). *Algorithm 1, in which line (1) represents a step of the NJ algorithm on the distance matrix D, and line (2) utilizes the dynamic programming algorithm provided in Lemma 5, has a time complexity of $O(k^3 + k \times n)$.*

Proof. The NJ algorithm has a time complexity of $O(k^3)$, where k is the number of input ortholog trees. For each pair of trees T_1 and T_2, computing $pMinEvolRec(T_1, T_2)$, $pMinEvolRec(T_2, T_1)$ and $MinEvolRec(T_1, T_2)$ is achieved in $O(n)$, where n the overall number of transcripts. Therefore, the overall time complexity of Algorithm 1 is $O(k^3) + O(k - 1) \times O(n)$, which is $O(k^3 + k \times n)$.   □

3.2   Algorithm for the L-MinEvolRec Problem

The heuristic algorithm for the MinEvolRec problem can be customized to provide a heuristic for the L-MinEvolRec problem. The main idea is to modify how $pMinEvolRec(T_1, T_2)$ is computed. It follows the same approach as Algorithm 1, except for the computation of $pMinEvolRec(T_1, T_2)$, where $B(T_1, T_2)$ is replaced by $LB(T_1, T_2)$, defined as follows.

Property 3 (Valid bipartitions at the root of a label-compatible SP supertree of two trees). Let T_1 and T_2 be two subtrees for two disjoint subsets of $\mathbb{T}$. Suppose T is an SP supertree of T_1 and T_2, label-compatible with T_1 and T_2, and which can be obtained by grafting T_1 onto an edge of T_2, without loss of generality. We have $l(T) = l(T_1) \cup l(T_2)$, and the root of T subdivides its set of leaves, $l(T)$, into a valid bipartition $(L_l, L_r) \in LB(T_1, T_2)$ such that:

1. $LB(T_1, T_2) = \{(L_l, L_r) \in B(T_1, T_2) \mid l(T_2) \subseteq L_l$ or $l(T_2) \subseteq L_r$ or $rec_{T_2}(r(T_2)) = rec(L_l, L_r)\}$,
   where $rec(L_l, L_r) = Cre$ if $lca_G(L_l \cup L_r) = lca_G(L_l)$ or $lca_G(L_l \cup L_r) = lca_G(L_r)$,
2. otherwise, $rec(L_l, L_r) = Dup$ if $rec_G(lca_G(L_l \cup L_r)) = Dup$, and $rec(L_l, L_r) = Spe$ if $rec_G(lca_G(L_l \cup L_r)) = Spe$.

The dynamic programming algorithm obtained by replacing $B(T_1, T_2)$ by $LB(T_1, T_2)$ in the algorithm $pMinEvolRec(T_1, T_2)$ (Lemma 5) is named $L - pMinEvolRec$. In other words, to compute $L{-}pMinEvolRec(T_1, T_2)$, all bipartitions (L_l, L_r) considered in the case (4) of Lemma 5 should be label-compatible with T_1 and T_2, i.e. if (L_l, L_r) subdivides $l(T_2)$ by placing $l(T_{2l})$ and $l(T_{2r})$ on different sides, then the root of a tree separating L_l and L_r should have the same LCA-reconciliation label as the root of T_2.

The time complexity of the heuristic Algorithm 1 to compute $L - MinEvolRec(\mathbb{T}, \Gamma, D, G)$ follows from Theorem 1.

Lemma 6. *Let T_1 and T_2 be two trees for two disjoint subsets of $\mathbb{T}$.*

$$L - MinEvolRec(T_1, T_2) = \min\{L{-}pMinEvolRec(T_1, T_2), L{-}pMinEvolRec(T_2, T_1)\}$$

such that:

(1) $L-pMinEvolRec(T_1, T_2) = 0$ *if* $l(T_1) = \emptyset$ *and* $l(T_2) = \emptyset$ *(Stop condition)*
(2) $L-pMinEvolRec(T_1, T_2) = cost(T_1)$ *if* $l(T_2) = \emptyset$ *(Stop condition)*
(3) $L-pMinEvolRec(T_1, T_2) = cost(T_2)$ *if* $l(T_1) = \emptyset$ *(Stop condition)*
(4) Otherwise, $L-pMinEvolRec(T_1, T_2) = \displaystyle\min_{(L_l, L_r) \in LB(\{T_1, T_2\})}$

$$\left\{ \begin{array}{l} cost(L_l, L_r) + \\ L - pMinEvolRec(T_{1|l_l}, T_{2|l_l}) + \\ L - pMinEvolRec(T_{1|l_r}, T_{2|l_r}) \end{array} \right\}$$

Proof. As long as all possible valid and label-compatible bipartitions are considered in the algorithm, the output of the algorithm will be a label-compatible SP supertree of T_1 and T_2 of minimum LCA-reconciliation cost.   □

Lemma 7. *The time complexity of the heuristic Algorithm 1 to compute* $L- MinEvolRec(\mathbb{T}, \Gamma, D, G)$ *follows from Theorem 1: Algorithm 1, in which line (1) represents a step of the NJ algorithm on the distance matrix* D, *and line (2) utilizes the dynamic programming algorithm provide in Lemma 6, has a time complexity of* $O(k^3 + k \times n)$.

Proof. Trivial, by following the proof of Theorem 1.   □

4 Results and Discussion

To the best of our knowledge, only two methods, other than the one presented in this paper, have been developed for the inference of transcript phylogenies: TrEvoR [7] and PhyloSofS [2]. These methods rely on a maximum parsimony framework based on a model of transcript evolution that includes three events: mutation (gain or loss of exons), fork or birth (creation of new transcripts), and death (loss of transcripts). Unfortunately, the source code for TrEvoR was no longer available at the time of writing this paper. As for PhyloSofS, the tool requires a specific format for the input data that should be generated using another tool, ThorAxe [24]. Therefore, PhyloSofS could not be included in the experimental comparison based on the simulated data either.

In Sect. 4.1, we provide the results of the comparison of our transcript tree inference method with true transcript trees obtained using SIMSPLICEEVOL [13]. In Sect. 4.2, we discuss the results of the comparison PhyloSofS and our method for two case studies on the transcripts of two real gene families.

4.1 Comparison with Simulated True Transcript Phylogenies

While we could not include TrEvoR and PhyloSofS in the comparison based on the simulated data, we included GIGA [23]. GIGA algorithm constructs gene trees agglomeratively from a distance matrix representation of gene sequences. It uses simple rules to minimize the cost of LCA-reconciliation and joins two subtrees at each step by gene speciation or gene duplication. GIGA can be adapted for transcript tree reconstruction by replacing the input species tree with a gene tree, the input distance matrix representation of gene sequences with a distance matrix of transcripts, and replacing the gene duplication event with a transcript creation event in the algorithm rules to construct the transcript tree agglomeratively.

We used 50 transcript phylogenies simulated with SIMSPLICEEVOL, as described in the Supplementary Material (S1-Table). We compared the 50 true transcript phylogenies to the transcript trees obtained using GIGA and the 6 versions of our method, considering the following variations: (1) the MINEVOLREC or the L_MINEVOLREC algorithm, and (2) the distance measures $dist_{min}$, $dist_{avg}$, or $dist_{max}$ between two ortholog trees. For the assessment, we considered the computation time of the methods, and three normalized distances between the inferred trees and the true transcript phylogenies: the normalized Robinson-Foulds (nRF) distance, the normalized Quartet (nQ) distance and the normalized triplet (nT) distance. Given two rooted trees T_1 and T_2 for a set $\mathbb{T}$, the nRF distance between T_1 and T_2 is defined as $d_{RF}^n(T_1, T_2) = (|C(T_1) \setminus C(T_2)| + |C(T_2) \setminus C(T_1)|)/(|C(T_1)| + |C(T_2)|)$, where, for any tree T, $C(T) = \{l(x)|x \in v(T)\}$. The nQ distance between T_1 and T_2 is defined as $d_Q^n(T_1, T_2) = (|F(T_1) \setminus F(T_2)| + |F(T_2) \setminus F(T_1)|)/(|F(T_1)| + |F(T_2)|)$, where for any tree T, $F(T)$ of size $\binom{|\mathbb{T}|}{4}$ represents the set of quartet topologies for all sets of four different leaves. The nT distance corresponds to the nQ distance when triplet topologies are considered instead of quartet topologies. These distances allow us to quantify the dissimilarity between the inferred trees and the true trees. Therefore, the smaller the distance is, the more accurate the phylogeny inference method is. Unlike our method, GIGA provides transcript trees that can contain polytomies, possibly resulting in unrooted trees if the polytomy is found at the root. To address this issue, we resolve the polytomies found in a given transcript tree by retaining a resolved topology that has the smallest nRF, nQ, or nT distance to the corresponding true tree. Note that among the 50 reconstructed GIGA trees in the dataset, only 11 have polytomies in their topology, and these are located at creation nodes.

Figure 2a provides details on the computational time for each method. It shows that GIGA exhibits higher computational efficiency, with faster processing times than both MINEVOLREC and L-MINEVOLREC. In Fig. 2b, Fig. 2c and Fig. 2d, comparisons are presented between repectively the nRF distance, the nT distance and nQ distance for the six versions of our methods using MACSE [18] sequence alignments or the true alignments, and the nRF distance for GIGA, across the 50 datasets. The results reveal that the two versions of our method, MIN_MINEVOLREC and MIN_L-MINEVOLREC, outperform all the other methods in the comparison. Notably, for these two methods, the average nRF distance of the reconstructed transcript trees to the true trees consistently remains lower than or equal to 0.2. Similarly, the nQ distance for the two methods remains lower than 0.1 (S4-Figure in Supplementary material). However, when using MACSE sequence alignments with MIN_L-MINEVOLREC and MIN_MINEVOLREC, the algorithm shows a slight performance increase compared to the other versions. This is surprising, as we expected to observe superior performance when using the true alignments. This arises because, in some cases,

MACSE's alignments of duplicated exons yield lower distances between recent paralogs and orthologous transcripts than the true alignments, leading to a better prediction of recent paralogs and orthologs. Indeed, MACSE's alignments

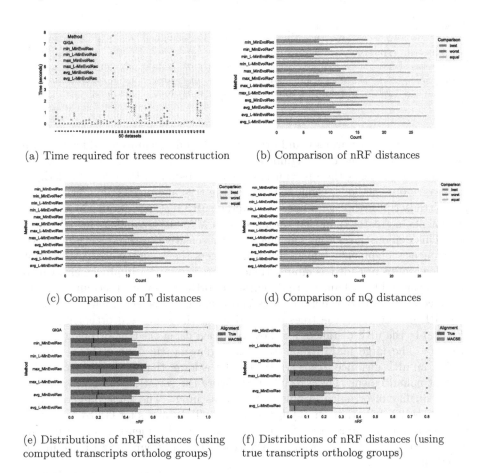

(a) Time required for trees reconstruction

(b) Comparison of nRF distances

(c) Comparison of nT distances

(d) Comparison of nQ distances

(e) Distributions of nRF distances (using computed transcripts ortholog groups)

(f) Distributions of nRF distances (using true transcripts ortholog groups)

Fig. 2. (a) Computing the time required for transcript tree reconstruction using GIGA and the six versions of our method. (b) Based on the nRF distance to the true transcript trees, the number of trees inferred by our method that are the best (lower nRF distance), worst (higher nRF distance), or have equal nRF distances compared to the trees inferred by GIGA. The asterisk (*) indicates the use of the true sequence alignments provided by SimSpliceEvol; otherwise, MACSE alignments are used. (c) Similar comparisons to those in (b) obtained using the nT distance to the true transcript trees. (d) Similar comparisons to those in (b) obtained using the nT distance to the true transcript trees. (e) Distributions of nRF distances between the inferred and the true transcript trees for GIGA and the six versions of our method using the computed ortholog groups [16]. (f) Distributions of nRF distances between the inferred and the true transcript trees for the six versions of our method when using the true ortholog groups as input. GIGA is not included because it does not take ortholog groups as input.

Table 1. Accuracy of pairwise homology relationships inferred by *GIGA* and MIN_MINEVOLREC using MACSE alignments. The asterisk indicates the use of the true transcript ortholog groups provided by SIMSPLICEEVOL; otherwise, they are pre-computed using the method proposed in [16]

Pairwise relations	True	MIN_MINEVOLREC				MIN_MINEVOLREC*				GIGA			
	#	#	Prec	Recall	F1-score	#	Prec	Recall	F1-score	#	Prec	Recall	F1-score
ortho-orthologs	1410	1192	0.878	0.742	0.804	1446	0.946	0.97	0.958	1175	0.872	0.727	0.793
para-orthologs	138	135	0.956	0.935	0.945	138	0.978	0.978	0.978	135	0.948	0.928	0.938
recent paralogs	75	76	0.461	0.467	0.464	79	0.937	0.987	0.961	42	0.667	0.373	0.478
ancient paralogs	2994	3214	0.871	0.936	0.902	2954	0.984	0.971	0.977	3265	0.865	0.943	0.902

predict some recent paralogy and orthology relations accurately, while they are missed when using the true alignments.

Figure 2e presents the distributions of nRF distances for GIGA and the 6 versions of our methods, using MACSE or the true alignments along with computed transcript ortholog groups. The average nRF distance obtained using true alignments is greater than that obtained with MACSE alignments for all methods. This demonstrates that our methods are sensitive to the quality of multiple sequence alignments, which, in turn, impacts the accuracy of the transcript ortholog groups provided as inputs. Hence, there is room for improvement, which can be achieved by enhancing the accuracy of ortholog groups precomputed and provided as input to our method. This improvement is evident when we compare the nRF distances for the trees obtained using the computed ortholog groups (Fig. 2e) to the nRF distances for those obtained using the true ortholog groups (Fig. 2f). The large majority of trees inferred using the true ortholog groups exhibit an nRF distance to the true trees that is lower than 0.1, indicating a significant enhancement. These results underscore the sensitivity of our approach to the quality of ortholog groups presented as input. The same observation is yield by the distributions of nQ distances and nT distances for GIGA and the 6 versions of our methods, using MACSE or the true alignments along with computed transcript ortholog groups or the true ortholog groups. (S4-Figure in the Supplementary material).

We compared the accuracy of the pairwise homology relations obtained from the reconciliation of the transcript trees with their corresponding gene trees. For this comparison, only the best version of our method is included, i.e., MIN_-MINEVOLREC. For MIN_MINEVOLREC and GIGA trees, we employed the LCA-reconciliation method to reconcile the trees with the gene tree. Table 1 presents the number of pairwise relations for each type of homoloy relation, including ortho-orthology, para-orthology, recent and ancient paralogy, over the 50-trees dataset, for SIMSPLICEEVOL (i.e., the true relations), MIN_MINEVOLREC, MIN_MINEVOLREC* (with the true ortholog groups as input), and GIGA. The table, also presents the precision, recall, and F1-score for each type of relation for MIN_MINEVOLREC, MIN_MINEVOLREC*, and GIGA. Precision is the ratio of the number of true relations predicted over the total number of predicted relations, while recall is the ratio of the number of true relations predicted over the total number of true relations for each type of relation. The F1-score is the harmonic mean of precision and recall. We observe that both MIN_MINEVOLREC and GIGA achieve significantly lower F1-scores than MIN_MINEVOLREC* for recent paralog relations. For MIN_MINEVOLREC, This arises primarily because the method tends to identify the best reconciliation cost, leading to the inference of recent paralogs that may not necessarily be true recent paralogs. As for GIGA, the method tends to group transcripts of the same gene under the same lowest common ancestor, and thus to also infer more false negative recent paralogs. On the other hand, both methods, MIN_MINEVOLREC and GIGA, exhibit good precision and recall for ortho-orthology, para-orthology, and ancient paralogy relations. MIN_MINEVOLREC* provides a good compromise between the two other methods and performs particularly well in inferring orthologous and paralogous transcripts, with precision, recall, and F1-scores near 1. This result once again emphasizes that our approach is highly sensitive to the quality of orthologs presented as input.

4.2 Comparison of Transcript Phylogenies of Families Inferred for Two Case Studies on TAF6 and PAX6 Gene Families

We compared the best version of our method (MIN_MINEVOLREC) with PHYLOSOFS and GIGA on two case studies involving the reconstruction of transcript phylogenies for two real gene families: the TAF6 and PAX6 gene families. The detailed results of the comparisons on the PAX6 gene family can be found in S1-Text in the Supplementary Material.

TAF6, also known as Transcription initiation factor TFIID subunit 6, along with other TAFs, is indispensable for assembling the pre-initiation complex and regulating gene transcription in various eukaryotic cellular processes. In this study, we investigate the TAF6 transcript family derived from ThorAxe, which consists of genes having a 1:1 orthology relationship with the human gene ENSG00000106290 across three species: Mouse (ENSMUSG00000036980), Zebrafish (ENSDARG00000102998), and Drosophila melanogaster (FBgn0010417). S2-Table in the Supplementary Material provides the details about these genes which comprise a total of 16 transcripts, distributed as follows: 10 in Human, 2 in Mouse, 3 in Zebrafish, and 1 in Drosophila melanogaster. Figure 3 shows the transcript phylogenies T_1^1, T_1^2 and T_1^3 inferred by PHYLOSOFS using three different configuration of parameter settings: the default setting (`Birth cost = 3, Death cost = 0, and Mutation cost = 2`); the second configuration (`Creation cost = 3, Death cost = 2, and Mutation cost = 2`); and the third configuration (`Creation cost = 3, Death cost = 3, and Mutation cost = 2`). The leaf nodes in these phylogenies represent the structure of transcripts, called isoform structure. If multiple transcripts share an identical structure, they are depicted by a unique leaf node in the phylogenies. For instance, the isoform structures 7#0, 7#1 and 6#0 correspond respectively to 2 ({h1, h2}), 3 ({h4, h5, h6}) and 2 ({m1, m2}) transcripts, while other isoform structures have only 1 transcript. In Fig. 3, T_2^* and T_2 represent the transcript tree reconstructed using MIN_MINEVOLREC, and T_3 the transcript tree reconstructed using GIGA.

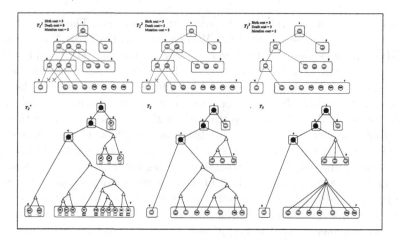

Fig. 3. (Top) The transcript phylogenies T_1^1, T_1^2, and T_1^3 inferred using PHYLOSOFS for the TAF6 family. T_1^1 is obtained using the default cost parameters (Birth cost=0, Death cost=0, and Mutation cost=2), while for T_1^2 and T_1^3, the death cost is increased to 2 and 3, respectively. The 3 trees are embedded in the gene tree where extant and ancestral genes are represented by rectangles. T_1^1 contains three trees: red, blue and green. White nodes represent extant transcripts that are not included in any tree. Any set of multiple recent paralogs is represented as a single leaf node, called an isoform structure (for instance, the isoform structure 7#1, 7#0, and 6#0 correspond respectively to 3, 2, and 2 transcripts). The red cross symbol at the termination of a branch symbolizes a loss, i.e., the death of a transcript. **(Bottom)** The transcript tree T_2^* represent the phylogeny of the TAF6 family inferred using MIN_MINEVOLREC. The inferred ortholog groups used to reconstruct T_2^* are indicated with colored surroundings at the leaves of the phylogeny. The isoform structure of leaves is also indicated below each leaf. Triangle internal nodes represent creation events, while black circle internal nodes represent speciations. T_2 represents the tree T_2^* where the leaves are grouped by isoform structure and T_3 represents the transcript tree of the TAF6 family reconstructed using GIGA. (Color figure online)

In terms of computing time, GIGA is faster than the other two methods, while PHYLOSOFS takes significantly more time.

T_1^1 contains three trees: red, blue, and green. The red tree originates from one ancestral transcript 1#0 at the root of the gene tree gene 1, while the other two, the blue and green trees, originate from two distinct ancestral transcripts, 2#1 and 2#2 respectively, at the gene 2. Twelve extant transcripts out of 16 are included in the three trees, while the remaining four transcripts correspond to orphan transcripts, i.e., not included in any tree. In the case of the two other configurations for PHYLOSOFS, namely T_1^2 and T_1^3, where the cost of transcript death is increased, the consequence is that the green tree disappears in T_1^2 resulting in 5#2 and 7#1 becoming orphan transcripts. In T_1^3, the blue tree also disappears resulting in 5#1 and 7#0 becoming orphan transcripts as well.

An observation of the multiple sequence alignment of TAF6 transcripts (see Fig. 4) shows that the transcripts 5#2 (z3) and 7#1 (h4,h5,h6) have very dis-

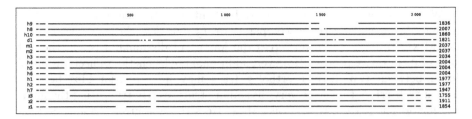

Fig. 4. Multiple sequence alignment of TAF6 transcripts (Figure generated using SeaView [10]).

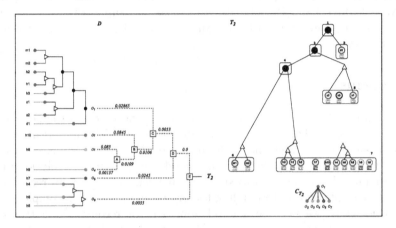

Fig. 5. (Left): The guide tree $\mathcal{D}$ with branch lengths, where the leaves (O_1, O_2, O_3, O_4, O_5, and O_6) represent the ortholog trees. The internal nodes (A, B, C, D, and E) represent the merging steps involving subtree grafting operations. **(Right)** The transcript phylogenies T_2 of the TAF6 family obtained after truncating $\mathcal{D}$. The connectivity tree C_{T_2} is also depicted.

similar structures. Similarly, the transcripts **5#1** (z1) and **7#0** (h1,h2) also have some how dissimilar structures. Thus, instead of having **5#2** and **7#1**, as well as **5#1** and **7#0** in the same tree in T_1^1, it would be more accurate to have them separated as in T_1^3.

T_2 contains 12 creation nodes, while $T3$ contains 2 creation nodes with polytomies in T_3. These polytomies indicate a difficulty of the GIGA method in resolving evolutionary relationships among the offspring of transcript creation events.

MIN_MINEVOLREC and PHYLOSOFS produce similar phylogenies when PHYLOSOFS is used with a parameter setting that penalizes transcript losses. The difference between the results of PHYLOSOFS and MIN_MINEVOLREC arises when PHYLOSOFS is used with default cost parameters (T_1^1) assigning a null cost to transcript losses, while MIN_MINEVOLREC considers a unitary cost for transcript losses. In this setting, PHYLOSOFS phylogenies tend to exhibit more losses.

We also observe that MIN_MINEVOLREC infers additional phylogenetic connections between transcripts compared to T_2^1 and T_3^1. This is because the MIN_MINEVOLREC algorithm forces the joining of dissimilar trees in the last steps of its iterative process. However, additional phylogenetic connections introduced between two distinct ortholog trees may infer relationships between orphan transcripts that have very dissimilar strutures. By truncating the guide tree for orthologs trees to join only those trees below a given distance threshold, we observe that MIN_MINEVOLREC and PHYLOSOFS provide similar phylogenies (see Fig. 5). More precisely, by truncating the guide tree, the results of MIN_MINEVOLREC align more closely with T_1^3. In summary, the results of MIN_MINEVOLREC and PHYLOSOFS are consistent, supporting their accuracy in the absence of ground truth, but MIN_MINEVOLREC is significantly faster than PHYLOSOFS.

5   Conclusion

Understanding transcript evolution is crucial for the functional annotation of genes and genomes. The study of transcript evolution is also important in phylogenetic analyses to infer the evolutionary relationships between genes and species, since transcript evolution is intricately related to the evolution of the genes from which they are produced. This is particularly crucial for reconstructing the tree of life and understanding the evolution of biological complexity.

In this study, we propose a novel approach for reconstructing transcript phylogenies based on given precomputed groups of orthologous transcripts. We present two methods, MINEVOLREC and L-MINEVOLREC, to combine a set of transcript ortholog trees into a supertree (or superforest) of transcripts. Given a set of transcript ortholog groups, the method relies on a guide tree defined for the set of ortholog groups to iteratively performs pairwise supertree construction in the internal nodes of the guide tree to contruct, at each iteration, a supertree having a minimum cost reconciliation with the gene tree. On simulated data, the MINEVOLREC method demonstrates the highest precision. Its application to the TAF6 and PAX6 gene families results in an effective prediction of the phylogenetic relationships between transcripts, corroborating the findings of previous studies, with fast computing times.

Future work will involve applying the method on a larger scale to multiple gene families and relating the results to known biological knowledge about transcript isoform functions. We will also extend the method to allow the prediction of unannotated transcript isoforms. Additionally, in this work, we assume that the gene trees are correct, while, in truth, they may contain errors. We will explore the possibility of using transcript homology information to correct gene trees.

References

1. Adami, C., Ofria, C., Collier, T.C.: Evolution of biological complexity. Proc. Natl. Acad. Sci. **97**(9), 4463–4468 (2000)
2. Ait-Hamlat, A., Zea, D.J., Labeeuw, A., Polit, L., Richard, H., Laine, E.: Transcripts' evolutionary history and structural dynamics give mechanistic insights into the functional diversity of the jnk family. J. Mol. Biol. **432**(7), 2121–2140 (2020)
3. Araújo, G.S., Telles, G.P., Walter, M.E.M., Almeida, N.F.: Distance-based live phylogeny. In: International Conference on Bioinformatics Models, Methods and Algorithms, vol. 4, pp. 196–201. SCITEPRESS (2017)
4. Black, D.L.: Protein diversity from alternative splicing: a challenge for bioinformatics and post-genome biology. Cell **103**(3), 367–370 (2000)
5. Chauve, C., El-Mabrouk, N.: New perspectives on gene family evolution: losses in reconciliation and a link with supertrees. In: Batzoglou, S. (ed.) RECOMB 2009. LNCS, vol. 5541, pp. 46–58. Springer, Heidelberg (2009). https://doi.org/10.1007/978-3-642-02008-7_4
6. Christinat, Y., Moret, B.M.E.: Inferring transcript phylogenies. BMC Bioinformatics **13**(9), S1 (2012)
7. Christinat, Y., Moret, B.M.E.: A transcript perspective on evolution. IEEE/ACM Trans. Comput. Biol. Bioinf. **10**(6), 1403–1411 (2013)
8. Desper, R., Gascuel, O.: Theoretical foundation of the balanced minimum evolution method of phylogenetic inference and its relationship to weighted least-squares tree fitting. Mol. Biol. Evol. **21**(3), 587–598 (2004)
9. Gascuel, O.: Mathematics of Evolution and Phylogeny. OUP Oxford, Oxford (2005)
10. Gouy, M., Tannier, E., Comte, N., Parsons, D.P.: Seaview version 5: a multiplatform software for multiple sequence alignment, molecular phylogenetic analyses, and tree reconciliation. In: Katoh, K. (ed.) Multiple Sequence Alignment. MMB, vol. 2231, pp. 241–260. Springer, New York (2021). https://doi.org/10.1007/978-1-0716-1036-7_15
11. Harrow, J., et al.: Gencode: the reference human genome annotation for the encode project. Genome Res. **22**(9), 1760–1774 (2012)
12. Kidd, K.K., Sgaramella-Zonta, L.A.: Phylogenetic analysis: concepts and methods. Am. J. Hum. Genet. **23**(3), 235 (1971)
13. Kuitche, E., Jammali, S., Ouangraoua, A.: Simspliceevol: alternative splicing-aware simulation of biological sequence evolution. BMC Bioinformatics **20**(20), 640 (2019)
14. Kuitche, E., Lafond, M., Ouangraoua, A.: Reconstructing protein and gene phylogenies using reconciliation and soft-clustering. J. Bioinform. Comput. Biol. **15**(06), 1740007 (2017)
15. Lafond, M., Chauve, C., El-Mabrouk, N., Ouangraoua, A.: Gene tree construction and correction using supertree and reconciliation. IEEE/ACM Trans. Comput. Biol. Bioinf. **15**(5), 1560–1570 (2017)
16. Ouedraogo, W.Y.D.D., Ouangraoua, A.: Inferring clusters of orthologous and paralogous transcripts. In: Jahn, K., Vinar, T. (eds.) Comparative Genomics. RECOMB-CG 2023. LNCS, vol. 13883, pp. 19–34. Springer, Cham (2023). https://doi.org/10.1007/978-3-031-36911-7_2
17. Pauplin, Y.: Direct calculation of a tree length using a distance matrix. J. Mol. Evol. **51**, 41–47 (2000)
18. Ranwez, V., Douzery, E.J., Cambon, C., Chantret, N., Delsuc, F.: MACSE V2: toolkit for the alignment of coding sequences accounting for frameshifts and stop codons. Mol. Biol. Evol. **35**(10), 2582–2584 (2018)

19. Rzhetsky, A., Nei, M.: Theoretical foundation of the minimum-evolution method of phylogenetic inference. Mol. Biol. Evol. **10**(5), 1073–1095 (1993)
20. Saitou, N., Nei, M.: The neighbor-joining method: a new method for reconstructing phylogenetic trees. Mol. Biol. Evol. **4**(4), 406–425 (1987)
21. Szollosi, G.J., Tannier, E., Daubin, V., Boussau, B.: The inference of gene trees with species trees. Syst. Biol. **64**(1), e42–e62 (2015)
22. Telles, G.P., Almeida, N.F., Minghim, R., Walter, M.E.M.: Live phylogeny. J. Comput. Biol. **20**(1), 30–37 (2013)
23. Thomas, P.D.: GIGA: a simple, efficient algorithm for gene tree inference in the genomic age. BMC Bioinformatics **11**(1), 1–19 (2010)
24. Zea, D.J., Laskina, S., Baudin, A., Richard, H., Laine, E.: Assessing conservation of alternative splicing with evolutionary splicing graphs. Genome Res. **31**(8), 1462–1473 (2021)
25. Zerbino, D.R., et al.: Ensembl 2018. Nucl. Acids Res. **46**(D1), D754–D761 (2018)

Gene-Adjacency-Based Phylogenetics Under a Stochastic Gain-Loss Model

Yoav Dvir[1,2]([✉]), Shelly Brezner[1]([✉]), and Sagi Snir[1]([✉])

[1] Department of Evolutionary and Environmental Biology, University of Haifa, Haifa, Israel
shellybrzner@gmail.com, ssagi@research.haifa.ac.il
[2] Tel-Hai Academic College, Tel Hai, Israel
yoavdvir@sci.haifa.ac.il

Abstract. A key task in molecular systematics is to decipher the evolutionary history of strains of a species. Standard markers are often too crude in this fine systematic resolution to provide a phylogenetic signal. However, among prokaryotes, events in genome dynamics (GD) such as gene gain in horizontal gene transfer (HGT) between organisms and gene loss seem to provide a quite sensitive signal. The synteny index (SI) marker captures differences between a pair of genomes in terms of both gene order and gene content. Recently, it was shown to be consistent under the Jump model, a simple model of GD where the only operation is a gene jump.

In this work, we extend the Jump model to a richer model, allowing for gene gain/loss events, the most prevalent GD events in prokaryotic evolution. Despite the increased model complexity, our new representation yields a significant reduction in the number of variables, leading to a simple equation to estimate the model parameter and, consequently, the consistency of the phylogenetic reconstruction. Additionally, with a more straightforward representation, we can easily calculate the asymptotic variance of the parameter estimation, allowing us to obtain a bound for the expected error. We tested the new model and its associated reconstruction approach on actual and simulated data, where the theoretical asymptotic assumptions do not hold. Our simulation results show a very high accuracy under short evolutionary distances. Applying the method to several families in the ATGC database resulted in relative agreement with other reconstruction approaches based on other signals. The code is on GitHub under the link: https://github.com/shellybre/indels_project.

Keywords: Prokaryotic Genome Dynamics · Phylogenetics · Markovian Processes · Birth-Death Theory · Statistical Consistency

1 Introduction

The unprecedented increase in the quality and quantity of publicly available genomic sequences creates opportunities to advance the long-standing quest for evolution: accurate tracking of delicate relationships at different levels of

© The Author(s), under exclusive license to Springer Nature Switzerland AG 2024
C. Scornavacca and M. Hernández-Rosales (Eds.): RECOMB-CG 2024, LNBI 14616, pp. 69–85, 2024.
https://doi.org/10.1007/978-3-031-58072-7_4

divergence from the deepest branches in the history of life to the shallowest branches that lead to strains within a species [8]. However, despite these technological achievements, the latter task, denoted *phylogeny reconstruction*, remains methodologically challenging [16]. One of the main obstacles to accurate phylogenetic analysis is horizontal gene transfer (HGT), the transfer of genetic data through different paths from vertical inheritance [9,18,27]. In prokaryotes, HGT is widespread, links distant lineages in the Tree of Life, and effectively turns it into a *network of life* [23,30,45]. However, since most of the genetic code is inherited from the parent to the child, it is widely accepted that even among prokaryotes the signal of vertical tree-like evolution is discernible [1,29,50].

As a result, the sought-after phylogenetically sensitive marker provides a clear and reliable signal, keeping systematists searching and devising alternative footprints and techniques to support and enrich existing knowledge. Standard phylogenetic markers, usually in the form of ribosomal genes, often do not provide a sufficiently sensitive signal. It became apparent that within prokaryotes, continuous gene mobility or genome dynamics (GD) can be harnessed to achieve this goal [19,20,28,33]. Under the umbrella of GD, the most dominant operations are mainly duplication, transfer, and loss (DTL) [4,10,25,39,43]. All these GD operations deteriorate genome synteny, that is, gene order conservation within a genome, during evolution. Phylogenetics under GD is divided into *order-based* [13,32,49] where the phylogenetic marker is gene order (or loss of) along the chromosome and *content-based* [12,38] where the order is ignored and the signal is drawn from the symmetric difference between the two gene sets.

Most modern phylogenetic approaches, or systematics in general, assume a stochastic model under which events and their probabilities are defined [11]. This framework allows one to associate a given observation, or simply data, with the likelihood of a series of events yielding this observation. Likelihood-based models were devised for GD events such as genome rearrangement [5,32,34,40], and gene gain/loss [15,42,48]. To the best of our knowledge, none of the above models deals with HGT.

A signal related to GD that captures both gene order and content is the *synteny index* (SI), suggested in [2,37]. In SI, a "neighborhood" is defined around every gene in a genome. The respected neighborhoods are compared for two orthologues, genes with common ancestry under speciation (as opposed to duplication or transfer). Although this approach has produced decent results in simulations and real data analysis [36,37], no rigorous analysis was provided. In [35], a first stochastic model was suggested, *the Jump* model that contains a single operation, in which a random gene "jumps" to a random location, a "slot" between two other genes in the genome. The jumps occur in poison rates. Subsequently, it was shown that SI is *consistent* and *additive/* in the Jump model, which means that under an infinite amount of information, SI reconstructs the correct edge lengths, allowing the reconstruction of the correct tree [17]. Despite the significant step in introducing the jump model as the first stochastic model for HGT, several limitations remain associated with this approach. First, the jump operation is not a prevalent event in microbial evolution. It also forces genomes to contain identical gene sets, an unrealistic situation, mainly when distantly related organisms are analyzed [46]. Next, the complexity of analyzing neighborhood genes implies complicated equations that prevent efficient calculations.

This work introduces a novel model that includes the two most common GD events, gene gain and loss [24, 44]. Under this innovation, it is possible to infer accurate rates of events during evolution. Consequently, we do not need to devise heuristics to analyze genomes over unequal gene sets; instead, we must solve the induced equations analytically. Furthermore, we also formulate novel statistics, conserved gene adjacency (CGA), that replace the gene neighborhood k used in the SI approach. We develop estimators for gene rates. Under the realistic assumption that both gain/loss rates are equal, we construct an estimator for the rate obtained as a solution for a nonlinear equation. We show that it is consistent and asymptotically linear and derive its variance from an error of $O(n_0^{-2})$, where n_0 is the initial length of the genome. Therefore, CGA eliminates complex expressions rendered by SI, leading to the analytical variance calculation. However, under the molecular clock assumption of rate constancy [51], we can derive and solve analytic expressions for the number of events using symmetry arguments.

We applied our method in software to simulated data and real data from the ATGC database. On several ATGC data sets representing various families, we observed a strong signal among closely related species that fades rapidly for moderate distances. Beyond resemblance to existing methods, the new model supplements existing models by providing competing explanations for observable data.

2   Preliminaries

In this section, we define the basic notions used in this work. We consider a model of the *genome* that evolves. The genome at time t is indicated by $\mathcal{G}(t)$. The genome at time t is considered a sequence of n_t different *genes*, $\mathcal{G}(t) = (g_1^t, g_2^t, ..., g_{n_t}^t)$, where g_i^t is the gene located in the $i - th$ location of the genome at time t. All genes come from an infinite *gene pool* $\mathcal{M} = \{m_1, m_2, ...\}$. Our starting genome at time $(t = 0)$ is $\mathcal{G}(0) = (g_1^0, g_2^0, ..., g_{n_0}^0) = (m_1, m_2, ..., m_{n_0})$ where n_0 is the number of genes in $\mathcal{G}(0)$.

Our model consists of two operations: a gene *gain* and a gene *loss*. In the event of a gain, a new gene is inserted into the genome at one of its ends or in a *slot* between two consecutive genes. The genes added to the genome come from $\mathcal{M}$. Thus, in the k-th gain event, the new gene added is denoted by m_{n_0+k}.

In a loss, a gene is removed from the genome while the other genes keep their order in the sequence. For example, if $\mathcal{G}(0) = (g_1^0, g_2^0, g_3^0, g_4^0, g_5^0) = (m_1, m_2, m_3, m_4, m_5)$ then at time t_1, in a loss event, the genome loses the gene m_2, thus $\mathcal{G}(t_1) = (g_1^{t_1}, g_2^{t_1}, g_3^{t_1}, g_4^{t_1}) = (m_1, m_3, m_4, m_5)$ and later in time t_2, in a gain event, the gene m_6 is added, so $\mathcal{G}(t_2) = (g_1^{t_2}, g_2^{t_2}, g_3^{t_2}, g_4^{t_2}, g_5^{t_2}) = (m_1, m_3, m_6, m_4, m_5)$.

We now introduce a stochastic framework to the model. Under this framework, gain and loss events proceed by a Poisson process with a characteristic rate. Specifically, at every time point t, each of the n_t genes in $\mathcal{G}(t)$ undergoes a loss process with rate μ. Similarly, each of the $n_t + 1$ slots undergoes a gain process with the event rate λ.

3   Gene Neighborhood Probabilities

We provide a basic definition that is central to this work. We say that a gene g in a genome $\mathcal{G}$ is at *gene distance* (or simply *distance*) d from another gene g' also in $\mathcal{G}$, or formally $d_\mathcal{G}(g, g') = d$, if there are d genes between them along $\mathcal{G}$. In particular, g and g' are at a distance zero iff they are adjacent.

Now, since our model is uniform across all genes in a genome, we can define a general probability space for any two genes.

Definition 1. *For time $t > 0$, any two genes g and g', and any integers i, j*

$$P_{i,j}(t) = \Pr\left[d_{\mathcal{G}(t)}(g, g') = j \mid d_{\mathcal{G}(0)}(g, g') = i\right]. \tag{1}$$

In other words, $P_{i,j}(t)$ denotes the probability that genes at a distance i at time zero will be at a distance j at time t. Equivalently, this is the probability that for any g_u^0, g_{u+i+1}^0 such that $1 \leq u, u + i + 1 \leq n_0$, there will be g_v^t, g_{v+j+1}^t such that $g_u^0 = g_v^t$ and $g_{u+i+1}^0 = g_{v+j+1}^t$. It is possible to obtain an expression for $P_{i,j}(t)$ in terms of λ, μ, and t.

We note that for two genes with a distance equal to i at time zero, $P_{i,j}(t)$ is the joint probability that these genes survive during time t, which equals $e^{-2\mu t}$, and the probability $p_{i,j}(t)$ that given that they survived, the distance between the same two genes will be j at time t. Then,

$$P_{i,j}(t) = e^{-2\mu t} p_{i,j}(t). \tag{2}$$

The idea of describing $p_{i,j}(t)$ as a Markov chain appeared in [35]. This is a birth-death-immigration model ([3], Chap. 3 page 96) describing a dynamic population; in our case, the population of genes that lies between the genes g_u, g_{u+i+1} at time 0. This population undergoes events of two types between time 0 and time t. Events of "birth" when a gene is added to the genome via HGT, or events of "death" when a gene is lost (see Fig. 1). Let the random variable $X_t, 0 \leq t$ describe this birth-and-death process where X_t holds the population size (gene distance) with the state space $0, 1, \ldots$.

Now, for $X_t = n$, there are $n+1$ places (slots) between the gene pair in which a gene could be added (see again Fig. 1(L)); thus, the birth rate is $\lambda_n = (n+1)\lambda$. Similarly, n genes between the gene pair are subjected to a loss event; thus, the death rate is $\mu_n = n\mu$. In a basic setting of a birth-death process, $\lambda_n = n\lambda$, and $\mu_n = n\mu$ in which λ_n is smaller by λ compared to our case. This general and well-known model is analyzed and solved in ([3], Chap. 3 page 108) Our case could be treated as a birth-and-death process with an "immigration" rate denoted by α, which is a constant independent of the size of the population that is added to the birth rate. Then $\lambda_n = n\lambda + \alpha$. The specific case described above

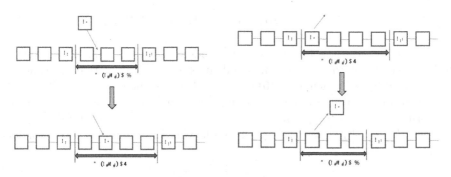

Fig. 1. Gain/Loss events: (L) A genome undergoes a gain event by receiving gene g_h, increasing $d(g_l, g_{\ell'})$ from 3 to 4. (R) A genome undergoes a loss event by losing the gene g_h, decreasing $d(g_l, g_{\ell'})$ from 4 to 3.

is solved by setting $\alpha = \lambda$. Note that this "immigration" does not refer to gene immigration. This is a common name for the parameter α that originates from a different model application. Here, the "births" simulate the gene immigration.

This work only deals with the simplest case where $\lambda = \mu$. This assumption fits the cases where the sizes of the genomes in question are similar. The instances where $\lambda \neq \mu$ will be dealt with separately in a different paper.

The case is relevant to our model assumptions, and the resulting transition probability is presented in the following proposition.

Proposition 1. *([3], Ch. 3, Eq. 2.38–2.39)*
    If $0 < \lambda = \mu$,

$$p_{i,j} = (\lambda t)^{i+j} \frac{(\lambda t)^{i+j}}{(1 + \lambda t)^{i+j+(\alpha/\lambda)}} \sum_{k=0}^{i \wedge j} \frac{i!}{k!(i-k)!} (-1)^k \left(1 - \frac{1}{\lambda^2 t^2}\right)^k \frac{(i + (\alpha/\lambda))_{j-k}}{(j-k)!}.$$

$$(3)$$

About the notations in the above equations:

1. *The notation* $i \wedge j$ *means the minimum between* i *and* j.
2. *If* $j > k$, *then* $(i+(\alpha/\lambda))_{j-k} = (i+(\alpha/\lambda))(i+(\alpha/\lambda)+1)...(i+(\alpha/\lambda)+j-k-1)$.

We can now state our primary results on gene adjacency based on Proposition 1. We obtain the following lemma from Eqs. 3.

Lemma 1. *Assume that a genome evolves* $t > 0$ *and let* λ *be the gain and loss rates. Then:*

$$P_{0,0}(t) = \frac{e^{-2\lambda t}}{1 + \lambda t}.$$

$$(4)$$

Proof. The equation of Lemma 1 is obtained by substituting in Eq. 3 $i = 0$ and $j = 0$ and applying Eq. (2).

4 Parameter Estimation from Observed Data

In this section, we show how to infer the model parameters from data assumed to be generated by the gain/loss model defined above. Assume that the genome $\mathcal{G}(t)$ evolved from the genome $\mathcal{G}(0)$ over time t. We define $n_{i,j} = n_{i,j}(\mathcal{G}(0), \mathcal{G}(t))$ as the number of pairs of genes with distance i in $\mathcal{G}(0)$ and distance j in $\mathcal{G}(t)$ (or formally $\{(g, g') : d_{\mathcal{G}(0)}(g, g') = i \ \wedge \ d_{\mathcal{G}(t)}(g, g') = j\}$). Therefore, $n_{i,j} = \left|\{(u, v) | g_u^0 = g_v^t, g_{u+i+1}^0 = g_{v+j+1}^t\}\right|$. The following is an unbiased estimator for $P_{i,j}(t)$:

$$\widehat{P}_{i,j}(t) = \frac{n_{i,j}}{n_0 - i - 1}. \tag{5}$$

where n_0 is the length of the genome at time 0. Then we plug $\widehat{P}_{0,0}(t) = \frac{n_{0,0}}{n_0-1}$ into Eq. 4 and obtain

$$\frac{n_{0,0}}{n_0 - 1} = \frac{e^{-2\bar{\lambda}t}}{1 + \bar{\lambda}t}. \tag{6}$$

By solving this equation for $\bar{\lambda}t$ we obtain the estimator we refer as CGA.

The variance (standard deviation squared) is used as a quality measure to compare estimators. We can estimate the variance in simulation or compute it theoretically. We can theoretically obtain the variance with an error of $O(n_0^{-2})$. To do this, we will prove the following proposition.

Proposition 2. *The expectancy and the variance of $n_{0,0}$ are given by*

$$E[n_{0,0}] = (n_0 - 1)\frac{e^{-2\lambda t}}{1 + \lambda t} \tag{7}$$

$$Var(n_{0,0}) = (n_0 - 1)\frac{e^{-2\lambda t}}{1 + \lambda t}\left(1 - \frac{e^{-2\lambda t}}{1 + \lambda t}\right) + 2(n_0 - 2)\frac{e^{-3\lambda t} - e^{-4\lambda t}}{(1 + \lambda t)^2}. \tag{8}$$

Proof. Denote

$$n_{0,0} = \sum_{i=1}^{n_0-1} I_{(i,i+1)},$$

where $I_{(i,i+1)}$ is an indicator random variable equal to 1 if the genes m_i and m_{i+1} are consecutive in $\mathcal{G}(t)$ and equal to 0 otherwise. Now, by its definition, $I_{(i,i+1)}$ attains its true value exactly with probability $P_{0,0}(t)$ (see Definition 1), and by Eq. 6 we have $P_{0,0}(t) = e^{-2\lambda t}/(1 + \lambda t)$. Since the expectation of an indicator variable is its probability, hence by the linearity of expectation,

$$n_{0,0} = \sum_{i=1}^{n_0-1} E[I_{(i,i+1)}]$$

$$= \sum_{i=1}^{n_0-1} \frac{e^{-2\lambda t}}{1 + \lambda t},$$

and Eq. 7 follows.

We also have,

$$Var(n_{0,0}) = Var(\sum_{i=1}^{n_0-1} I_{(i,i+1)})$$

$$= \sum_{i=1}^{n_0-1} Var(I_{(i,i+1)}) + 2\sum_{i<j} Cov(I_{(i,i+1)}, I_{(j,j+1)}).$$

Clearly,

$$Var(I_{(i,i+1)}) = P_{0,0}(t)(1 - P_{0,0}(t))$$

and if $i + 1 < j$, then due to mutual independence

$$Cov(I_{(i,i+1)}, I_{(j,j+1)}) = 0.$$

If $i + 1 = j$ then

$$Cov(I_{(i,i+1)}, I_{(j,j+1)}) = E[I_{(i,i+1)}I_{(i+1,j+2)}] - E[I_{(i,i+1)}]E[I_{(i+1,j+2)}]$$

and

$$E[I_{(i,i+1)}I_{(i+1,i+2)}] = \frac{e^{-3\lambda t}}{(1 + \lambda t)^2}.$$

This completes the proof.

In probability theory, one of the main questions about an estimator is whether it is asymptotically normal. Showing that an estimator is asymptotically normal implies that, for large genome sizes, its distribution could be approximated by a normal distribution. This means that it is consistent and can approximate its confidence intervals. The central limit theorem (CLT) is essential for proving asymptotic normality. It states the conditions for which a mean of i.i.d. variables is an asymptotic average and its asymptotic distribution. Although $\widehat{P}_{0,0}(t) = \frac{I_1 + \cdots + I_{n_0-1}}{n_0-1}$ is indeed a mean, it is not possible to use the basic CLT to prove its asymptotic normality because the random indicator variables of the form $I_{i,i+1}$ are dependent. The following specialized version of the CLT is suitable for our case. For this theorem, the following definitions are needed.

Definition 2. *Let $X_1, X_2, \ldots$ be a sequence of random variables, then,*

1. *If the distribution of the random vector $(X_n, X_{n+1}, \ldots, X_{n+j})$ does not depend on n, the sequence is said to be stationary.*
2. *The sequence is said to be m-dependent if the random vectors $(X_1, \ldots, X_k)$ and $(X_{k+n}, \ldots, X_{k+n+l})$ are independent whenever $n > m$.*

With these definitions, we have the following specialized version of the CLT. We have adopted a version of this theorem that fits the needs of this paper.

Theorem 1. *([6], pages 363–364) Suppose that the sequence $X_1, X_2, \ldots$ is non-negative, stationary and m dependent and that $E[X_n = 0]$ and $E[X^{12}] < \infty$. If $S_n = X_1 + \ldots + X_n$ then*

$$n^{-1}Var[S_n] \to \sigma^2 = E[X_1^2] + 2\sum_{k=1}^{m} E[X_1 X_{1+k}], \tag{9}$$

where the series converges. If $\sigma > 0$, then $n^{1/2}S_n \to_d N(0, \sigma^2)$.

Note that the variance we have calculated in Eq. 8 agrees to the variance implicated by this version of CLT.

Theorem 2. *The estimator $\widehat{P}_{0,0}(t)$ has expectation and variance*

$$E[\widehat{P}_{0,0}(t)] = \frac{e^{-2\lambda t}}{1 + \lambda t} \tag{10}$$

$$Var(\widehat{P}_{0,0}(t)) = \frac{1}{(n_0 - 1)}\left[\frac{e^{-2\lambda t}}{(1 + \lambda t)} + \frac{(2 - \frac{2}{n_0-1})e^{-3\lambda t} - (3 - \frac{2}{n_0-1})e^{-4\lambda t}}{(1 + \lambda t)^2}\right], \tag{11}$$

and it is asymptotically normal, i.e., when $n_0 \to \infty$, then

$$n_0^{1/2}\left(\widehat{P}_{0,0}(t) - \frac{e^{-2\lambda t}}{1 + \lambda t}\right) \to_d N(0, \mathcal{AV}) \tag{12}$$

where the asymptotic variance is

$$\mathcal{AV} = \frac{e^{-2\lambda t}}{(1 + \lambda t)}\left(1 - \frac{e^{-2\lambda t}}{1 + \lambda t}\right) + 2\frac{e^{-3\lambda t} - e^{-4\lambda t}}{(1 + \lambda t)^2} \tag{13}$$

Proof. The sequence $I_{1,2}, I_{2,3}, ..I_{n_0-2,n_0-1)}$ is 1-dependent and the sequence is stable. Therefore, by theorem 1 as n_0 tends to infinity $\widehat{P}_{0,0}(t)$ is asymptotically normal.

Corollary 1. *The variance of the estimator $\overline{\lambda t}$ is,*

$$V(\overline{\lambda t}) = \frac{V(\widehat{P}_{0,0}(t))}{D^2} + O\left(n_0^{-2}\right). \tag{14}$$

Proof. By Eq. 4, $\widehat{P}_{0,0}(t) - P_{0,0}(t) = D(\overline{\lambda t} - \lambda t) + O(\overline{\lambda t} - \lambda t)^2$, where $D = -\frac{e^{-2\lambda t}}{1+\lambda t}\left(2 - \frac{1}{1+\lambda t}\right)$ is the derivative of the right side of Eq. 4 concerning λt.

As a result of the delta method ([22], page 58, Theorem 8.12), the estimator $\overline{\lambda t}$ is also asymptotically normal:

Corollary 2. *The estimator $\overline{\lambda t}$ is asymptotically normal, and*

$$n_0^{1/2}(\overline{\lambda t} - \lambda t) \to_d N\left(0, \frac{\mathcal{AV}}{D^2}\right). \tag{15}$$

The asymptotic normality of the estimators shown in corollary 2 yields the following desired property:

Corollary 3. *The estimators $\widehat{P}_{0,0}(t)$ and $\overline{\lambda t}$ are consistent.*

Proof. See ([22], page 54 Theorem 8.2).

5   Experimental Results

In this section, we provide the results of applying the methods of the previous section to genomic data.

5.1   Simulation over a Single Branch

we have used (assuming $\lambda = 1$), $t = \lambda t = 0.02, 0.03, .., 0.08$, each experiment was repeated 10,000 times. and we presented the results for the CGA estimator defined by Eq. 6. The results are presented in Fig. 2. We see that there is no significant bias. On the right graph, we present sample standard deviations. We also calculated the theoretical approximation to the standard deviation of the CGA method with an error of $O(n_0^{-2})$, and it can be seen that they are almost equal to the sample standard deviation. Because we have shown that the CGA estimator is asymptotically normal, approximately 95% of the CGA method's estimations fall within the two standard deviation margins.

We also compared the running times of the SI and CGA methods. The results are summarized in Table 1. Comparisons were made for edge lengths of 0.02, 0.2, and 2. For each length, we averaged the running time in 100 runs. The CGA has some advantages in run-time. In both methods, each gene in the first genome is searched in the second genome. This part appears in both algorithms and is dominant in running time. Given that the gene is found, there is a difference between the methods. While the CGA needs to check whether it has the same consecutive gene in both genomes, SI counts how many genes appear at the intersection of the k Neihborwoods of the two genomes, which takes more time. It can be seen that when the edge length is 2, the advantage of CGA is reduced. Because fewer genes survive, SI must make fewer comparisons between neighborhoods.

Table 1. Running time comparisons between SI and CGA for edge lengths 0.002,0.02 and 0.2. Mean for 100 runs in seconds.

Edge length	0.02	0.2	2
SI	0.168	0.168	0.158
CGA	0.147	0.148	0.150

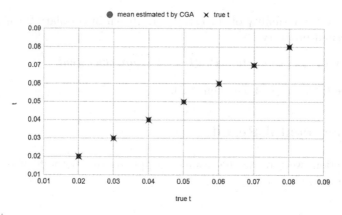

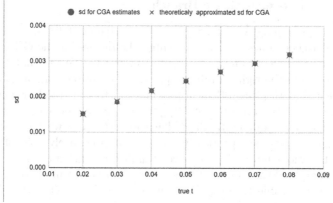

Fig. 2. Simulation Results, Single edge, genome size $n_0 = 5000$. (T): Mean estimated t (B): Standard deviation.

5.2   Simulation over a Tree

We extend the experiment above from a single edge to an entire tree. The test here is different, focusing on tree reconstruction (topology) and not on the model parameters. The procedure is as follows: For a given parameter ℓ, a random tree model is generated with an edge length exponentially distributed with the mean ℓ. A genome evolves from the root genome with 5000 genes down the tree, where for every edge e, we set $\mu = \lambda = \ell_e$ where the same procedure as was taken for the single edge experiment is used. At the end of this procedure, there are "genomes" in the form of ordered lists on the leaves of the tree. For every pair of these genomes, we applied the following methods for measuring the distance:

1. Symmetric CGA- we needed a symmetric estimation of the distance between two genomes, so instead of solving Eq. 6, we solved for $\widehat{\lambda t}$ the equation,

$$\frac{2n_{0,0}}{n_a + n_b - 2} = \frac{e^{-2\widehat{\lambda t}}}{1 + \widehat{\lambda t}}, \tag{16}$$

where n_A and n_B are the two genomes' lengths (numbers of genes).

2. CGC (Corrected Gene Content)- The distance between two genomes is

$$- 2log \left(\frac{2c(A, B)}{min(n_A, n_B)} \right), \tag{17}$$

where $c(A, B)$ are the number of genes common between genomes A and B [47].

3. DCJ-indel distance from UniMoG software [7].

4. SI is adapted to handle unequal genome length. Instead of the genome length, we used the mean lengths of the two genomes.

The reconstructed tree was compared to the original model tree, using the Robinson–Foulds (RF) symmetric difference [31]. In terms of the normalized error rate (incorrect edges) versus the average length of the edges ℓ, the results are shown in Fig. 3 for three intervals of edge length. We see results for trees with very short edges at the top, from 10^{-3} to 10^{-2}. As can be seen, the SI and CGA have similar graphs in all three ranges. They perform well with a tiny error rate from 10^{-3} to 10^{-2} (top) and from 10^{-2} to 10^{-1} (mid). Their error rate increases from 10^{-1} to 1 (bottom), reaching an error rate of 0.5. for edge length 1.

The DCJ starts with a tiny edge length with rates similar to those of SI and CGA; however, its rate starts to rise fast, with an edge length of 0.005, and from there it has the highest error rate compared to the other methods. The CGC begins with a tiny edge length, on the left graph, with a higher error rate compared to the other methods; however, its error rate decreases, and then it is similar to the rates of the SI and CGA on edge lengths 10^{-2} to 10^{-1}. In the range of edge lengths from 10^{-1} to 1, the CGC error rate slowly increases compared to the SI and CGA and reaches an error rate of 0.25 for edge length 1, which is half the error rate of the SI and CGA.

The CGC is based on the probability of gene survival, expressed as $e^{-\lambda t}$, while the CGA and SI rely on information on gene pairs and their probability of survival, which is $e^{-2\lambda t}$. As a result, the data utilized by SI and CGA degrade at a faster rate. This difference highlights the advantage of SI and CGA for short edge lengths. For example, in short lengths, the rapid decrease in the number of consecutive pairs of surviving genes in CGA creates a strong signal that improves the estimate of λt, where in GCG, the number of survived genes decreases slower, hence creating a weaker signal.

It is also worth noting that in scenarios with long connections, there may be instances where there are no common adjacent gene pairs, leading to the application of a heuristic solution $n_{0,0} = 1$ in Eq. (6). Similarly, when the SI index is 0, a heuristic solution SI $= 1$ is adopted. In conclusion, both approaches demonstrate satisfactory performance in phylogenetic reconstruction.

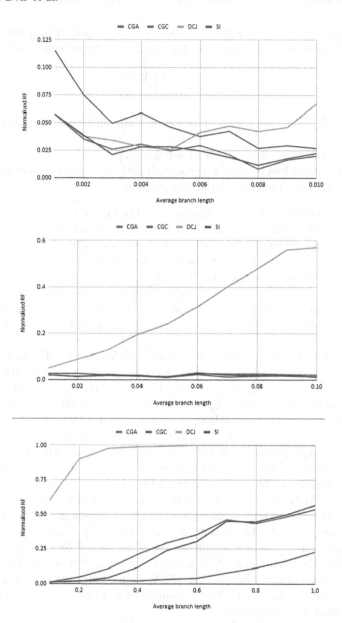

Fig. 3. Simulation Normalized RF Result for tree reconstruction:. Genome
size $n = 5000$ (T): Edge length in the Very Short Edge of 0.001–0.01. (M): Edge length
in the middle range of 0.01–0.1. (B) Edge length in the Long range of 0.1–1

5.3   Real Data Analysis

We report our results applying the CGA approach to real microbial data. The main resource we use is the Alignable Tight Genomic Clusters (ATGC) database [21,26], a collection of closely related bacterial and archaeal genomes. The ATGC has several precomputed features that facilitate gene order analysis, which is mandatory here. In particular, Clusters of Orthologous Genes (COGs) were identified in all ATGCs [41], allowing us to apply our approaches according to gene order. The current version of ATGC contains 410 groups of bacterial and archaeal genomes that jointly encompass approximately 3,700 genomes that encode 12.4 million proteins that are classified into nearly 50,000 COGs. Additionally, each ATGC is accompanied by a mutation-based tree that we compared with our gene order-based approaches.

The first ATGC data set we analyzed was ATGC007 *Lactococcus lactis* in 14 taxa. Figure 4 shows three trees created from this data set. On the left, the original protein-based tree is shown; the CGA-based tree is shown in the middle; and on the right, the SI-based tree. To facilitate visual inspection, we selected the cherry of NZ9000 and MG1363, which occurs in the three trees, as an outgroup.

We measured the RF symmetric difference between the original protein sequence-based tree for the data set and each GD-based tree (SI and CGA). For the two comparisons, we obtained an RF score of eight, which means that the protein-based tree shared seven edges with each of these GD-based trees (since all trees are binary unrooted trees and hence contain eleven internal edges, which implies that an RF score of eight means seven common edges).

We consider this result interesting, as genome dynamics and point mutations are seemingly unrelated processes. However, we observe a significant agreement between the trees (the SI and CGA versus the protein tree).

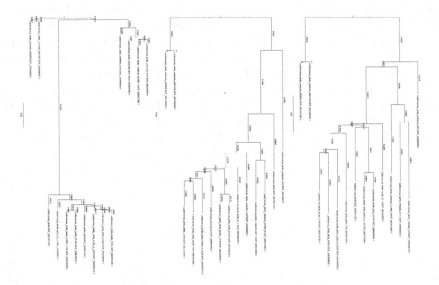

Fig. 4. Real Data Analysis, The ATGC007 *Lactococcus lactis*: Data Set over 14 taxa. (L) The original tree, (M) The CGA-based tree and (R) the SI-based tree.

6   Conclusions

This work presents a novel stochastic model for the most prevalent genome dynamics (GD) operations, gene gain, and loss. Alongside this model, we have devised a gene order-based phylogenetic reconstruction approach based on conserved gene adjacency (CGA) among evolving genomes undergoing gene-gain/loss events. The model assumes a Poissonian process that acts independently on each gene of the genome. To the best of our knowledge, this is the first time such a stochastic, Poisson-based model has been used for this type of events.

The work resembles the Synteny Index (SI) approach by extracting a signal from the gene order along the genome. However, the CGA employs a significantly simpler signal, allowing additional quantitative properties, such as simple equations, to be solved to obtain the estimator, a more straightforward statistical consistency proof, and theoretically computed variance.

The CGA requires slightly less computational time than the SI because of its simplicity. However, the significance of the CGA lies in its ability to indicate the location of the signal within the data. When considering each gene present in $\mathcal{G}(0)$ that persists in $\mathcal{G}(t)$, the CGA only verifies if the consecutive gene in both genomes is the same, while the SI searches for the overlap between the two k-neighborhoods. Although the CGA processes significantly less data, it produces a comparable outcome. This may imply that the data utilized by the CGA capture most, if not all, of the signal that the SI captures.

We tested CGA in three settings: simulation on a single edge corresponding to a parent-child relationship, simulation along a tree, and testing on real microbial data from the Alignable Tight Genomic Clusters (ATGC) database. The CGA's results are similar to the SI's in tests with simulated trees. The CGC has poorer results than the CGA on short edges $0.001 - 0.01$, similar results in the range $0.01 - 0.1$, and better results than the CGA on comparatively long edges $0.1 - 1$.

SI and CGA count the number of neighboring genes that survived, while CGC counts the total number of genes that survived. This difference explains the variations in the results, as explained in more detail in Sect. 5.2.

Subsequent studies will focus on improving the accuracy of estimating long edge lengths. Additionally, efforts will be made to integrate the jump model with gain/loss mechanisms to create a more adaptable framework.

Regarding the analysis of real microbial data, although we analyzed only a small number of data sets, the results allude to a general trend of basic agreement between the mutation, the protein-based tree, and the genome dynamics trees (SI and CGA). This finding has implications beyond the scope of this paper for the correlation of these processes. We also note that in several ATGCs, the SI-based tree was found to be more similar to the protein tree than that of the CGA, whereas in others, the opposite was found. This does not point to any advantage to one of the methods since the protein tree is not a "golden standard," let alone with these very short edges of several thousandths of length. In the future, we will apply the indel approach to the EggNOG orthology database with wide divergence [14] with genomes in a wide range of gene sets.

In principle, devising accurate models and methods that faithfully describe actual biological events is highly important, as such accurate modeling allows subsequent reconstruction of the history of these events. We expect the theory we developed here to serve as the basis for subsequent microbial genomic research, with a richer set of events, a more significant number of genomes, or other applications rather than phylogenetic reconstruction.

References

1. Abby, S.S., Tannier, E., Gouy, M., Daubin, V.: Lateral gene transfer as a support for the tree of life. Proc. Natl. Acad. Sci. **109**(13), 4962–4967 (2012)
2. Adato, O., Ninyo, N., Gophna, U., Snir, S.: Detecting horizontal gene transfer between closely related taxa. PLOS comp. Biol. **11**, e1004408 (2015). https://doi.org/10.1371/journal.pcbi.1004408
3. Anderson, W.J.: Continuous-Time Markov Chains: An Applications-oriented Approach. Springer, New York (2012)
4. Bansal, M.S., Kellis, M., Kordi, M., Kundu, S.: RANGER-DTL 2.0: rigorous reconstruction of gene-family evolution by duplication, transfer and loss. Bioinformatics **34**(18), 3214–3216 (2018)
5. Biller, P., Guéguen, L., Tannier, E.: Moments of genome evolution by double cut-and-join. BMC Bioinform. **16**(14), S7 (2015)
6. Billingsley, P.: Probability and Measure. Wiley, Hoboken (2008)
7. Braga, M.D., Willing, E., Stoye, J.: Double cut and join with insertions and deletions. J. Comput. Biol. **18**(9), 1167–1184 (2011)
8. Delsuc, F., Brinkmann, H., Philippe, H.: Phylogenomics and the reconstruction of the tree of life. Nat. Rev. Genet. **6**(5), 361–375 (2005). https://doi.org/10.1038/nrg1603
9. Doolittle, W.F.: Phylogenetic classification and the universal tree. Science **284**(5423), 2124–9 (1999)
10. Doyon, J.P., Scornavacca, C., Gorbunov, K.Y., Szöllősi, G.J., Ranwez, V., Berry, V.: An efficient algorithm for gene/species trees parsimonious reconciliation with losses, duplications and transfers. In: Tannier, E. (ed.) RECOMB-CG 2010. LNCS, vol. 6398, pp. 93–108. Springer, Heidelberg (2010). https://doi.org/10.1007/978-3-642-16181-0_9
11. Felsenstein, J.: Evolutionary trees from DNA sequences: a maximum likelihood approach. J. Mol. Evol. **17**(6), 368–376 (1981)
12. Fitz Gibbon, S.T., House, C.H.: Whole genome-based phylogenetic analysis of free-living microorganisms. Nucleic Acids Res. **27**(21), 4218–4222 (1999)
13. Hannenhalli, S., Pevzner, P.A.: Transforming cabbage into turnip: polynomial algorithm for sorting signed permutations by reversals. J. ACM (JACM) **46**(1), 1–27 (1999)
14. Huerta-Cepas, J., et al.: eggNOG 5.0: a hierarchical, functionally and phylogenetically annotated orthology resource based on 5090 organisms and 2502 viruses. Nucleic Acids Res. **47**(D1), D309–D314 (2018). https://doi.org/10.1093/nar/gky1085
15. Huson, D.H., Steel, M.: Phylogenetic trees based on gene content. Bioinformatics **20**(13), 2044–2049 (2004)
16. Kapli, P., Yang, Z., Telford, M.J.: Phylogenetic tree building in the genomic age. Nat. Rev. Genet. **21**(7), 428–444 (2020). https://doi.org/10.1038/s41576-020-0233-0

17. Katriel, G., et al.: Gene transfer-based phylogenetics: analytical expressions and additivity via birth-death theory. Syst. Biol. **72**, syad060 (2023). https://doi.org/10.1093/sysbio/syad060

18. Koonin, E.V., Makarova, K.S., Aravind, L.: Horizontal gene transfer in prokaryotes: quantification and classification. Annu. Rev. Microbiol. **55**, 709–42 (2001)

19. Koonin, E.V., Makarova, K.S., Wolf, Y.I.: Evolution of microbial genomics: conceptual shifts over a quarter century. Trends Microbiol. **29**(7), 582–592 (2021). https://doi.org/10.1016/j.tim.2021.01.005

20. Koonin, E.V., Puigbo, P., Wolf, Y.I.: Comparison of phylogenetic trees and search for a central trend in the "forest of life". J. Comput. Biol. **18**(7), 917–924 (2011)

21. Kristensen, D.M., Wolf, Y.I., Koonin, E.V.: ATGC database and ATGC-COGs: an updated resource for micro- and macro-evolutionary studies of prokaryotic genomes and protein family annotation. Nucleic Acids Res. **45**(D1), D210–D218 (2017). https://doi.org/10.1093/nar/gkw934

22. Lehmann, E.L., Casella, G.: Theory of Point Estimation. Springer, New York (2006)

23. Martin, W.: Mosaic bacterial chromosomes: a challenge en route to a tree of genomes. BioEssays **21**, 99–104 (1999)

24. McInerney, J., McNally, A., O'Connell, M.: Why prokaryotes have pangenomes. Nat. Microbiol. **2**(4) (2017). https://doi.org/10.1038/nmicrobiol.2017.40, https://eprints.whiterose.ac.uk/113972/. 2017 Macmillan Publishers Limited, part of Springer Nature. This is an author produced version of a paper published in Nature Microbiology. Uploaded in accordance with the publisher's self-archiving policy

25. Nakhleh, L., Ruths, D., Wang, L.S.: RIATA-HGT: a fast and accurate heuristic for reconstructing horizontal gene transfer. In: Wang, L. (ed.) Computing and Combinatorics. LNCS, vol. 3595, pp. 84–93. Springer, Heidelberg (2005). https://doi.org/10.1007/11533719_11

26. Novichkov, P.S., Ratnere, I., Wolf, Y.I., Koonin, E.V., Dubchak, I.: ATGC: a database of orthologous genes from closely related prokaryotic genomes and a research platform for microevolution of prokaryotes. Nucleic Acids Res **37**((Database issue)), D448-54 (2009)

27. Ochman, H., Lawrence, J.G., Groisman, E.A.: Lateral gene transfer and the nature of bacterial innovation. Nature **405**(6784), 299–304 (2000)

28. Pang, T.Y., Lercher, M.J.: Each of 3,323 metabolic innovations in the evolution of E. coli arose through the horizontal transfer of a single DNA segment. Proc. Nat. Acad. Sci. U.S.A **116**(1), 187–192 (2019). https://doi.org/10.1073/pnas.1718997115

29. Puigbò, P., Wolf, Y.I., Koonin, E.V.: The tree and net components of prokaryote evolution. Genome Biol. Evol. **2**, 745–756 (2010)

30. Ragan, M.A., McInerney, J.O., Lake, J.A.: The network of life: genome beginnings and evolution. introduction. Philos. Trans. R. Soc. Lond B Biol. Sci. **364**(1527), 2169–2175 (2009)

31. Robinson, D., Foulds, L.: Comparison of phylogenetic trees. Math. Biosci. **53**, 131–147 (1981)

32. Sankoff, D., Nadeau, J.H.: Conserved synteny as a measure of genomic distance. Discret. Appl. Math. **71**(1–3), 247–257 (1996)

33. Schönknecht, G., Weber, A.P.M., Lercher, M.J.: Horizontal gene acquisitions by eukaryotes as drivers of adaptive evolution. BioEssays **36**(1), 9–20 (2014). https://doi.org/10.1002/bies.201300095, iSBN: 1521-1878 (Electronic)\ 265-9247 (Linking)

34. Serdoz, S., et al.: Maximum likelihood estimates of pairwise rearrangement distances. J. Theor. Biol. **423**, 31–40 (2017)

35. Sevillya, G., Doerr, D., Lerner, Y., Stoye, J., Steel, M., Snir, S.: Horizontal gene transfer phylogenetics: a random walk approach. Mol. Biol. Evol. **37**(5), 1470–1479 (2019). https://doi.org/10.1093/molbev/msz302

36. Sevillya, G., Snir, S.: Synteny footprints provide clearer phylogenetic signal than sequence data for prokaryotic classification. Mol. Phylogenet. Evol. **136**, 128–137 (2019)

37. Shifman, A., Ninyo, N., Gophna, U., Snir, S.: Phylo SI: a new genome-wide approach for prokaryotic phylogeny. Nucleic Acids Res. **42**(4), 2391–2404 (2013)

38. Snel, B., Bork, P., Huynen, M.A.: Genome phylogeny based on gene content. Nat. Genet. **21**(1), 108–110 (1999). https://doi.org/10.1038/5052

39. Stolzer, M., Lai, H., Xu, M., Sathaye, D., Vernot, B., Durand, D.: Inferring duplications, losses, transfers and incomplete lineage sorting with nonbinary species trees. Bioinformatics **28**(18), i409–i415 (2012)

40. Sumner, J.G., Jarvis, P.D., Francis, A.R.: A representation-theoretic approach to the calculation of evolutionary distance in bacteria. J. Phys. A: Math. Theor. **50**(33), 335601 (2017)

41. Tatusov, R.L., et al.: The cog database: new developments in phylogenetic classification of proteins from complete genomes. Nucleic Acids Res. **29**(1), 22–28 (2001)

42. Terauds, V., Sumner, J.: Maximum likelihood estimates of rearrangement distance: implementing a representation-theoretic approach. Bull. Math. Biol. **81**(2), 535–567 (2019)

43. Tofigh, A., Hallett, M., Lagergren, J.: Simultaneous identification of duplications and lateral gene transfers. IEEE/ACM Trans. Comput. Biol. Bioinform. (TCBB) **8**(2), 517–535 (2011)

44. Vanchurin, V., Wolf, Y.I., Koonin, E.V., Katsnelson, M.I.: Thermodynamics of evolution and the origin of life. Proc. Nat. Acad. Sci. **119**(6), e2120042119 (2022). https://doi.org/10.1073/pnas.2120042119, https://www.pnas.org/doi/abs/10.1073/pnas.2120042119

45. Wolf, Y., Rogozin, I., Grishin, N., Koonin, E.V.: Genome trees and the tree of life. Trends Genet. **18**(9), 472–479 (2002)

46. Wolf, Y.I., Makarova, K.S., Lobkovsky, A.E., Koonin, E.V.: Two fundamentally different classes of microbial genes. Nat. Microbiol. **2**, 16208 (2016). https://doi.org/10.1038/nmicrobiol.2016.208

47. Wolf, Y.I., Rogozin, I.B., Grishin, N.V., Koonin, E.V.: Genome trees and the tree of life. Trends Genet. **18**(9), 472–479 (2002)

48. Woodhams, M., Steane, D.A., Jones, R.C., Nicolle, D., Moulton, V., Holland, B.R.: Novel distances for Dollo data. Syst. Biol. **62**(1), 62–77 (2012)

49. Yancopoulos, S., Attie, O., Friedberg, R.: Efficient sorting of genomic permutations by translocation, inversion and block interchange. Bioinformatics **21**(16), 3340–3346 (2005)

50. Zhaxybayeva, O., Gogarten, J.P., Charlebois, R.L., Doolittle, W.F., Papke, R.T.: Phylogenetic analyses of cyanobacterial genomes: quantification of horizontal gene transfer events. Genome Res. **16**(9), 1099–1108 (2006)

51. Zuckerkandl, E., Pauling, L.: Molecules as documents of evolutionary history. J. Theor. Biol. **8**(2), 357–66 (1965)

Tools for Evolution Reconstruction

REvolutionH-tl: Reconstruction of Evolutionary Histories tool

José Antonio Ramírez-Rafael[1,2,6], Annachiara Korchmaros[2],
Katia Aviña-Padilla[1], Alitzel López Sánchez[3],
Andrea Arlette España-Tinajero[4], Marc Hellmuth[5],
Peter F. Stadler[2,6,7,8,9], and Maribel Hernández-Rosales[1(✉)]

[1] Center for Research and Advanced Studies of the National Polytechnic Institute,
Irapuato Unit, Irapuato, Guanajuato, Mexico
maribel.hr@cinvestav.mx
[2] University of Leipzig, Leipzig, Saxony, Germany
[3] University of Sherbrooke, Sherbrooke, QC, Canada
[4] Universidad Autónoma de San Luis Potosí, Instituto de Física,
San Luis Potosí, Mexico
[5] Stockholm University, Stockholm, Sweden
[6] Max Planck Institute for Mathematics in the Sciences, Leipzig, Saxony, Germany
[7] University of Vienna, Vienna, Austria
[8] Santa Fe Institute, Santa Fe, NM, USA
[9] Universidad Nacional de Colombia, Bogotá, Colombia

Abstract. Orthology detection from sequence similarity remains a difficult and computationally expensive problem for gene families with large numbers of gene duplications and losses. REvolutionH-tl implements a new graph-based approach to identify orthogroups, orthology, and paralogy relationships first, and it uses this information in a second step to infer event-labeled gene trees and their reconciliation with an inferred species tree. It avoids using gene trees and species trees upon input and settles for a maximal subtree reconciliation in cases where noise or horizontal gene transfer precludes a global reconciliation. The accuracy of the tool is comparable to competing tools at substantially reduced computational cost. REvolutionH-tl is freely available at https://pypi.org/project/revolutionhtl/.

Keywords: best matches · orthology · evolutionary scenarios

1 Introduction

Phylogenetic trees describe the history of the "taxa" at their leaves such that inner nodes depict the emergence of new lineages while edges correspond to their preservation over time. In the context of species evolution, branching points denote speciation events, while in gene evolution, internal nodes represent speciation, duplication, or other significant gene-related events. Combining the phylogeny

© The Author(s), under exclusive license to Springer Nature Switzerland AG 2024
C. Scornavacca and M. Hernández-Rosales (Eds.): RECOMB-CG 2024, LNBI 14616, pp. 89–109, 2024.
https://doi.org/10.1007/978-3-031-58072-7_5

of genes with that of the species in which the gene resides, "gene family histories" or "evolutionary scenarios" describe the evolution of genes as gene trees embedded into species trees. In this setting, orthologous genes are those that arise from speciation events, while paralogs originate from gene duplications along the edges of the species tree [7, 29]. The reconstruction of evolutionary relationships between genomic entities is a prerequisite for understanding morphological evolution, demographic changes in closely related species, and the functional annotation of newly discovered genes. Functional gene annotation relies on identifying orthologous genes since these tend to retain ancestrally functions [9].

Orthology detection has been studied extensively. Tree-based methods reconcile gene tree models with species trees to identify orthologous genes (orthologs). Graph-based methods, on the other hand, use sequence similarity to infer relationships between genes and then extract orthology from the corresponding graphs. These methods are computationally efficient and scale well with large datasets compared to tree-based approaches [18,19,22,32]. Recent mathematical interest has focused on detecting orthology through best match graphs (BMGs), which group closely related genes between species in directed graphs. We introduce REvolutionH-tl, a bioinformatics tool that uses BMGs to detect orthologs and reconstruct gene and species phylogenies in the presence of gene duplication and loss events [13]. Our method uses best hit data from sequence similarity searches to construct BMGs. Gene and species phylogenies are reconstructed by maximizing the number of informative triples with no assumptions regarding evolutionary models. Moreover, our tool performs tree reconciliations by placing duplication and loss events along the branches of the species tree [6,18]. Comparisons with ProteinOrtho [18] and OrthoFinder [6], two widely-used tools for orthology inference, and with Astral-Pro [31] for species tree reconstruction demonstrate that REvolutionH-tl is not only faster but also offers performance that is on par with these established tools, thereby positioning it as a more powerful tool for evolutionary analyses.

2   REvolutionH-tl Workflow

The workflow of REvolutionH-tl is divided into 5 steps: (1) *Computation of alignment hits*, computation of all-against-all pairwise alignments p (2) *BMG estimation and orthogroup detection*, (3) *Gene tree reconstruction and orthology assignment*, (4) *Species tree reconstruction*, and (5) *Tree reconciliation*. This methodology is illustrated in Fig. 1. Each step in REvolutionH-tl can be executed individually or simultaneously using a single command. In the subsequent subsections, we explain the methods employed for each step. All mathematical concepts referenced herein are thoroughly described in Appendix A.1.

Computation of Alignment Hits. The aim of this step is to compute sequence similarity between genes of different species. The input is a collection of fasta files, each of them corresponding to one species and containing a list of sequences. Alignments are computed using Diamond [4] for amino acid sequences or BLAST [2,17] for DNA sequences. As output, REvolutionH-tl generates a directory containing the alignment hits for each pair of species.

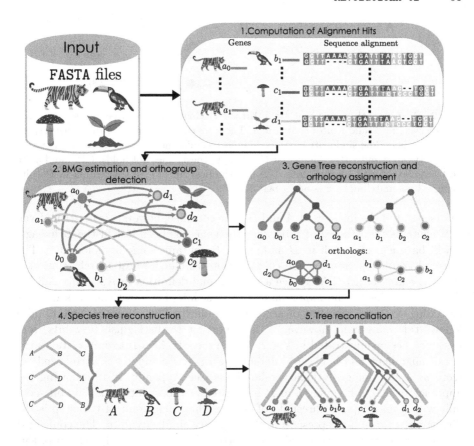

Fig. 1. REvolutionH-tl methodology. (1) Computation of alignments using BLAST for DNA sequences or Diamond for protein sequences. (2) Estimation of the best match graphs (BMG) from best hits. Each connected component of the BMG defines an orthogroup, which is a set of genes descending from a common ancestral gene for a given set of species. (3) A gene tree is inferred from each BMG, duplication ■ and speciation • events are assigned to its internal nodes. Two genes are orthologs if the last common ancestor of the corresponding leaves in the gene tree is a speciation node; this step provides all orthology assignments. (4) A species tree is reconstructed based on color triples from the set of inferred gene trees. (5) Gene trees are mapped to species trees via the reconciliation map, inferring gene losses and editing gene trees inconsistent with the species tree. The reconciled tree helps to detect how many duplications and losses occur in each branch and the number of genes at speciation nodes. (Color figure online)

BMG Estimation and Orthogroup Detection. For each gene x in species A, a gene y in species B is a *best match* if y is evolutionary most closely related to x among all genes y' in species B [10]. Note that best matches are not necessarily unique. Moreover, if B contains an ortholog z of x, then z is necessarily a best match of x since (ignoring horizontal transfer) no gene in A can be more closely related to a gene in B than gene pairs that originated in the speciation event

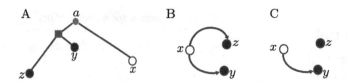

Fig. 2. Best matches are not best hits. (A) Gene tree on three genes x, y, z and two species (black and white) rooted in a. The least common ancestor of x with respect to y and z is a. The similarity score (weighted tree distance) between x and y is larger than between x and z (the Molecular Clock Hypothesis does not hold). (B) z and y are best matches of x in species black. (C) y is the best hit of x in species black. Despite being a best match of x, z is not a best hit of x.

that separated the lineages of A and B. The converse, however, is not true. Even if B contains no ortholog of x because the gene has been lost in the B lineage, x will have a best match in B as long as B contains some homolog of x. The directed graph representing the (usually not symmetric) best match relation is the *best match graph* (BMG); see a more formal definition in Appendix A.3.

Best matches can be *approximated* by best hits, i.e., most similar sequences according to some measure of sequence similarity [29]. REvolutionH-tl also follows this paradigm. However, it is important to distinguish best hits from best matches, as illustrated in Fig. 2. More precisely, REvolutionH-tl takes the list of alignment hits between different species and computes a normalized bit score as a clue for evolutionary relatedness; see Appendix A.4. Following the approach in [22], a homolog y in B is counted as a best hit for x if its bit score falls within an adaptive threshold of the best hit. The result of this step is the directly *best hit graph* representing the sequence similarity data. The connected components of the *best hit graph* are reported as *orthogroups*. The output of this step is composed of two *.tsv files. One is the list of the estimated best hits represented as ordered gene pairs, and the other is the list of the orthogroups.

Gene Tree Reconstruction and Orthology Assignment. Gene trees can be inferred from BMGs as their so-called least resolved trees (LRTs) [10]. To this end, we extract the set of informative triples from the best hit graph as described in [26]. Since the best hit graph is only an approximation of the true BMG, the estimated triple set $\mathfrak{R}$ is usually not consistent. Thus, REvolutionH-tl employs a heuristic to extract a maximum set of consistent triples $\widehat{\mathfrak{R}}$ [26,30]. The reconstruction of the gene tree T from $\widehat{\mathfrak{R}}$ is a well-understood problem that can be solved efficiently using the BUILD algorithm [28]. Although the gene tree T generated at this step is not fully resolved in general; it already contains all the orthology relationships between the genes within the orthogroup.

Finally, we assign event labels $t(v)$ to internal nodes v of the gene tree using the species overlap method [11,16], which assigns a speciation event to a node v if the species of the genes below v do not overlap and a duplication event otherwise. The result is an event-labeled gene tree (T, t, σ), where σ denotes the map that assigns to each gene (leaf x of T) the species $\sigma(x)$ in which it resides. A formal description of this step can be found in Appendix A.5. The list of orthologous genes is obtained from (T, t, σ) as the list of all gene pairs x, y whose least common ancestor in T is labeled as a speciation event. A key advantage of the relation-based approach is that a correct orthology assignment requires only a correct *discriminating tree* obtained by collapsing all adjacent pairs of inner nodes with the same event label. Therefore, it is important that tree T has few false branches, while we need only a moderate level of resolution.

The outputs of this step are three `tsv files`, namely (1) the list of gene trees in NHX format and the ID of the corresponding orthogroup, (2) a list of best matches recomputed from (T, t, σ), and (3) a list of orthologs. The last two files are reported as lists of pairs of genes.

Species Tree Reconstruction. The event-labeled gene trees (T, t, σ) contain information on the species tree S in the form of gene triples $((u, v), w)$, where u, v and w are genes from three different species such that their common last common ancestor $\text{lca}(u, w) = \text{lca}(v, w)$ is labeled as speciation event [12, 14, 15]. In this case, S displays $((\sigma(u), \sigma(v)), \sigma(w))$. For noiseless data, it suffices to run BUILD on this set of triples. A heuristic extension is used to account for errors in the previous steps, as described in [26], this heuristic forces a partition of the auxiliary Aho graph whenever BUILD encounters a step in which this graph is connected. For the technical details, refer to Appendix A.6. The output of this step is a single file containing the inferred species tree in NHX format. To accommodate duplication events that precede the last common ancestor v of all taxa in the species tree, we consider planted trees in which an extra parent of v is added, see Appendix A.1 for more details.

Tree Reconciliation. An *evolutionary scenario* is an embedding μ of the gene tree (T, t, σ) into the species tree S such that (i) every leaf (gene) x of T is mapped to the leaf $\mu(x) = \sigma(x)$, (ii) every node of T labeled as a speciation is mapped to a node (speciation event) in S, (iii) every duplicate node of T is mapped to an edge of S, and (iv) if two nodes u and v in T are comparable w.r.t. to the ancestor partial order of T, this is also the case for their embeddings $\mu(u)$ and $\mu(v)$ in the species tree S, see Fig. 3. A more formal definition is given in Appendix A.7.

To reconcile a gene tree with a species tree, we first need to test if the trees are consistent by identifying if the color triples of the gene tree are a subset of the triples in the species tree; see Appendix A.2. If this is not the case, we proceed by pruning the leaves of the gene tree appearing in most inconsistent triples and in the least consistent triples, as described in Appendix A.6-Algorithm 1. Once we ensure consistency, the reconciliation of a gene tree with a species tree is computed by mapping every node of the gene tree to a node or edge of the species tree. Leaves in the gene tree will map to the species they belong to in the species tree. Internal nodes in the gene tree whose label is a speciation event will map to an internal node of the species tree and those with a duplication label

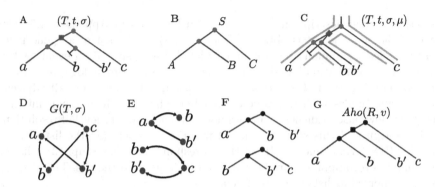

Fig. 3. Representation of evolutionary histories and gene tree reconstruction. (A) Event labeled gene tree (T, t, σ) with speciations • and duplications ■. We assume $\sigma(a) = A$, $\sigma(b) = \sigma(b') = B$, $\sigma(c) = C$. (B) Species tree S. (C) Reconciliation of the gene tree (T, t, σ) and the species tree S. The reconciliation map μ is shown implicitly by drawing the nodes of the gene tree in the corresponding nodes and edges of the species tree. (D) Best match graph $G(T, \sigma)$ for the gene tree (T, t, σ). (E) Informative triples $R(G, \sigma)$ obtained from the best match graph G. (F) Aho-graph for the informative triples $R(G, \sigma)$. (G) Three obtained by the algorithm BUILD when applied to the set of informative triples $R(G, \sigma)$. (Color figure online)

will be mapped to the edges of the species tree, always conserving the order of divergences.

At this point, the inferred gene tree might not be fully resolved; this is, there might be nodes in the gene tree with more than two children. Those nodes corresponding to a speciation event will be resolved when applying the reconciliation map, which will use the topology of the species tree to be reflected in those nodes in the gene tree; see Appendix A.8 for further details.

Furthermore, we can infer gene loss events once we determine that a speciation node u of the gene tree maps to an internal node v in the species tree, we check if all the children of v are mapped by a descendant of u, in the case this is not true, we infer gene losses; see Appendix A.9.

The input for this step is a list of gene trees and a species tree in NHX format. The user can provide a species tree; otherwise, REvolutionH-tl will compute it as described in the previous section. The output is a *.tsv file containing evolutionary scenarios, consisting of the orthogroup identifier, a reconciled gene tree in NHX format, and the map of the nodes of the gene tree to the nodes and edges of the species tree. As only duplication nodes will be mapped to edges in the species tree, we report the node v when the duplication is mapped to edge uv in the species tree. Therefore, the reconciliation map is a list of pairs of nodes. The reconciled gene tree contains evolutionary events, including gene losses.

3   Benchmarking Methodology

To show the correctness of our approach, we run the REvolutionH-tl workflow to predict orthologs and reconstruct evolutionary histories, then we contrast REvolutionH-tl to Proteinortho and Orthofinder for evaluating the orthology prediction and to Astral-Pro for comparing the species tree reconstruction. We use two datasets. The first dataset is a collection of simulated genomes and evolutionary histories, which we refer to as the synthetic dataset. The second dataset was selected from the gold standard database PhylomeDB [8]; in particular, we use the collection called (PhyId 500) [QfO] Reference Model Species Metaphylome *H. sapiens* and refer to it as QfO-PhylomeDB. The following parts describe the datasets and how we run the tools on such data. Furthermore, we explain how to evaluate the correctness of the predictions.

Simulation of the Synthetic Dataset. We use SaGePhy [20] to simulate evolutionary histories under three main parameters: the number of species $N \in \{5, 10, 15, ..., 50\}$, duplication rate d, and loss rate l; d and l are randomly and independently chosen in $\{i/10 : i \in \mathbb{N} \text{ and } 0 \leq i \leq 10\}$. This yields 1210 triples of parameters (N, d, l). We simulate five evolutionary histories of a species tree S for each, obtaining five gene trees evolving inside S. The simulated trees have a length for each branch; this real number indicates a rate of evolution. These trees were simulated, assuming a constant mutation rate for all the genes. Protein sequences are then simulated by choosing the *LG Substitution* model [21] among others offered by SaGePhy. Furthermore, we create the set F_N of N fasta files, one for each species; these fasta files are our simulated genomes. Similarly, we generated a small data set of evolutionary scenarios containing horizontal gene transfer events, for this dataset we generated species trees with 5 and 15 species, duplication and loss rates of 0.25, and HGT rates of 0.25 and 1. For each combination of transference rate and number of species, we simulated 10 gene trees.

This data set is the ground-truth for the following part of our analysis; the simulated trees represent the true evolutionary history of the simulated sequences, in consequence we can obtain the *true* orthogroups and orthology graphs. To measure the performance of a tool, we contrast this ground truth against the inferences of such a tool.

Correctness of Orthogroup Identification. Recognizing homology is the first and crucial step for evolution reconstruction. To test the correctness of the predicted orthogroups of the synthetic dataset, we compare the predictions made by REvolutionH-tl, ProteinOrtho, and Orthofinder against the ground-truth orthogroups generated with SaGePhy. To measure performance for this task, we classify the ground-truth orthogroups based on its inferences. In this context, we say that a true orthogroup X is *fully recovered* if there is an inferred orthogroup X' such that $X = X'$; otherwise, the inference is wrong. There are two types of wrong inferences; first, when an inferred orthogroup is a proper subset of the true orthogroup $X' \subset X$, in this scenario, we say that the true orthogroup X is *splitted*. On the other hand, if for the real orthogroups X_0 and X_1 there is an inferred orthogroup X' containing elements of both true orthogroups $X' \cap X_0 \neq \emptyset \neq X' \cap X_1$, then we say that the real orthogroups X_0

and X_1 are *merged*. We calculate the percentage of correct inferences and split and merged orthogroups for each tool in the analysis.

Standard Performance Metrics. To evaluate the performance of a tool for orthology prediction and gene tree inference, we first classify the individual predictions as *true positive* (*tp*), *false positive* (*fp*), *true negative* (*tn*), and *false negative* (*fn*). Here, an individual prediction may regard an orthology relation or a triple in a gene or species tree. Once the individual predictions have been classified, we compute the standard performance metrics *precision* $= tp/(tp + fp)$, *recall* $= tp/(tp + fn)$, *false positive rate* $= fp/(fp + tn)$, and *accuracy* $= (tp + tn)/(tp + tn + fp + fn)$. Given an orthogroup X and its corresponding set of predictions, we can compute the performance A_X for each metric. We create collections of synthetic orthogroups by selecting simulations with fixed values of duplication and loss rates. Thus, for the collection of orthogorups $Y = \{X_0, X_1, ...\}$ we compute the performance of the collection as $A_Y = (A_{X_0} + A_{X_1} + ...)/|Y|$. For the QfO-PhylomeDB dataset, we run the tools for each orthogroup individually; for the synthetic dataset, we run the tools over complete genomes.

Performance for Orthology Inference. We compute the performance of orthology prediction per orthogroup. Precision is computed with respect to an orthogroup, which is a set of genes X; for this set, there is a set of true orthology relations $\Theta_X \subseteq X \times X$, and a set of inferred orthology relations $I_X \subseteq X \times X'$, where $X \subseteq X'$, since X' might include genes that do not belong to orthogroup X. Next, we compute the values for performance metrics using the following classification of predictions $fp = |I_X \backslash \Theta_X|$, $fn = |\Theta_X \backslash I_X|$, $tp = |I_X \cap \Theta_X|$, and $tn = |X \times X'| - |I_X \cup \Theta_X|$.

Performance for Tree Reconstruction. Given a real tree T and an inferred tree T', we compute the precision false positive rate, and recall performance over their induced rooted triples. In this case we have the values $fp = |R(T') \backslash R(T)|$, $fn = |R(T) \backslash R(T')|$, $tp = |R(T) \cap R(T')|$, and $tn = 3\binom{n}{3} - fp - fn - tp$ for n taxa.

4   Results

Our main contribution is `REvolutionH-tl`, an open-source, cross-platform Python software [23] for fast and accurate orthology prediction and the inference of gene and species trees with promising applications to whole-genome comparative studies. In Fig. 4, we illustrate the results obtained by `REvolutionH-tl` for the genomes of four vertebrates.

Exact Orthogroups. Table 1 shows the metrics of the orthogroup performance running the tools on the simulated genomes. Our analysis revealed that for

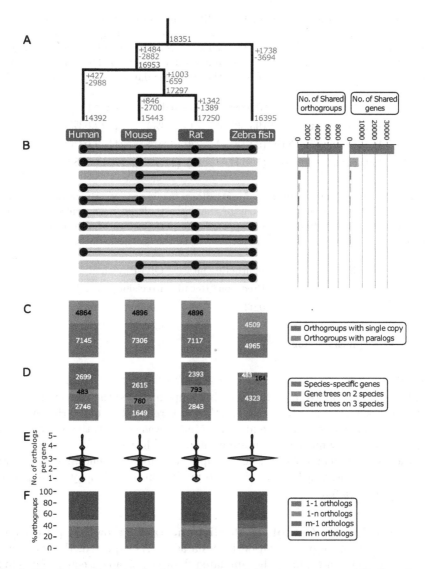

Fig. 4. REvolutionH-tl analysis on four vertebrate genomes. (A) Overview of the reconciliation map of gene trees across four species with the species tree. Numbers denote duplication events with a $+$, while gene losses are indicated by $-$. Red numbers at internal nodes signify the count of genes observed in the ancestral species. The numbers on the leaves represent the total number of genes analyzed for each species. (B) Distribution of orthogroups and genes shared among different species sets, with a significant portion shared by all species. (C) Analysis of orthogroups containing single-copy genes versus those with paralogs. (D) Enumeration of species-specific genes and gene trees containing genes only present in two or three species. (E) Representation of the distribution of orthologs per gene within each genome. (F) Percentage breakdown of orthologs 1-to-1, 1-to-many, many-to-1, and many-to-many across species. (Color figure online)

Table 1. Orthogroups recovered by different platforms and tools. A real orthogroup is *fully recovered* when there is an inferred orthogroup that coincides with it, it is *splitted* when its genes are contained in more than one inferred orthogroup, it is *merged* when it is strictly contained in an inferred orthogroup. The true orthogroups used for this analysis were obtained from the synthetic dataset.

Method	Fully Recovered	Splitted	Merged
REvolutioH-tl	5786 (95.63%)	20 (0.33%)	244 (4.03%)
ProteinOrtho	5576 (92.16 %)	230 (3.80%)	244 (4.03%)
Orthofinder	3613 (59.71 %)	691 (11.42%)	1746 (28.85%)

Table 2. Global performance of orthology assignment. Precision and recall were computed for all the predictions of the three tools on both synthetic and the PhylomeDB collection datasets. As expected, the performance of all the tools decays when dealing with real sequences, in contrast with the synthetic dataset.

Tool	Synthetic Dataset		QfO-PhylomeDB	
	Precision	Recall	Precision	Recall
REvolutionH-tl	0.93	0.97	0.78	0.36
Proteinortho	0.96	0.97	0.84	0.37
Orthofinder	0.85	0.92	0.66	0.47

REvolutionH-tl, almost all inferred orthogroups matched the true orthogroups, resulting in a high percentage of correctly recovered orthogroups. However, the performance of Proteinortho and Orthofinder was comparatively lower, with a decrease in the number of accurately inferred orthogroups. We also observed that the most common error was the grouping of unrelated genes, i.e., the merging of orthogroups, which resulted in the prediction of homology among genes that were not related.

High Accuracy for Orthology Prediction. Table 2 shows the performance metrics for orthology prediction on both datasets. As expected, the tools have higher precision and recall in the simulated dataset , as a consequence of simulating a non-ultrametric dataset. In particular, REvolutionH-tl maintains high performance. These numbers prove that the efficacy of the three tools is comparable. Figure 5 shows that REvolutionH-tl has a very good performance despite high gene duplication and loss rates. In addition, we evaluated the performance of orthology prediction under the influence of horizontal gene transfer events. Table 3 shows a noticeable decline in performance with an increasing number of HGT events. However, we observed that this decline is mitigated when more species are included in the analysis. It is noteworthy that in the absence of HGT events, REvolutionH-tl has the highest accuracy, moreover, as the number of transfers increases, the three tools reach a similar level of performance.

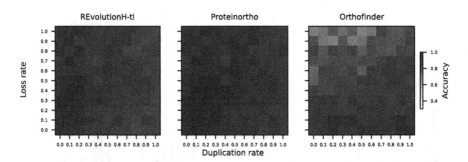

Fig. 5. Distribution of orthology performance. Heatmaps show tool accuracy for orthology assignment in the synthetic dataset. Each cell displays the average accuracy for orthogroups simulated with a specific duplication and loss rate combination.

Table 3. Accuracy of orthology prediction in the presence of horizontal gene transferences. This table presents the performance results for three rates of HGT. The ground-truth evolutionary scenarios for this analysis were simulated with gene duplication and loss rates equal to 0.25.

Tool	5 species			15 species		
	0 hgt	1 hgt	>1 hgt	0 hgt	1 hgt	>1 hgt
REvolutionH-tl	1.000	0.778	0.415	0.996	0.800	0.618
Proteinortho	1.000	0.803	0.393	0.996	0.800	0.615
Orthofinder	0.875	0.770	0.430	0.885	0.804	0.649

Accuracy for Gene and Species Tree Reconstruction. In the relation-based approach, orthology inference depends only on the accuracy of best matches and the correctness of the event labels. It does not require that either the gene tree or the species is fully resolved. For our task, therefore, we are primarily concerned with false-positive triples, while false negatives have little impact.

We assess the accuracy of gene and species tree reconstruction by evaluating the precision, recall, and false positive rate of their triple inference. While REvolutionH-tl exhibits commendable performance in reconstructing gene trees before reconciliation, its efficacy notably improves after post-reconciliation, as shown in Fig. 6(A) and Fig. 7. This enhancement stems from the resolution of multifurcated speciation nodes present before reconciliation, which can be effectively disentangled afterward. Additionally, during reconciliation, gene trees incongruent with the species tree are edited to achieve consistency by selectively eliminating sources of noise or erroneous orthology predictions. Regrettably, a comparative analysis with Orthofinder output was unfeasible, as we could only procure a comprehensive gene tree encompassing all genes studied rather than individual orthogroups.

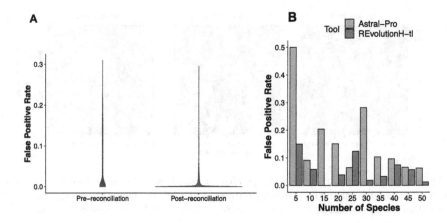

Fig. 6. False positive rate of gene and species tree reconstruction. The violin plot (A) shows the gene triple false positive rate pre-reconciliation and post-reconciliation. The barplot (B) shows the false positive rate for species tree reconstruction for `REvolutionH-tl` and `Astral-Pro` on the synthetic dataset.

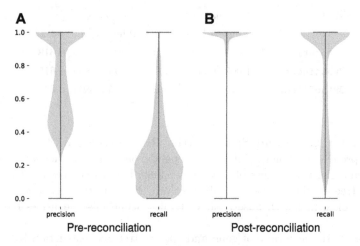

Fig. 7. Recall and precision for gene tree reconstruction. The violin plots show the triple performance for gene tree reconstruction pre-reconciliation (A) and post-reconciliation (B).

The violin plots in Fig. 6(A) show that, overall, our tool exhibits a low false positive rate for the gene triples. This indicates that the inferred phylogenetic information generally avoids incorrect groupings, albeit it may sometimes lack resolution. The barplot in Fig. 6(B) illustrates the performance metrics for the set of species trees inferred from the synthetic dataset; more precisely, for each number of species, all available (about 600) gene trees were considered to infer one single species tree. The observable fluctuation of the performance metrics

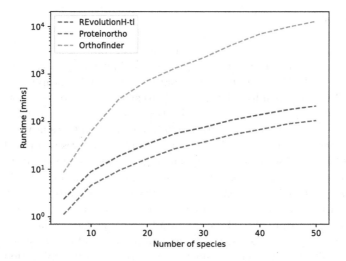

Fig. 8. Running times on synthetic genomes. For 50 species, `Orthofinder` takes nine days, while `Proteinortho` takes less than two hours and `REvolutionH-tl` less than four hours. Compared to `Proteinortho`, `REvolutionH-tl` also computes gene trees, species tree, and reconciled gene trees.

primarily stems from variations in the number of inferred gene trees utilized for reconstructing the species trees. This variability partially arises from generating gene trees from split or merged orthogroups. However, if the gene tree topology exhibits noise, such as due to long-branch attraction (LBA), the associated triples will likewise be noisy. Consequently, this leads to poor species tree inference when constructing the maximal species tree displaying all gene tree triples. Nonetheless, the method employed for species tree reconstruction stands on its own merits, as precision and recall are 1 without these confounding factors.

Moreover, `REvolutionH-tl` outperformed `Astral-Pro` in terms of the inferred species trees derived from the same set of gene trees. The barplot in Fig. 6(B) shows that, in general, our tool exhibits a lower false positive rate than `Astral-Pro`.

Fast Reconstruction of Evolution. The running times of the three tools are shown in Fig. 8. `REvolutionH-tl` finished analyzing a dataset of 50 species in several hours, while `Orthofinder` took 9 days to complete the same computation. The time of execution of `REvolutionH-tl` and `Proteinortho` are similar; the main difference in this case, is that `REvolutionH-tl` returns event-labeled gene tree, species trees, reconciled trees, orthogroups, and orthology assignments, while `Proteinortho` only yields the latter two.

5   Discussion

`REvolutionH-tl` is a tool for the fast and accurate inference of orthology relationships, as well as the reconstruction of evolutionary scenarios. It relies on approximating BMGs with efficiently computable best hit graphs and exploits recent mathematical results relating BMGs to gene trees, species trees, and their reconciliation. Importantly, only similarities obtained from pairwise sequence comparisons are required as a starting point. The method takes advantage of two features of best match theory: first, ancient duplications result in best matches only in the presence of extensive complementary gene loss. Hence, it usually separates such old paralog groups into distinct connected components. This avoids the expensive reconstruction of large gene trees. Second, incompletely resolved gene trees are sufficient for orthology inference. Empirically, we observe that the trees obtained from the relational data tend to have fewer false triples compared to other approaches. Resource-demanding computational procedures, in particular multiple sequence alignments and probabilistic explorations of gene tree topologies, can be avoided altogether.

The comparison of `REvolutionH-tl` vs. `Proteinortho` and `Orthofinder` orthology prediction shows that `REvolutionH-tl` has a remarkable performance for orthology assignments, particularly in terms of recall. In addition, `REvolutionH-tl` is computationally efficient. Although `Proteinortho` exhibits slightly superior orthology performance and running times compared to `REvolutionH-tl`, the former solely provides orthology results. In contrast, the latter generates evolutionary scenarios, i.e., gene trees embedded into a species tree, thereby validating orthology predictions using an evolutionary hypothesis. This is a relevant difference because providing an explicit evolutionary scenario is the only way to ensure the biological feasibility of orthology inference.

In addition, the comparison of `REvolutionH-tl` and `Astral-Pro` species tree reconstructions from the same set of (unlabeled) gene trees demonstrates `REvolutionH-tl` efficiency in estimating species trees.

`REvolutionH-tl` is a versatile tool designed to enhance research in comparative genomics. It supports various applications, from the reconstruction of evolutionary histories to the use of orthology for functional annotation of genes. Understanding the evolutionary history of genes and species is crucial for deciphering the relationships between different organisms and tracking changes in genomic complexity. Moreover, the ability to identify orthologs and paralogs serves as an initial step in hypothesizing the functions of newly discovered genes.

Given its capabilities, `REvolutionH-tl` is particularly effective in contributing to integrating whole genomes and annotations within an evolutionary framework, making it a valuable resource for studies aiming to uncover the intricacies of genetic evolution and function. In general, `REvolutionH-tl` maximizes parsimony; the estimation of evolutionary relatedness is based on the bit score, which is prone to LBA artifacts; furthermore, when we assign evolutionary events and compute reconciliation, we aim to minimize the number of gene duplication and loss events.

In the face of the poor performance of previous parsimonious methodologies, REvolutionH-tl stands out in reconstructing evolutionary histories with high precision, this is avoiding the inference of false positive triples in both gene and species trees. On the other hand, REvolutionH-tl has a high recall for phylogeny reconstruction, meaning that our tool successfully recovers most of the true triples, returning nearly fully resolved phylogenies. Despite REvolutionH-tl outputs trees with polytomies, the high triple precision suggests that we can trust these trees, and they can serve to constrain the search space of other methodologies, such as maximum likelihood or Bayesian optimization.

The source of noise in our predictions comes mainly from scenarios where gene duplication immediately precedes speciation events; in this case, REvolutionH-tl methodology is susceptible to incorrectly identifying out-paralogous genes as in-paralogous. In our framework, noise is handled by graph editing, which eliminates contradictions between informative triples from the best hit graph. Furthermore, tree reconciliation was demonstrated to be useful for distinguishing the true topology of gene trees.

Currently, refining speciation events in gene trees depends on having a species tree for reconciliation. REvolutionH-tl can infer species trees with an acceptable level of accuracy. The information about gene and species tree topologies is obtained using the best match theory; thus, the more duplicated genes are in the analysis, the more phylogenetic signal is obtained. This may represent a problem when dealing with close species or when the whole genome of the species of interest is unavailable.

Despite the good performance of REvolutionH-tl, there are several avenues for improvement. Future upgrades to the tool will refine (i) the estimations of evolutionary relatedness, (ii) the weights used for informative triples, (iii) the resolving of multifurcated nodes in the inferred gene trees, and (iv) reconstruction of species tree from gene trees lacking duplication events.

Acknowledgments. This work was supported in part by CINVESTAV-University of California (UC Alianza MX) joint project and by the German Research Foundation (DFG, STA 850/49-1). KAP (CVU:227919) and JARR (CVU:1147711) received financial support from CONAHCyT. We express our gratitude to Marisol Navarro Miranda, Erika Viridiana Cruz Bonilla, and Luis Fernando Flores Lopez for their valuable contributions to the design of the methodology figure for REvolutionH-tl.

Disclosure of Interests. Authors have no competing interests to declare that are relevant to the content of this article.

A   Appendix

A.1   Notation

A *graph* $G = (V(G), E(G))$ consists of two sets: a non-empty set of objects $V(G)$, called *nodes*, and a set $E(G)$ of *edges*. Each edge, noted as $e = uv$, connects a pair of nodes $u, v \in V(G)$. The edge is called an *arrow* when this connection has a direction. In such cases, v is an *out-neighbor* of u. When we count the number of connections to a node v, we refer to this count as the *degree* of the node, denoted as $\deg_G(v)$. Furthermore, the *out-degree* $\deg_G^+(v)$ of a node v is the number of its out-neighbors. Based on this concept, graphs are divided into two main families depending on the nature of their connections: those with edges, known as *undirected graphs*, and those with arrows, known as *directed graphs*. A *subgraph* H of G is also a graph where $V(H) \subseteq V(G)$ and $E(H) \subseteq E(G)$. Moreover, the *subgraph of G induced by* $V' \subseteq V(G)$ denoted as $G[V']$ is a subgraph where $V(H) = V'$ and its set of edges consists of all edges in $E(G)$ that connect the nodes in V'.

In a graph G, a *path* from node u to node v is a sequence of nodes starting at u and ending at v, with consecutive nodes connected by edges. A graph is termed *connected* if there is a path linking every pair of its nodes.

A *tree* T is a connected undirected graph that becomes disconnected by removing any edge. In this context, every tree is *rooted*, meaning it has a designated root node ρ_T, with the structure visualized such that all other nodes fall hierarchically beneath the root (refer to Fig. 3(A)). The *leaves* of the tree, $L(T)$, are nodes with zero out-degree. The *inner nodes*, $V^0(T)$, are those nodes that are neither leaves nor the root of T.

Although rooted trees are considered undirected, the convention $uv \in E(T)$ indicates u as the unique *parent* of v and v as a *child* of u, with $\mathrm{ch}_T(u)$ representing children of u. Also, u is an *ancestor* of v, and v a *descendant* of u, if u lies on the unique path from v to ρ_T. We express this as $v \preceq_T u$, or more strictly as $v \prec_T u$ if $v \neq u$. Nodes $u, v \in V(T)$ are *non-comparable*, noted as $x \parallel_T y$, if neither is an ancestor or descendant of the other; they are *comparable* otherwise. The *last common ancestor* of a set $X \subseteq V(T)$, $\mathrm{lca}_T(X)$, is the most distant node u from ρ_T that is an ancestor of all nodes in X. For individual nodes $x, y \in V(T)$, we denote $\mathrm{lca}_T(x, y)$ as their last common ancestor. Furthermore, in [15] the $\preceq_T$ relationship has been extended to consider edges within T; for edges $e_0 = uv, e_1 = xy$ and a node z, $e_0 \preceq_T e_1$ if $v \preceq_T y$, $z \prec_T e_0$ if $z \preceq_T v$, and $e_0 \prec_T z$ if $u \preceq_T z$.

For any node v in $V(T)$, the expression $T(v)$ denotes the *subtree* rooted at v, encompassing all descendants of v. The restriction $T_{|L'}$ of T to a leaf subset $L' \subseteq L(T)$ is its minimal subtree connecting all leaves in L', excluding degree-two inner nodes. A tree T *displays* another tree T' with leaves L', denoted $T' \leq T$, if T' arises from contracting inner edges of $T_{|L'}$. If $L(T) = L(T')$, T is a *refinement* of T'. The *cluster* $C_T(v)$ includes all leaves in the subtree $T(v)$.

All trees in this paper are *phylogenetic*, meaning each inner node $v \in V^0(T)$ has an out-degree $\deg_T^+(v) > 1$, except the root. In some cases, like in Fig. 3(C), we examine *planted trees* formed by adding a new node 0_T and edge $0_T \rho_{T'}$ to a phylogenetic, rooted tree T'.

A *triple* $xy|z$ is a tree on three leaves x, y and z where x and y share a closer common ancestor than either does with z, triples are pivotal for supertree construction [3,5,28]. Each tree T corresponds to rooted triples $R(T)$. A triple set R is *consistent* if it's part of $R(T)$ for some tree T that displays R. The

BUILD algorithm [1,28] checks this, returning a supertree for consistent R or noting inconsistency. It uses the *Aho-graph* $[R, L']$ (with $L' = \bigcup(L(R))$ and edge xy for each triple $xy|z \in R$) to assess consistency; a disconnected graph confirms consistent triples.

A.2   Evolutionary Scenarios

In a species tree S, leaves symbolize extant species, and inner nodes indicate speciations. Conversely, a gene tree (T, t, σ) depicts genes as leaves $L(T)$. The function $\sigma : L(T) \to L(S)$ maps each gene to its residing species. The function $t : V(T) \to \{\bullet, \Box, \odot, \times\}$ classifies nodes in the gene tree based on evolutionary processes: $t(x) = \bullet$ for speciation, $t(x) = \Box$ for duplication, $t(x) = \odot$ for extant genes, and $t(x) = \times$ for gene loss, as detailed in [15] and illustrated in Fig. 3B.

An *evolutionary scenario* (S, T, t, σ) merges a gene tree (T, t, σ) with a species tree S via the reconciliation map μ, as introduced in [11,15] and exemplified in Fig. 3. Detailed mathematical constraints of such scenarios are elaborated in Appendix A.7.

Constructing an evolutionary scenario requires consistency between gene and species trees, assessed using color triples. Let $\mathfrak{R}(T) = \{r \in R(T) \mid t(\text{lca}_T(L(r))) = \bullet \text{ and } |\sigma(L(r))| = 3\}$ be the set of speciation triples of the gene tree. Given a triple $ab|c \in R(T)$, the corresponding *color triple* is $\sigma(ab|c) = \sigma(a)\sigma(b)|\sigma(c)$. Finally, let $\mathfrak{R}_\sigma(T) = \{\sigma(r) \text{ for all } r \in \mathfrak{R}(T)\}$ be the set of *color triples* of the gene tree. Here, the gene tree (T, t, σ) and a species tree S are consistent whenever $\mathfrak{R}_\sigma(T) \subseteq R(S)$ [12,15]. Consistency is required to ensure that a reconciliation between (T, t, σ) and S exists.

A.3   Best Match Graphs

The concept of *best match graphs* (BMGs) [10,11,24,25,27] outlines that a gene y is a *best match* for x if, x and y reside in distinct species and $\text{lca}_T(x, y) \preceq \text{lca}_T(x, y')$ for all genes y' in the species $\sigma(y)$, i.e., y is one of the genes in $\sigma(y)$ that is evolutionary most closely related to x.

The best match graph $G(T, \sigma)$, a directed graph, represents these relationships, with an arrow xy indicating y is the best match of x. The tree (T, t, σ) *explains* $G(T, \sigma)$.

For any directed graph G and a node-coloring map $\sigma : V(G) \to M$, the *informative triples* $\mathcal{R}(G, \sigma)$ ascertain if G is a BMG. A triple $r \in \mathcal{R}(G, \sigma)$ exists with $L(r) = x, y, y' \in V(G)$ and $\sigma(x) \neq \sigma(y) = \sigma(y')$ if $xy \in E(G)$, $xy' \notin E(G)$, and, if T is binary, $yy'|x$ for both $xy, xy' \in E(G)$. (G, σ) is a BMG if and only if $(G, \sigma) = G(\text{aho}(\mathcal{R}(G, \sigma)), \sigma)$ [10,25]. Figure 3AD-G depicts the interplay between gene trees, best match graphs, and informative triples.

A.4   Selection of Best Hits

Each alignment hit $\overrightarrow{xy}$ is associated with a *bit score* $\omega(\overrightarrow{xy})$, we estimate the evolutionary relatedness between two genes x and y as the *normalized bit-score* $\omega_{xy} = (\omega(\overrightarrow{xy})/\text{length}(y) + \omega(\overrightarrow{yx})/\text{length}(x))/2$.

For each gene x, we identify the most closely related genes in a different species $Y \neq \sigma(x)$. A gene y from species $\sigma(y) = Y$ is considered a *best hit* of x if its alignment hit score ω_{xy} meets or exceeds an *adaptive threshold* defined as $f \cdot \omega_{x|Y}$, where $\omega_{x|Y} = \max(\omega_{xy}$ where $\sigma(y) = Y)$. Here, f is a factor between zero and one. This threshold, aimed at identifying paralogous best hits, was introduced in [22], and we set $f = 0.95$.

A.5   From Best Hits to Gene Trees

We start by constructing a *best hit graph* (G, σ), which is a directed graph where nodes are the genes of the orthogroup, and there is an arrow xy if y is best hit of x. Then we proceed to find a least resolved gene tree (T^*, σ) that maximizes the similarity of the best hit graph (G, σ) and the best match graph $G(T^*, \sigma)$. To do so, we use the heuristic introduced in [26], which consists of finding the maximum set of consistent, informative triples $\mathcal{R}(G, \sigma)$.

The three (T^*, σ) are further refined into an *augmented tree* (T, σ), which allows us to assign evolutionary events in such a way that duplication events are minimized while maintaining the same best match graph, this is $G(T^*, \sigma) = G(T, \sigma)$ [27].

Now, we create the evolutionary events map $t : V(T) \rightarrow \{\bullet, \square, \odot\}$ in such a way that for a node $v \in V(T)$, if such a node is a leaf then we set $t(v) = \odot$, on the contrary, we set $t(v) = \bullet$ if $\sigma(C_T(v')) \cap \sigma(C_T(v'')) = \emptyset$, otherwise $t(v) = \square$.

Finally, having the event-labeled gene tree (T, t, σ), we compute the orthology relation underling this tree as the relation that comprises all pairs (x, y) and (y, x) of genes x and y for which $t(\mathrm{lca}_T(x, y)) = \bullet$.

A.6   Consistency of Triple Sets

To reconcile a gene tree (T, t, σ) inconsistent with the species tree S, we modify (T, t, σ) to a consistent tree (T', t, σ). We differentiate between consistent triples $R_C = \{r \in \mathfrak{R}(T) : \sigma(r) \in R(S)\}$ and inconsistent triples $R_I = \mathfrak{R}(T) \backslash R_C$. The aim is to eliminate triples in R_I while retaining those in R_C. Removing a leaf $a \in L(T)$ also removes all triples $r \in R(T)$ with $a \in L(r)$. Utilizing this, we can select a subset of inconsistent leaves $L_I \subseteq L(R_I)$, set $L' = L(R_T) \backslash L_I$, and construct a consistent tree $T' = T_{|L'}$. The steps for this tree editing are outlined in Algorithm 1.

A.7   Tree Reconciliation

Once we ensure consistency between the gene tree (T, t, σ) and species trees S, we perform a reconciliation map as follows.

Algorithm 1: Prune-L

---

input : A gene tree T, and a species tree S.

output: A gene tree T' consistent with the species tree S.

1 $R_C \leftarrow$ set of consistent triples;
2 $R_I \leftarrow$ set of inconsistent triples;
3 $L \leftarrow L(T)$;
4 $L \leftarrow \emptyset$;
5 Let w_x^C be the number of triples $r \in R_C$ where $x \in L(r)$;
6 Let w_x^I be the number of triples $r \in R_I$ where $x \in L(r)$;
7 **while** $R_I \neq \emptyset$ **do**
8 $\quad$ $m \leftarrow \max(\{w_x^I \text{ for } x \in L\})$;
9 $\quad$ $X \leftarrow \{x \in L \text{ such that } w_x^I = m\}$;
10 $\quad$ $x \leftarrow \underset{x \in X}{\mathrm{argmin}}\,(w_x^C)$;
11 $\quad$ Pop x from L and add it to L';
12 $\quad$ Delete the triples $r \in R_C \cup R_I$ where $x \in L(r)$;
13 $\quad$ Update w_x^C and w_x^I ;
14 **end**
15 $T' \leftarrow T_{|L'}$;
16 **return** T'

Lets assume that $x, y \in V(T)$, then the reconciliation map $\mu : V(T) \to V(S) \cup E(S)$ from the gene tree (T, t, σ) to the species tree S satisfies:

(U0) *Root Constrain.* $\mu(0_T) = 0_S$

(U1) *Gene Constrain.* If $t(x) = \odot$, then $\mu(x) = \sigma(x) \in L(S)$

(U2) *Speciation Constrain.* If $t(x) = \bullet$, then $\mu(x) = \mathrm{lca}_S(\sigma(C_T(x))) \in V^0(S)$, and
$\quad \mu(y_0) \parallel_S \mu(y_1)$ for any two distinct children $y_0, y_1 \in \mathrm{ch}_T(x)$

(U3) *Duplication Constrain.* If $t(x) = \square$, then $\mu(x) = e \in E(S)$ and $\mathrm{lca}_S(\sigma(C_T(x))) \prec e$

(U4) *Ancestor Constrain.* If $x \prec_T y$, then $\mu(x) \preceq_S \mu(y)$

The reconciliation map $\mu : V(T) \to V(S) \cup E(S)$ is computed in linear time [15]. Given a node $v \in V(T)$ such that $t(v) \neq \square$, it is straightforward to determine which element of $V(S) \cup E(S)$ corresponds to $\mu(v)$ by just looking at constraints $U0 - 2$, in the case when $t(v) = \square$ we set $\mu(v) = xy \in E(S)$ such that $y = \mathrm{lca}_S(\sigma(C_T(x)))$, this assignation minimizes the gene-loss events.

A.8 Resolving Speciation Nodes

To refine a node $x \in V(T)$ with more than two children via a map $f : \mathrm{ch}_T(x) \to y_0, y_1$, perform: (i) add nodes y_0, y_1 to the tree, (ii) remove edges xy for each $y \in \mathrm{ch}_T(x)$, and (iii) add edges xy_0, xy_1, and $f(y)y$ for each $y \in \mathrm{ch}_T(x)$. When u is a speciation node, use the reconciliation map μ and map $f : \mathrm{ch}_T(u) \to \mathrm{ch}_S(\mu(u))$ to resolve u. For $v \in \mathrm{ch}_T(u)$ and $v' \in \mathrm{ch}_S(\mu(u))$, set $f(v) = v'$ iff $\mu(v) \preceq v'$.

A.9   Inferring Gene Loss

For a speciation node $x \in V(T)$ in the gene tree, the reconciliation map μ helps detect gene losses by mapping x to a node $y = \mu(x)$ in the species tree. If we find a node $y' \in \mathrm{ch}_S(y)$ for which all nodes $x' \in V(T)$ satisfying $x' \prec x$ also fulfill $\mu(x') \parallel_S y'$, a gene loss at y' is inferred.

References

1. Aho, A.V., Sagiv, Y., Szymanski, T.G., Ullman, J.D.: Inferring a tree from lowest common ancestors with an application to the optimization of relational expressions. SIAM J. Comput. **10**(3), 405–421 (1981). https://doi.org/10.1137/0210030
2. Altschul, S.F., et al.: Gapped BLAST and PSI-BLAST: a new generation of protein database search programs. Nucleic Acids Res. **25**(17), 3389–3402 (1997). https://doi.org/10.1093/nar/25.17.3389
3. Bininda-Emonds, O.: Phylogenetic Supertrees: Combining Information to Reveal the Tree of Life. Computational Biology, Springer, Dordrecht (2004). https://doi.org/10.1007/978-1-4020-2330-9
4. Buchfink, B., Reuter, K., Drost, H.G.: Sensitive protein alignments at tree-of-life scale using DIAMOND. Nat. Methods **18**, 366–368 (2021). https://doi.org/10.1038/s41592-021-01101-x
5. Dress, A., Huber, K.T., Koolen, J., Moulton, V., Spillner, A.: Basic Phylogenetic Combinatorics. Cambridge University Press, Cambridge (2011). https://doi.org/10.1017/CBO9781139019767
6. Emms, D.M., Kelly, S.: OrthoFinder: phylogenetic orthology inference for comparative genomics. Genome Biol. **20**, 1–14 (2019)
7. Fitch, W.: Homology: a personal view on some of the problems. Trends Genet. **16**, 227–231 (2000). https://doi.org/10.1016/S0168-9525(00)02005-9
8. Fuentes, D., Molina, M., Chorostecki, U., Capella-Gutiérrez, S., Marcet-Houben, M., Gabaldón, T.: PhylomeDB V5: an expanding repository for genome-wide catalogues of annotated gene phylogenies. Nucleic Acids Res. **50**(D1), D1062–D1068 (2021). https://doi.org/10.1093/nar/gkab966
9. Gabaldón, T., Koonin, E.V.: Functional and evolutionary implications of gene orthology. Nat. Rev. Genet. **14**(5), 360–366 (2013)
10. Geiß, M., et al.: Best match graphs. J. Math. Biol. **78**(7), 2015–2057 (2019). https://doi.org/10.1007/s00285-019-01332-9
11. Geiß, M.: Best match graphs and reconciliation of gene trees with species trees. J. Math. Biol. **80**(5), 1459–1495 (2020)
12. Hellmuth, M.: Biologically feasible gene trees, reconciliation maps and informative triples. Algorithms Mol. Biol. **12**(1), 23 (2017). https://doi.org/10.1186/s13015-017-0114-z
13. Hellmuth, M., Stadler, P.F.: The theory of gene family histories. arXiv preprint arXiv:2304.11826 (2023)
14. Hellmuth, M., Wieseke, N., Lechner, M., Lenhof, H.P., Middendorf, M., Stadler, P.F.: Phylogenomics with paralogs. Proc. Natl. Acad. Sci. U.S.A. **112**, 2058–2063 (2015). https://doi.org/10.2307/2412448
15. Hernandez-Rosales, M., Hellmuth, M., Wieseke, N., Huber, K.T., Moulton, V., Stadler, P.F.: From event-labeled gene trees to species trees. BMC Bioinform. **13**(19), S6 (2012). https://doi.org/10.1186/1471-2105-13-S19-S6

16. Huerta-Cepas, J., Dopazo, H., Dopazo, J., Gabaldón, T.: The human phylome. Genome Biol. **8**, R109 (2007)

17. Kerfeld, C.A., Scott, K.M.: Using BLAST to teach "E-value-tionary" concepts. PLoS Biol. **9**(2), e1001014 (2011). https://doi.org/10.1371/journal.pbio.1001014

18. Klemm, P., Stadler, P.F., Lechner, M.: Proteinortho6: pseudo-reciprocal best alignment heuristic for graph-based detection of (co-) orthologs. Front. Bioinform. **3**, 1322477 (2023)

19. Kristensen, D., Wolf, Y., Mushegian, A., Koonin, E.: Computational methods for gene orthology inference. Brief. Bioinform. **5**(12), 399–420 (2019)

20. Kundu, S., Bansal, M.S.: SaGePhy: an improved phylogenetic simulation framework for gene and subgene evolution. Bioinformatics **35**(18), 3496–3498 (2019). https://doi.org/10.1093/bioinformatics/btz081

21. Le, S.Q., Gascuel, O.: An improved general amino acid replacement matrix. Mol. Biol. Evol. **25**, 1307–1320 (2008). https://doi.org/10.1093/molbev/msn067

22. Lechner, M., Findeiß, S., Steiner, L., Marz, M., Stadler, P.F., Prohaska, S.J.: **Proteinortho**: detection of (co-)orthologs in large-scale analysis. BMC Bioinform. **12**(1), 124 (2011). https://doi.org/10.1186/1471-2105-12-124

23. Python Software Foundation: Python language reference (2023). http://www.python.org

24. Schaller, D., et al.: Corrigendum to "Best match graphs". J. Math. Biol. **82**(6), 47 (2021). https://doi.org/10.1007/s00285-021-01601-6

25. Schaller, D., Geiß, M., Hellmuth, M., Stadler, P.F.: Best match graphs with binary trees. In: Martín-Vide, C., Vega-Rodríguez, M.A., Wheeler, T. (eds.) AlCoB 2021. LNCS, vol. 12715, pp. 82–93. Springer, Cham (2021). https://doi.org/10.1007/978-3-030-74432-8_6

26. Schaller, D., Geiß, M., Hellmuth, M., Stadler, P.F.: Heuristic algorithms for best match graph editing. Algorithms Mol. Biol. **16**(1), 19 (2021). https://doi.org/10.1186/s13015-021-00196-3

27. Schaller, D., Geiß, M., Stadler, P.F., Hellmuth, M.: Complete characterization of incorrect orthology assignments in best match graphs. J. Math. Biol. **82**(3), 20 (2021). https://doi.org/10.1007/s00285-021-01564-8

28. Semple, C., Steel, M., Steel, B.: Phylogenetics. Oxford Lecture Series in Mathematics and Its Applications, Oxford University Press, Oxford (2003)

29. Stadler, P.F., et al.: From pairs of most similar sequences to phylogenetic best matches. Algorithms Mol. Biol. **15**(1), 1–20 (2020). https://doi.org/10.1186/s13015-020-00165-2

30. Wu, B.Y.: Constructing the maximum consensus tree from rooted triples. J. Comb. Optim. **8**(1), 29–39 (2004). https://doi.org/10.1023/B:JOCO.0000021936.04215.68

31. Zhang, C., Mirarab, S.: ASTRAL-Pro 2: ultrafast species tree reconstruction from multi-copy gene family trees. Bioinformatics **38**(21), 4949–4950 (2022)

32. Zmasek, C.M., Eddy, S.R.: A simple algorithm to infer gene duplication and speciation events on a gene tree. Bioinformatics **17**(9), 821–828 (2001)

Gene Tree Parsimony in the Presence of Gene Duplication, Loss, and Incomplete Lineage Sorting

Prottoy Saha, Md. Shamiul Islam, Tasnim Rahman, Adiba Shaira,
Kazi Noshin, Rezwana Reaz, and Md. Shamsuzzoha Bayzid[✉]

Department of Computer Science and Engineering, Bangladesh University
of Engineering and Technology, Dhaka 1205, Bangladesh
shams_bayzid@cse.buet.ac.bd

Abstract. Inferring species trees from multi-locus data needs to account
for gene tree discordance due to various biological processes, including
incomplete lineage sorting (ILS) and gene duplication and loss (GDL).
Gene tree parsimony (GTP) is a popular approach for estimating species
trees that seeks to minimize the number of evolutionary events required
to reconcile the species tree with gene trees. Minimizing gene duplication and loss (MGD and MGDL) are GTP approaches that typically
make the simplifying assumption that population-related effects such as
incomplete lineage sorting (ILS) are negligible. However, this assumption is problematic for denser phylogenies, where ILS is more prominent.
Here, we extend the existing GTP methods to account for both GDL
and ILS by minimizing a weighted sum of the GDL and "deep" coalescence events required for a given collection of gene trees. We provide
a graph-theoretic characterization and present a dynamic programming
algorithm for this problem. Through an extensive evaluation study on
a collection of simulated and empirical datasets, we compared our proposed GTP approaches with the leading methods in the field.

Keywords: Gene tree · species tree · gene tree discordance · gene
duplication and loss · incomplete lineage sorting

1 Introduction

Estimating species trees from genes sampled throughout the whole genome is
now common due to the rapid growth rate of newly sequenced genomes. However, biological processes can result in different loci having different evolutionary
histories (known as gene tree discordance/heterogeneity), making species tree
estimation from multi-locus data challenging. Incomplete lineage sorting (ILS),
and gene duplication and loss (GDL) are two dominant reasons for gene tree
heterogeneity.

The standard approach for estimating species trees is known as concatenation (or combined analysis), which concatenates multiple sequence alignments of

C. Scornavacca and M. Hernández-Rosales (Eds.): RECOMB-CG 2024, LNBI 14616, pp. 110–128, 2024.
https://doi.org/10.1007/978-3-031-58072-7_6

different genes into a single super-alignment and then estimates a tree from this alignment. However, this approach cannot be used when the species' genomes contain multiple copies of some genes, which can result from gene duplication. Given the prevalent occurrence of gene duplication and loss, species tree estimation requires a different type of approach in this case. Given a collection of gene trees, gene tree parsimony (GTP) is an optimization problem that seeks a species tree that implies the minimum number of evolutionary events, such as gene duplications and gene losses. A natural optimization problem in this context is the *Minimize Gene Duplications* (MGD), which seeks a species tree by minimizing the total number of duplications required to explain the observed gene trees under the maximum parsimony reconciliation [15]. Related to MGD is the *Minimize Gene Duplication and Loss* (MGDL) problem, which considers both duplications and losses. DupTree [32], iGTP [7,10], DynaDup [3–5] and earlier similar dynamic programming based methods [18] are popular tools to tackle these NP-hard problems. Some powerful approaches, on the other hand, are less dependent on maximum parsimony reconciliation and are more agnostic regarding the causes of gene tree discordance. In other words, they do not look for a species tree that suggests the fewest number of evolutionary events, such as gene duplications and gene losses. PHYLDOG [7] and Guenomu [14] are two examples of such approaches, which co-estimate gene trees and species trees. These methods are computationally intensive and can only be applied to small datasets. Another strategy that employs a GDL-agnostic optimization criterion is MulRF [11]; however, it was demonstrated to be statistically consistent in a general duplication-only model of gene evolution [24]. ASTRAL-Pro [36] is a recent development, which is an enhanced version of the widely adopted species tree estimation method ASTRAL, to handle multiple-copy gene trees by accounting for both orthology and paralogy.

Non-parametric GTP approaches, MGD and MGDL, are commonly used alternatives that explicitly take GDL into account and are scalable to large datasets with hundreds of taxa and genes. These methods are suitable for eukaryotic organisms at sufficiently large evolutionary distances, where gene tree discordance typically arises due to duplication and loss events. However, at smaller evolutionary distances, ILS (also known as deep coalescence), modeled by the multi-species coalescent (MSC) [19], is a highly probable phenomenon that arises when gene copies fail to coalesce at the speciation point, meaning that gene copies at a single locus extend deeper than the speciation events [21]. Maddison [21] introduced the minimizing deep coalescence (MDC) problem as a parsimonious criterion for inferring the species tree from a set of gene trees, assuming the discordance is exclusively due to ILS. The MDC problem was later proved to be NP-hard [38]. Existing GTP approaches do not jointly model duplication, loss, and deep coalescence, thus limiting their applicability and accuracy. While unified probabilistic and discrete models exist for gene tree-species tree reconciliation to infer the most parsimonious (MP) history of a gene family in the presence of duplications, losses, and ILS [1,2,9,25,27,34], developing species tree estimation methods under a unified model is less explored. This paper addresses the GTP problem to consider both GDL and ILS events. Unlike the unified probabilistic models like DLCoal [25], we introduce a simpler parsimony problem MGDLX (minimize gene duplication, loss, and "extra lineages" resulting from deep coalescence events) which seeks a species tree that minimizes the summation of gene duplication, loss, and deep coalescent events. The key contributions of this study are as follows.

- We formalize the MGDLX problem under the GTP framework to jointly consider a weighted summation of different types of evolutionary events.
- We formulate MGDLX as a minimum weight maximum clique problem (see Theorem 4), and show how to solve this problem efficiently using dynamic programming. We show that this optimal clique can be found in polynomial time in the number of vertices of the graph, because of the special structure of the graphs. We also show that a constrained version of this problem, where the subtree-bipartitions of the species tree are drawn from the subtree-bipartitions of the input gene trees, can be solved in time that is polynomial in the number of gene trees and taxa.
- We evaluated MGD, MGDL, MGDLX with DupTree, MulRF, and ASTRAL-Pro in an extensive evaluation study containing a collection of simulated and empirical datasets.

2 Prior Terminology and Theory

2.1 Notations and Definitions

We now define some general terminology we will use throughout this paper; other terminology will be introduced as needed. We denote a gene tree by gt and a species tree by ST. The nodes in T are denoted by $V(T)$ and the set of taxa that appear in leaves is denoted by $L(T)$. We will assume that gene trees and species trees are rooted binary trees, with leaves drawn from the set of n taxa, and we allow the gene trees to have multiple copies of the taxa, and even to miss some taxa. Note that since T can have multiple copies of some taxa, it is possible for $|L(T)|$ to be smaller than the number of leaves in T. We consider the *restriction-based* approach, which was used by most of the literature [3,4,16,17,20,38], where an incomplete gene tree gt is reconciled with a species tree ST by taking the homeomorphic subtree ST' of ST (i.e., ST is restricted to the taxon set of gt, denoted by $ST|_{L(gt)}$, and then nodes with in-degree and out-degree 1 are suppressed). Other approaches for handling incomplete gene trees have also been proposed (see [4,6]). A subtree of T rooted at a node v in T is called a *clade*. The set of taxa that appear in the leaves in a clade rooted at v is called a *cluster* and is denoted by $c_T(v)$.

The most recent common ancestor (MRCA) of a set A of leaves in T is denoted by $MRCA_T(A)$. Given a gene tree gt and a species tree ST, where $L(gt) \subseteq L(ST)$, we define $\mathcal{M} : V(gt) \rightarrow V(ST)$ by $\mathcal{M}(v) = MRCA_{ST}(c_{gt}(v))$. The optimal embedding for various criteria we discuss (e.g., MDC, MGD, MGDL) is obtained using $\mathcal{M}$ [12,21,37,38]. For a rooted gene tree gt and a rooted species tree ST, where $L(gt) \subseteq L(ST)$, an internal node v in gt is called a *duplication node* if $\mathcal{M}(v) = \mathcal{M}(v')$, for some child v' of v, and otherwise v is a *speciation node*. The minimum number of losses required to reconcile a gene tree with a species tree is obtained under the MRCA mapping and standard formula based on the MRCA mapping has been extensively studied [3,4,16,17,20]. *Extra lineages* (resulting from deep coalescence) is defined by embedding the gene tree gt into the species tree ST, and then counting the number of lineages on each edge of the species tree. The number of extra lineages in a branch of T is the number of lineages on it minus 1 [21,30].

We denote by $Dup(gt, ST)$ and $loss(gt, ST)$ the number of duplication nodes and the number of losses, respectively. $Duploss(gt, ST)$ is defined as the sum of $Dup(gt, ST)$ and $loss(gt, ST)$. $XL(gt, ST)$ denotes the number of extra lineages of a species tree ST with respect to a gene tree gt. For a set $\mathcal{G}$ of gene trees, the number of duplications is denoted by $Dup(\mathcal{G}, ST) = \sum_{gt \in \mathcal{G}} Dup(gt, ST)$, the number of losses is denoted by $loss(\mathcal{G}, ST) = \sum_{gt \in \mathcal{G}} loss(gt, ST)$, and the number of extra lineages is denoted by $XL(\mathcal{G}, ST) = \sum_{gt \in \mathcal{G}} XL(gt, ST)$. $Duploss(\mathcal{G}, ST)$ is the sum of the number of duplications and losses, i.e., $Duploss(\mathcal{G}, ST) = Dup(\mathcal{G}, ST) + loss(\mathcal{G}, ST)$.

Subtree-Bipartition, Domination, Compatibility Graph. The material in this section is from [3–5]. We have included this material to make this paper self-contained and easy to follow.

Bayzid et al. [3,4] introduced the concept of "subtree-bipartition" of an internal node u in a rooted, binary tree T, denoted by $\mathcal{SBP}_T(u)$, which is the bipartition $c_T(l)|c_T(r)$, where l and r are the two children of u. The set of subtree-bipartitions of a tree T is denoted by $\mathcal{SBP}_T = \{\mathcal{SBP}_T(u) : u \in V_{int}(T)\}$. Given a pair of subtree-bipartitions $BP_i = P_{i_1}|P_{i_2}$ and $BP_j = P_{j_1}|P_{j_2}$, where P_{i_1}, P_{i_2}, P_{j_1} and P_{j_2} represent clusters; BP_i is *dominated by* BP_j (and conversely that BP_j dominates BP_i) if either of the following two conditions holds: (1) $P_{i_1} \subseteq P_{j_1}$ and $P_{i_2} \subseteq P_{j_2}$, or (2) $P_{i_1} \subseteq P_{j_2}$ and $P_{i_2} \subseteq P_{j_1}$. A subtree-bipartition $A|B$ is dominated by a species tree T if one of T's subtree-bipartitions dominates $A|B$. Given a set $\mathcal{G}$ of rooted binary gene trees on the set $\mathcal{X}$ of n taxa, the compatibility graph $CG(\mathcal{G})$ has one vertex for each possible subtree-bipartition defined on $\mathcal{X}$ and there is an edge between two vertices if and only if the associated subtree-bipartitions are compatible, i.e., they can co-exist in a single tree. Note that if two subtree-bipartitions are compatible, then their associated clusters (produced by unioning the two parts of the bipartition) are also either disjoint or one contains the other [3].

3   Solving MGDLX

The input to $MGDLX$ is a set $\mathcal{G} = \{gt_1, gt_2, \ldots, gt_k\}$ of rooted gene trees such that $L(gt_i) \subseteq \mathcal{X}$, where $i \in 1, 2, \ldots, k$. The output is a rooted and binary species tree ST on $\mathcal{X}$ such that the summation of duplication, loss, and extra lineage cost $DLX(\mathcal{G}, ST) = Dup(\mathcal{G}, ST) + loss(\mathcal{G}, ST) + XL(\mathcal{G}, ST)$ is minimized. We can also incorporate different weights for different evolutionary events. Let c_d, c_l, and c_{xl} be the cost of a duplication, loss, and extra lineage, respectively. In that case, the weighted version of MGDLX seeks a species tree that minimizes $c_d Dup(\mathcal{G}, ST) + c_l loss(\mathcal{G}, ST) + c_{xl} XL(\mathcal{G}, ST)$.

3.1   Prior Results on MGD

Definition 1 (From [3]). *An internal node u in a gene tree gt is a speciation node with respect to a species tree ST if $\mathcal{SBP}_{gt}(u)$ is dominated by ST. Otherwise, this is a duplication node.*

Let $\mathcal{SBP}_{dom}(gt, ST)$ be the set of subtree bipartitions in gt that are dominated by ST. That means, $\mathcal{SBP}_{dom}(gt, ST) = \{bp : bp \in \mathcal{SBP}_{gt}$ and bp is dominated by $ST\}$. Note that, by Definition 1, $|\mathcal{SBP}_{dom}(gt, ST)|$ is the number of speciation nodes in gt with respect to ST. Subsequently, the number of duplication nodes in gt is $(n-1) - |\mathcal{SBP}_{dom}(gt, ST)|$.

Bayzid *et al.* [3] proposed a graph theoretic characterization of the MGD problem and proposed a dynamic programming solution accordingly. In the context of MGD, the weight of a node v (corresponding to a subtree-bipartition $X|Y$) in a compatibility graph $CG(\mathcal{G})$ is defined as follows [3].

$$W_{dom}(v) = \sum_{gt \in \mathcal{G}} |\{bp : bp \in \mathcal{SBP}_{gt} \text{ and } bp \text{ is dominated by } v\}| \qquad (1)$$

Subsequently, the weight of a clique $\mathcal{C}$ in $CG(\mathcal{G})$ is defined as follows.

$$W_{dom}(\mathcal{C}) = \sum_{v \in \mathcal{C}} W_{dom}(v)$$

Theorem 1 (From [3]). *Let $\mathcal{G} = \{gt_1, gt_2, \ldots, gt_k\}$ be a set of k rooted binary gene trees on the n taxa in $\mathcal{X}$. Let $\mathcal{C}$ be an $(n-1)$-clique in $CG(\mathcal{G})$ maximizing $W_{dom}(\mathcal{C})$, and let ST be the species tree defined by the clique (so that SBP_{ST} corresponds to $\mathcal{C}$). Then ST is a binary species tree that optimizes MGD with respect to $\mathcal{G}$.*

3.2   Prior Results on Minimizing Deep Coalescence (MDC)

Given a collection of gene trees $\mathcal{G} = \{gt_1, gt_2, \ldots, gt_k\}$, MDC seeks a species tree T such that the total number of extra lineages (EL) required for $\mathcal{G}$, denoted by $XL(T, \mathcal{G}) = \sum_i (XL(T, gt_i))$, is minimized.

For any cluster A in gt and a cluster B in ST, A is B-maximal if (1) $A \subseteq B$, and (2) for any cluster A' in gt, if $A \subseteq A'$, then $A' \not\subseteq B$ (meaning that no other cluster of gt containing A is a subset of B). We define $k_B(gt)$ to be the number of B-maximal clusters within gt.

Yu *et al.* [35] proposed a graph theoretic characterization of the MDC problem and showed how to compute the extra lineage cost of a gene tree gt with respect to a species tree ST.

Theorem 2 (From [35]). *Let gt be a rooted binary gene tree and ST a species tree on the same set of leaves. Then $MDC(gt, ST) = \sum(k_B(gt) - 1)$, where B ranges over the clusters of $ST|_{L(gt)}$.*

Definition 2 (From [3]). *Let v be a node with subtree-bipartition $p|q$, and $B = p \cup q$. We set $W_{xl}(v) = 0$ if $B \cap L(gt) = \emptyset$, and otherwise $W_{xl}(v) = k_B(gt) - 1$.*

3.3   Graph Theoretic Characterization for MGDLX

To assign appropriate weights to the vertices in a compatibility graph $CG(\mathcal{G})$ so that the total weight of a $(n-1)$-clique in $CG(\mathcal{G})$ can be used to compute the cost $DLX(\mathcal{G}, ST) = Dup(\mathcal{G}, ST) + loss(\mathcal{G}, ST) + XL(\mathcal{G}, ST)$ of a species tree ST, we leverage the relationship between different costs established in [38].

Theorem 3 (From [38]). *Let gt be a rooted binary gene tree and ST a rooted binary species tree such that $L(gt) \subseteq L(ST)$. Then, $loss(gt, ST) = MDC(gt, ST) + 2Dup(gt, ST)$. For an arbitrary gene tree where there may be more than one gene copy from a species, or missing taxa (i.e., $L(gt) \neq L(ST)$), gt is mapped to the restriction $ST|_{L(gt)}$ of ST. Then, $loss(gt, ST) = MDC(gt, ST) + 2Dup(gt, ST) + (|V(gt)| - |V(ST|_{L(gt)})|)$.*

We now prove in Theorem 4 that the minimum weight clique of size $n-1$ in an appropriately weighted compatibility graph corresponds to an optimal solution to the MGDLX problem.

Theorem 4. *Let $\mathcal{G} = \{gt_1, gt_2, \ldots, gt_k\}$ be a set of binary, rooted gene trees on set $\mathcal{X}$ of n species, and let $CG(\mathcal{G})$ be the compatibility graph with vertex weights defined by $W_{MGDLX}(v) = 2W_{xl}(v) - 3W_{dom}(v)$, where $W_{xl}(v)$ is as defined in Definition 2 and $W_{dom}(v)$ is the weight as defined in Eq. 1. Then, the set of bipartitions in an $(n-1)$-clique of minimum weight in $CG(\mathcal{G})$ defines a binary species tree ST that optimizes MGDLX, meaning that it minimizes $DLX(\mathcal{G}, ST)$.*

Proof. Let $\mathcal{C}$ be a clique of size $n-1$ and ST be the tree defined by the subtree-bipartitions represented by the nodes in $\mathcal{C}$. Let n_i be the number of leaves in gt_i, and $\sum_{i=1}^{k} n_i = N$. Also, note that $|V(gt_i)| = 2n_i - 1$. Then, by Theorems 1, 2, 3, and Definition 2,

$$DLX(\mathcal{G}, ST) = \sum_{i=1}^{k} (Dup(gt_i, ST)) + loss(gt_i, ST) + XL(gt_i, ST) + (|V(gt_i)| - |V(ST|_{L(gt_i)})|)$$

$$= \sum_{i=1}^{k} (Dup(gt_i, ST)) + [2(Dup(gt_i, ST)) + XL(gt_i, ST)] + XL(gt_i, ST)$$
$$+ (|V(gt_i)| - |V(ST|_{L(gt_i)})|)$$

$$= \sum_{i=1}^{k} 3Dup(gt_i, ST) + 2XL(gt_i, ST) + (|V(gt_i)| - |V(ST|_{L(gt_i)})|)$$

$$= \sum_{i=1}^{k} 3(n_i - 1) - 3 \sum_{v \in C} W_{dom}(v) + 2 \sum_{v \in C} W_{xl}(v) + \sum_{i=1}^{k}(2n_i - 1) - \sum_{i=1}^{k} |V(ST|_{L(gt_i)})|$$

$$= \sum_{v \in C} (2W_{xl}(v) - 3W_{dom}(v)) + \sum_{i=1}^{k} 3(n_i - 1) + \sum_{i=1}^{k}(2n_i - 1) - \sum_{i=1}^{k} |V(ST|_{L(gt_i)})|$$

$$= \sum_{v \in C} W_{MGDLX}(v) + \sum_{i=1}^{k}(5n_i - 4) - \sum_{i=1}^{k} |V(ST|_{L(gt_i)})|$$

$$= W_{MGDLX}(\mathcal{C}) + 5N - 4k - \sum_{i=1}^{k} |V(ST|_{L(gt_i)})|. \tag{2}$$

Note that $V(ST|_{L(gt_i)})$ does not depend on the topology of the species tree ST. Therefore, the clique with the minimum weight defines a tree ST that minimizes $DLX(\mathcal{G}, ST)$.

3.4   Dynamic Programming Approach for MGDLX

The graph-theoretic characterization of the optimal solution for MGDLX, given in Sect. 3.3, suggests an algorithm for finding the optimal solution, in which a max weight clique is sought in a compatibility graph. We present a DP algorithm similar to the ones proposed for MGD and MGDL in [3]. Let $\mathcal{SBP}$ be a set of input subtree-bipartitions. We compute $score(A)$ in order, from the smallest cluster to the largest cluster $\mathcal{X}$, corresponding to the subtree-bipartitions in $\mathcal{SBP}$. If $|A| = 2$, we set $score(A) = W_{MGDLX}(a_1|a_2)$, where $A = \{a_1, a_2\}$. When A contains more than two taxa, $score(A)$ is computed as follows.

$$score(A) = min\{score(A_1) + score(A_2) + 2W_{xl}(A_1|A_2) - 3W_{dom}(A_1|A_2) : A_1|A_2 \in \mathcal{SBP}\}$$

The score (duplication+loss+XL) of the optimal species tree ST is given by $score(\mathcal{X}) + 5N - 4k - \sum_{i=1}^{k} |V(ST|_{L(gt_i)})|$, by Theorem 4 and Eq. 2. $\mathcal{SBP}$ contains all possible subtree-bipartitions if an exact solution is desired. Otherwise, if $\mathcal{SBP}$ contains only those subtree-bipartitions from the input gene trees, then the algorithm finds the optimal constrained species tree in time that is polynomial in the number of gene trees and taxa.

We now have the following Theorem.

Theorem 5. *Let $\mathcal{G}$ be a set of k rooted, binary gene trees and $\mathcal{SBP}$ a set of subtree-bipartitions. Then the DP algorithm finds the species tree ST minimizing the total number of duplications, losses, and extra lineages subject to the* constraint that $\mathcal{SBP}_{ST} \subseteq \mathcal{SBP}$ in $O(n|\mathcal{SBP}|^2)$ time. Therefore, if $\mathcal{SBP}$ is all possible subtree-bipartitions, we have an exact but exponential time algorithm. However, if $\mathcal{SBP}$ contains only those subtree-bipartitions from the input gene trees, then the DP algorithm finds the optimal constrained species tree in $O(n^3k^2)$.

Proof. As a preprocessing step, for every subtree-bipartition $A_1|A_2 \in \mathcal{SBP}$, we compute $W_{xl}(A_1|A_2)$ - $W_{dom}(A_1|A_2)$ in $n|\mathcal{SBP}|^2$ time. Consequently, for a cluster A, $score(A)$ can be computed in $|\mathcal{SBP}|$ time since at worst we need to look at every subtree-bipartition in $\mathcal{SBP}$. Hence, the theorem follows.

4   Experimental Studies

4.1   Dataset

Simulated Dataset. The relative performance of various GTP methods compared to ASTRAL-Pro and MulRF has been extensively tested on previously used simulated datasets, utilizing a 25-taxa dataset analyzed in the ASTRAL-Pro study [36]. The dataset was produced using SimPhy [22] starting from a default model condition and adjusting different parameters (e.g., duplication rate, duploss rate, ILS level and number of genes). True gene trees were simulated under the DLCoal model [25] which is a combined model of ILS and gene duplication and loss. Next, gene sequence alignments were simulated from true gene family trees and finally gene trees were estimated from gene sequence alignments. The default setting, containing 25 ingroup species and an outgroup and 1000 genes, has a duplication rate (λ_+), which is also equal to the loss rate (λ_-) of 4.9×10^{-10}. Gene tree estimation error was controlled by varying sequence lengths (100 bp and 500 bp). The default ILS level is $[60\%, 80\%]$. We performed experiments with varying duplication rates from $0\sim5\lambda_+$ and loss/duplication ratios from $0\sim1$ (0, 0.1, 0.5, 1) with all other parameters kept the same as the default. We also performed some experiments with varying duplication rates and ILS levels.

Moreover, we assessed the GTP methods alone on a collection of single-copy datasets: the 48-taxon avian and 37-taxon mammalian simulated datasets from [23], and the 11-taxon dataset from [13]. These datasets contain different levels of ILS (from moderately low to extremely high levels of ILS), and range in terms of gene tree estimation errors (controlled by sequence lengths) and numbers of genes.

Empirical Dataset. We assessed the performance of GTP methods on two popular biological datasets: (1) the transcriptome dataset of land plant species (Plants83), which was first analyzed by [33], and (2) a dataset of 16 yeast species (Fungi16) with 7,280 multicopy gene trees from [8].

4.2   Methods and Measurements

The DP solutions to MGD and MGDL problems [3] were implemented in the DynaDup software package. This study introduces DP solutions for the Maximum Gene Duplication with Lineage Constraints (MGDLX) problem, which have also been integrated into the DynaDup software package. These methods are denoted as DynaDup-D, DynaDup-DL, and DynaDup-DLX. In our experiments, we set equal cost/weight to duplication, loss and extra lineage events, i.e., $c_d = c_l = c_{xl} = 1$. We compared these GTP methods with the leading species tree estimation methods that can handle multi-copy gene trees, such as ASTRAL-Pro (maximizing a refined measure of quartet similarity, accounting for both orthology and paralogy), DupTree [32] (minimizing the duplication reconciliation cost, and thus is a solution to the MGD problem), and MulRF [11] (optimizing an extension of the RF distance [26] to multi-labeled trees).

On the simulated datasets, we compared the estimated trees with the model species tree using normalized Robinson-Foulds (RF) distance [26]. For the biological dataset, we compared the estimated species trees with established evolutionary relationships. We assessed support values in the estimated trees using local posterior probabilities [29] computed by ASTRAL-Pro.

5   Results and Discussion

We performed two separate experiments to assess: (1) the relative performance of different GTP methods (MGD, MGDL, and MGDLX) considering various reasons for gene tree discordance, and (2) the performance of GTP methods compared to the best existing species tree estimation methods (e.g., ASTRAL-Pro, MulRF, DupTree) that can handle both orthology and paralogy with multi-copy gene trees.

5.1   Experiment 1: Comparing MGD, MGDL and MGDLX

Results on 11-Taxon Dataset. Figure 1 shows the results on 11-taxon dataset with varying levels of ILS and numbers of genes for both estimated and true gene trees (and thus accounting for the impact of gene tree estimation error (GTEE)). As expected, error rates of the methods decrease as we increase the number of genes, decrease the amount of ILS, and use true gene trees with no GTEE. On true gene trees, all the GTP methods perform well and there is no statistically significant differences ($P > 0.05$) between them. However, DynaDup-DLX is better than other on two model conditions, albeit the differences are small. For the estimated gene trees, DynaDup-DLX is notably better than other methods on five (out of 10) model conditions. These differences are statistically significant ($P < 0.05$) on three conditions. DynaDup-DL is the second best method followed by DynaDup-D and DupTree. These results suggest that, on this particular dataset, minimizing the summation of duplication, loss and deep coalescence events lead to better species trees than minimizing only the number of duplications or duplications and losses. DupTree was worse than other methods on most of the model conditions, especially with relatively small numbers of genes.

Results on 48-Taxon Dataset. The performance of the GTP methods on the 48-taxon dataset with varying ILS levels and numbers of gene trees is shown in

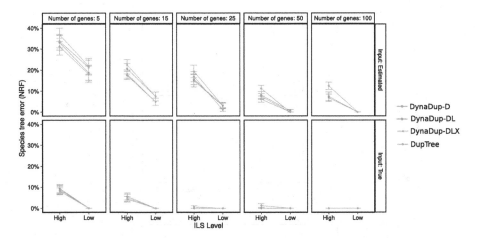

Fig. 1. Average RF rates on the 11-taxon dataset varying the number of genes and ILS levels for estimated and simulated gene trees over 50 replicates.

Fig. 2(a). On the estimated gene trees with 500 bp sequences, DynaDup-DLX and DynaDup-DL are more accurate than DynaDup-D and DupTree (as was observed on the 11-taxon dataset). Moreover, similar to the 11-taxon dataset, DynaDup-D was more accurate than DupTree. Interestingly, however, when considering the true gene trees, the trend was exactly the opposite – DupTree outperformed DynaDup-D, and DynaDup-D and DupTree are consistently and significantly better than DynaDup-DL and DynaDup-DLX.

Results on 37-Taxon Dataset. We evaluated the performance of the GTP methods on the 37-taxon dataset under various model conditions (Fig. 2(b)). For both estimated and true gene trees, DupTree and DynaDup-D performed comparably (with no statistically significant differences between them) and they substantially outperformed DynaDup-DL and DynaDup-DLX. DynaDup-DLX produced the least accurate results. Notably, while the error rate for DupTree and DynaDup-D dramatically reduced with increasing numbers of genes, the error rates of DynaDup-DL and DynaDup-DLX did not improve much beyond 100 gene trees – empirical evidence for the solutions to MGDL and MGDLX problems to be not statistically consistent.

5.2   Experiment 2: Comparing GTP Methods with the Best Alternative Species Tree Estimation Methods

Results on S25 Taxon Dataset. We compared MGD, MGDL, and MGDLX with ASTRAL-Pro, MulRF, and DupTree. The estimated gene trees in this dataset are unrooted. While ASTRAL-Pro, DupTree and MulRF can handle unrooted trees, DynaDup requires rooted gene trees. Therefore, we rooted them

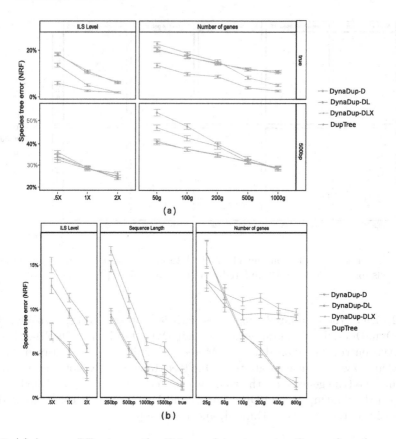

Fig. 2. (a) Average RF rates on the 48-taxon dataset varying the number of genes and ILS level for estimated and true gene trees over 20 replicates. The ILS level has been varied from 0.5X (highest) to 2X (lowest) while fixing the number of genes at 1000. The number of genes has been varied from 50 to 1000 with fixed 1X ILS. (b) Average RF rates on the 37-taxon dataset varying ILS level, the number of genes and sequence length over 20 replicates. The ILS level has been varied fixing the sequence length at 500bp and the number of genes at 200. The sequence length has been varied from 250bp to 1500bp with 200 gene trees and 1X ILS level (moderate ILS). The number of genes has been varied from 25 to 800, with 500bp sequence length and 1X ILS.

using the outgroup to run DynaDup on this dataset. We used the MulRF-estimated trees, which were provided by the ASTRAL-Pro study (instead of re-running it).

Varying Duplication Rate and Loss/Duplication Ratio. We assessed the performance of various methods under varying duplication rates and loss/duplication rates while fixing ILS level at 70% (Fig. 3(a)). On true gene trees, DynaDup, DupTree, and ASTRAL-Pro exhibited the highest accuracy, with no statistically significant differences among them. These three methods significantly outperformed others (DynaDup-DL, DynaDup-DLX, and MulRF), where MulRF produced the least accurate trees. As we increased the gene tree estimation error (GTEE) by varying gene sequence lengths (500bp and 100bp), the relative accuracy of DupTree and DynaDup-D drastically declined compared to ASTRAL-Pro. With a 500bp sequence length, DupTree and DynaDup-D were comparable to ASTRAL-Pro, but ASTRAL-Pro significantly outperformed the others on model conditions with 100bp sequence (i.e., higher GTEE). Notably, increasing the duplication rate λ_+ decreases the error, possibly due to the additional copies offering more information, similar to augmenting the number of loci. Higher loss/dup rate tends to increase tree errors due to the increased amount of missing data. However, the methods are reasonably robust to loss rates, especially on model conditions with low gene tree estimation errors (i.e., 500bp and true model conditions). MulRF performed poorly across all model conditions, especially with higher duplication rates.

Varying Duplication Rate and ILS Level. We then repeat this experiment in the case where ILS level also varies with the duplication rate, where loss/duplication rate is kept as 1 (Fig. 3(b)). All methods perform well in conditions with no ILS. However, the performance of MulRF, DynaDup-DLX, and DynaDup-DL degrades drastically with increasing ILS levels. ASTRAL-Pro, DynaDup-D, and DupTree, on the other hand, are much more robust to the increasing levels of ILS. Interestingly, although DupTree and DynaDup-D do not consider ILS into account, they are much more tolerant to ILS than DynaDup-DL, DynaDup-DLX, and MulRF. Increasing the duplication rate notably improves the performance of the methods as additional copies provide more information. These improvements are more prominent on challenging model conditions with higher levels of ILS.

5.3   Results on Empirical Dataset

Plant (1KP) Dataset. [33] analyzed the 103 plant species dataset with 424 single-copy gene trees using ASTRAL. The original analysis had also inferred 9,683 multicopy gene trees with up to 2,395 leaves for 80 of the 103 species and three additional genomes (a total of 83). These gene trees were not used in the original study because there were no appropriate species tree techniques. Later, ASTRAL-Pro and DupTree were used to evaluate these multicopy gene trees [36]. In this study, We reanalyzed this dataset with DynaDup-D, DynaDup-DL, and DynaDup-DLX (Fig. 4).

DynaDup-D successfully recovered major clades (e.g., *Gymnosperms, Magnoliids, Monocots, Gnepine, Lycophytes, Mosses, Liverworts, Hornworts, Zygnematophyceae* and *Coleochaetales*) as monophyletic groups, with the exception of *Eudicots* and *Monilophytes*. It incorrectly placed Petridium as a sister to Hibiscus, albeit with a very low local posterior probability-based branch support (~ 0), thereby relocating it from *Monilophytes* to *Eudicots*. While DynaDup-D achieved monophyly for various clades, the internal resolutions within these clades differed at multiple places compared to the ASTRAL tree reported in [33], resulting in 25 edges varying from ASTRAL. DynaDup-D placed Amborella as

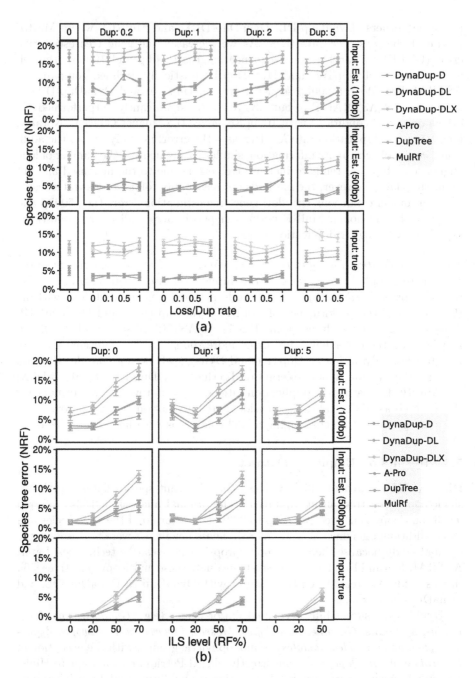

Fig. 3. Average RF rates on the S25 dataset for n = 25 taxa and k = 1000 gene trees. The performance of six methods have been evaluated over 50 replicates for true gene trees and estimated gene trees from 100 bp and 500 bp alignments (a) varying duplication rates (box column) and loss/duplication ratio (x-axis) and (b) varying duplication rates (box column) and ILS level (x-axis).

a sister to the rest of the *Angiosperms*, making it consistent with the ASTRAL and ASTRAL-Pro-estimated tree. Both ASTRAL-Pro and DynaDup-D support the *GnePine* hypothesis, where Gnetales is sister to Pinaceae, nested within the Coniferales. ASTRAL, on the other hand, supports the Gnetifier hypothesis (i.e., Gnetales is sister to Coniferales as a whole). DynaDup-D placed *Chara* as the sister to land plants, differing from ASTRAL-Pro and ASTRAL which put Zygnamatales as the sister to land plants.

We also applied DupTree to this dataset which differed from ASTRAL-tree in 33 branches. It did not recover *Eudicots* and *Monilophytes*. DynaDup-DL and DynaDup-DLX, on the other hand, differ from the ASTRAL-estimated tree in 24 and 28 branches, respectively.

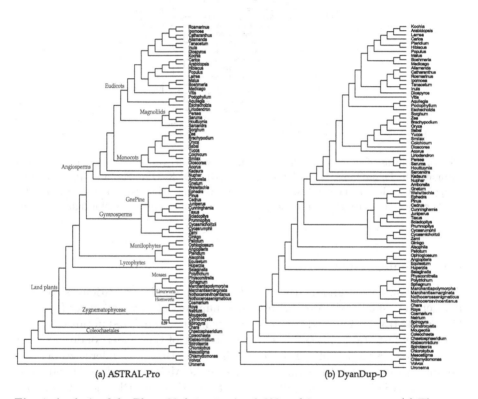

Fig. 4. Analysis of the Plants83 dataset using 9,683 multi-copy gene trees. (a) The tree estimated by ASTRAL-Pro, and (b) the tree estimated by DynaDup-D.

16 Species Yeast Dataset. [8] reconstructed a species tree with concatenation (combined analysis) using only single-copy gene trees based on 706 one-to-one orthologs. We re-analyzed this dataset with GTP methods using all the 7,280 multicopy gene trees. Notably, DynaDup-D, DynaDup-DL, DynaDup-DLX, DupTree, and ASTRAL-Pro returned an identical tree (Fig. 5). This tree differs from the tree reported in the original study [8] using single-copy gene trees on only one branch, regarding the relative position of *Saccharomyces castellii* and *Candida glabrata*. The original study placed *Saccharomyces castellii* as the sister to *Candida glabrata* and the *Saccharomyces* group by imposing a constraint during the maximum likelihood (ML) search. In contrast, the GTP approaches (DynaDup-D, DynaDup-DL, DynaDup-DLX, and DupTree), and ASTRAL-Pro reconstructed the sister relationship of *Candida glabrata* and *Saccharomyces*, which is aligned with the unconstrained ML search in the concatenated analysis and the tree reported by [28].

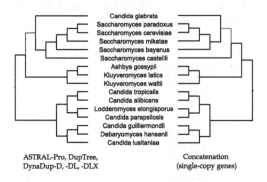

Fig. 5. Analysis of the Fungi16 dataset using 7,280 multicopy gene trees. Left: The tree estimated by DynaDup-D, DynaDup-DL, DynaDup-DLX, DupTree, and ASTRAL-Pro using 7,280 multicopy gene trees. Right: The tree estimated by concatenation using 706 single-copy genes with the red branch enforced as a constraint [8] (Color figure online).

5.4   Running Time

Table 1 shows the running time of different methods on various simulated datasets. On smaller datasets with limited numbers of taxa and genes (e.g. 11-, 25-, and 37-taxon simulated datasets), all the methods analyzed in this study are reasonably fast (take only a few seconds for most of the model conditions) and there is no substantial differences in running time. However, differences become notable as we increase the number of taxa/genes. MGD (DyanDup-D and Dup-Tree) and ASTRAL-Pro are significantly faster than MGDL and MGDLX on larger datasets.

Table 1. Running time of different methods on various datasets analyzed in this study. For simulated datasets, we show the average running time (over 50 replicates) on different model conditions.

Dataset	Model condition	Running time				
		MGD	MGDL	MGDLX	DupTree	ASTRAL-Pro
11-taxon	Estimated, 100gt, High ILS	0.102 s	0.131 s	0.142 s	0.033 s	0.029 s
37-taxon	500bp, 800gt, 1X ILS	1.805 s	2.867 s	2.918 s	14.852 s	5.990 s
48-taxon	2X, 1000gt, 500bp	38.960 s	55.245 s	55.396 s	55.410 s	17.055 s
S25 taxon	0 Dup: 0, ILS: 50	2.469 s	4.759 s	4.792 s	5.211 s	3.133 s
Fungi16		3.159 s	96.036 s	99.147 s	5.193 s	5.821 s
Plants83		3.436 h	51.161 h	51.162 h	52.327 min	17.369 min

6   Conclusions

In this work, we have presented a new algorithm for inferring the most parsimonious species tree from a collection of gene family trees that minimizes the total number of duplications, losses, and extra lineages required to reconcile the species tree with the gene trees. Previously, gene tree parsimony methods for inferring species trees were developed by separately taking deep coalescence (e.g., Phylonet-MDC [35]) or gene duplication and loss (e.g., DupTree and DynaDup) into account. However, no methods are available that minimize the summation of various evolutionary events. These parsimonious frameworks are easy yet effective approaches to take the reasons for gene tree heterogeneity into account while estimating species trees from a collection of gene trees.

Our experimental study underscores scenarios where MGDLX (DynaDup-DLX) and MGDL (DynaDup-DL) exhibit better performance than MGD (DynaDup-D and DupTree), as observed in 11- and 48-taxon datasets. However, our findings also revealed that MGD is better than MGDL and MGDLX under many model conditions (e.g., 37- and 25-taxon dataset). Especially, with highly accurate gene trees, MGD is as good as or even better than ASTRAL. This nuanced outcome indicates that the relative efficacy of GTP approaches is diverse, necessitating further investigations to identify conditions favoring specific GTP methodologies. Future studies should investigate if prior domain knowledge about the relative prevalence of various evolutionary events within particular datasets can be utilized to assign weights to different events, thereby aiming to minimize a weighted summation of number of duplication, loss, and extra lineages. Moreover, since the number of losses is usually substantially higher than the number of duplications or extra lineages, appropriate weights should be assigned to duplication and extra lineage events relative to loss events to more accurately reflect the relative impact of each type of event on the evolutionary history of the species. While DynaDup-DLX incorporates the capability to minimize such a weighted summation, further investigations are required to determine appropriate weights that contribute to improved species tree inference.

Overall, the performance of GTP approaches on various simulated and real biological datasets is promising. Although MGD, MGDL, and MGDLX are likely to be statistically inconsistent as MDC [31], these GTP approaches are not agnostic to the gene tree heterogeneity as they take into account the specific nature of the way incomplete lineage sorting or gene duplication and losses occur. Thus, these methods can be applied to both orthologous genes and gene families that by definition will include both paralogs and orthologs (in which case the gene trees could be multi-copy). In addition, with its parsimony approach, GTP is more applicable and scalable to a broad range of species and large datasets compared to sophisticated probabilistic approaches.

Data Availibility Statement. DynaDup-DLX is freely available in open source form at https://github.com/prottoy99/DynaDup. All the datasets analyzed in this paper are from previously published studies and are publicly available.

References

1. Ansarifar, J., Markin, A., Górecki, P., Eulenstein, O.: Integer linear programming formulation for the unified duplication-loss-coalescence model. In: Cai, Z., Mandoiu, I., Narasimhan, G., Skums, P., Guo, X. (eds.) ISBRA 2020. LNCS, vol. 12304, pp. 229–242. Springer, Cham (2020). https://doi.org/10.1007/978-3-030-57821-3_20

2. Arvestad, L., Berglund, A.C., Lagergren, J., Sennblad, B.: Bayesian gene/species tree reconciliation and orthology analysis using MCMC. Bioinform. Oxford **19**(1), 7–15 (2003)

3. Bayzid, M.S., Mirarab, S., Warnow, T.: Inferring optimal species trees under gene duplication and loss. In: Proceedings of Pacific Symposium on Biocomputing (PSB), vol. 18, pp. 250–261 (2013)

4. Bayzid, M.S., Warnow, T.: Gene tree parsimony for incomplete gene trees: addressing true biological loss. Algorithms Mol. Biol. **13**, 1 (2018)

5. Bayzid, M.S.: Inferring optimal species trees in the presence of gene duplication and loss: beyond rooted gene trees. J. Comput. Biol. **30**(2), 161–175 (2023)

6. Bayzid, M.S., Warnow, T.: Estimating optimal species trees from incomplete gene trees under deep coalescence. J. Comput. Biol. **19**(6), 591–605 (2012)

7. Boussau, B., Szöllősi, G.J., Duret, L., Gouy, M., Tannier, E., Daubin, V.: Genome-scale coestimation of species and gene trees. Genome Res. **23**(2), 323–330 (2013)

8. Butler, G., et al.: Evolution of pathogenicity and sexual reproduction in eight Candida genomes. Nature **459**, 657–662 (2009). https://doi.org/10.1038/nature08064

9. Chan, Y.B., Ranwez, V., Scornavacca, C.: Inferring incomplete lineage sorting, duplications, transfers and losses with reconciliations. J. Theoret. Biol. **432**, 1–13 (2017)

10. Chaudhary, R., Bansal, M.S., Wehe, A., Fernández-Baca, D., Eulenstein, O.: iGTP: a software package for large-scale gene tree parsimony analysis. BMC Bioinform. **11**, 1–7 (2010)

11. Chaudhary, R., Burleigh, J.G., Fernández-Baca, D.: Inferring species trees from incongruent multi-copy gene trees using the Robinson-Foulds distance. Algorithms Mol. Biol. **8**(1), 1–12 (2013). https://doi.org/10.1186/1748-7188-8-28

12. Chauve, C., Doyon, J.P., El-Mabrouk, N.: Gene family evolution by duplication, speciation, and loss. J. Comp. Biol. **15**(8), 1043–1062 (2008)

13. Chung, Y., Ané, C.: Comparing two Bayesian methods for gene tree/species tree reconstruction: a simulation with incomplete lineage sorting and horizontal gene transfer. Syst. Biol. **60**(3), 261–275 (2011)
14. De Oliveira Martins, L., Mallo, D., Posada, D.: A Bayesian supertree model for genome-wide species tree reconstruction. Syst. Biol. **65**(3), 397–416 (2016)
15. Goodman, M., Czelusniak, J., Moore, G., Romero-Herrera, E., Matsuda, G.: Fitting the gene lineage into its species lineage: a parsimony strategy illustrated by cladograms constructed from globin sequences. Syst. Zool. **28**, 132–163 (1997)
16. Górecki, P.: Reconciliation problems for duplication, loss and horizontal gene transfer. In: Proceedings of 8th Annual International Conference on Computational Molecular Biology, pp. 316 – 325 (2004)
17. Guigo, R., Muchnik, I., Smith, T.F.: Reconstruction of ancient molecular phylogeny. Mol. Phylogenet. Evol. **6**(2), 189–213 (1996)
18. Hallett, M.T., Lagergren, J.: New algorithms for the duplication-loss model. In: Proceedings of ACM Symposium on Computer Biology RECOMB2000, pp. 138–146. ACM Press, New York (2000)
19. Kingman, J.F.C.: The coalescent. Stoch. Processes Appl. **13**(3), 235–248 (1982)
20. Ma, B., Li, M., Zhang, L.: From gene trees to species trees. SIAM J. Comput. **30**(3), 729–752 (2000)
21. Maddison, W.P.: Gene trees in species trees. Syst. Biol. **46**, 523–536 (1997)
22. Mallo, D., de Oliveira Martins, L., Posada, D.: Simphy: phylogenomic simulation of gene, locus, and species trees. Syst. Biol. **65**(2), 334–344 (2016)
23. Mirarab, S., Bayzid, M.S., Boussau, B., Warnow, T.: Statistical binning enables an accurate coalescent-based estimation of the avian tree. Science **346**(6215), 1250463 (2014)
24. Molloy, E.K., Warnow, T.: FastMulRFS: fast and accurate species tree estimation under generic gene duplication and loss models. Bioinformatics **36**(Supplement–1), i57–i65 (2020)
25. Rasmussen, M.D., Kellis, M.: Unified modeling of gene duplication, loss, and coalescence using a locus tree. Genome Res. **22**(4), 755–765 (2012)
26. Robinson, D., Foulds, L.: Comparison of phylogenetic trees. Math. Biosci. **53**, 131–147 (1981)
27. Rogers, J., Fishberg, A., Youngs, N., Wu, Y.C.: Reconciliation feasibility in the presence of gene duplication, loss, and coalescence with multiple individuals per species. BMC Bioinform. **18**(1), 1–10 (2017)
28. Salichos, L., Rokas, A.: Inferring ancient divergences requires genes with strong phylogenetic signals. Nature **497**, 327–331 (2013). https://doi.org/10.1038/nature12130
29. Sayyari, E., Mirarab, S.: Fast coalescent-based computation of local branch support from quartet frequencies. Mol. Biol. Evol. **33**(7), 1654–1668 (2016)
30. Than, C.V., Nakhleh, L.: Species tree inference by minimizing deep coalescences. PLoS Comp. Biol. **5**(9), e1000501 (2009)
31. Than, C.V., Rosenberg, N.A.: Consistency properties of species tree inference by minimizing deep coalescences. J. Comp. Biol. **18**, 1–15 (2011)
32. Wehe, A., Bansal, M.S., Burleigh, J.G., Eulenstein, O.: Duptree: a program for large-scale phylogenetic analyses using gene tree parsimony. Am. J. Bot. **24**(13), 1540–1541 (2008)
33. Wickett, N.J., et al.: Phylotranscriptomic analysis of the origin and early diversification of land plants. Proc. Natl. Acad. Sci. **111**(45), E4859–E4868 (2014)

34. Wu, Y.C., Rasmussen, M.D., Bansal, M.S., Kellis, M.: Most parsimonious reconciliation in the presence of gene duplication, loss, and deep coalescence using labeled coalescent trees. Genome Res. **24**(3), 475–486 (2014)
35. Yu, Y., Warnow, T., Nakhleh, L.: Algorithms for MDC-based multi-locus phylogeny inference: beyond rooted binary gene trees on single alleles. J. Comput. Biol. **18**(11), 1543–1559 (2011)
36. Zhang, C., Scornavacca, C., Molloy, E.K., Mirarab, S.: ASTRAL-Pro: quartet-based species-tree inference despite paralogy. Mol. Biol. Evol. **37**(11), 3292–3307 (2020). https://doi.org/10.1093/molbev/msaa139
37. Zhang, L.: On a Mirkin-Muchnik-Smith conjecture for comparing molecular phylogenies. J. Comput. Biol. **4**(2), 177–188 (1997)
38. Zhang, L.: From gene trees to species trees II: species tree inference by minimizing deep coalescence events. IEEE/ACM Trans. Comput. Biol. Bioinf. **8**(9), 1685–1691 (2011)

Assessing the Potential of Gene Tree Parsimony for Microbial Phylogenomics

Samson Weiner[1], Yutian Feng[2], J. Peter Gogarten[2,3],
and Mukul S. Bansal[1,3(✉)]

[1] School of Computing, University of Connecticut, Storrs, USA
{samson.weiner,mukul.bansal}@uconn.edu
[2] Department of Molecular and Cell Biology, University of Connecticut, Storrs, USA
{yutian.feng,gogarten}@uconn.edu
[3] Institute for Systems Genomics, University of Connecticut, Storrs, USA

Abstract. A key challenge in microbial phylogenomics is that microbial gene families are often affected by extensive horizontal gene transfer (HGT). As a result, most existing methods for microbial phylogenomics can only make use of a small subset of the gene families present in the microbial genomes under consideration, potentially biasing their results and affecting their accuracy. One well-known approach for truly genome-scale phylogenomics is gene tree parsimony (GTP), which takes as input a collection of gene trees and finds a species tree that most parsimoniously reconciles with the input gene trees. While GTP based methods are widely used for phylogenomic studies of non-microbial species, their underlying reconciliation models are not designed to handle HGT and, therefore, they cannot be meaningfully applied to microbes. No GTP based methods have yet been developed for microbial phylogenomics.

In this work, we (i) design and implement the first GTP based approach, *PhyloGTP*, for microbial phylogenomics, (ii) use an extensive simulation study to systematically assess the accuracies of PhyloGTP and two other recently developed methods, SpeciesRax and ASTRAL-Pro-2, under a range of different conditions, and (iii) analyze two real microbial datasets with different characteristics. We find that PhyloGTP and SpeciesRax are more accurate than ASTRAL-Pro-2 across nearly all tested conditions, that PhyloGTP and SpeciesRax have similar accuracies overall, but there are conditions under which PhyloGTP consistently outperforms SpeciesRax, and that both PhyloGTP and Species-Rax can sometimes yield incorrect, misleading phylogenies on complex real datasets.

Keywords: Phylogenomics · Microbial evolution · Gene tree parsimony

1 Introduction

The accurate inference of phylogenetic relationships between different microbes is an important problem in evolutionary biology. A key difficulty in estimating

C. Scornavacca and M. Hernández-Rosales (Eds.): RECOMB-CG 2024, LNBI 14616, pp. 129–149, 2024.
https://doi.org/10.1007/978-3-031-58072-7_7

such phylogenies is the presence of extensive horizontal gene transfer (HGT) in microbial evolutionary histories. This can result in markedly different evolutionary histories for different gene families, obfuscating the underlying species-level or strain-level phylogeny. As a result, the traditional approach for reconstructing microbial phylogenies is to use only "well-behaved" gene families resistant to HGT. This includes the use of small-subunit ribosomal RNA genes (e.g., [48,65]) or of a concatenated alignment of a few core genes from the genomes of interest (e.g., [14,35,41]). Both these approaches, however, are known to be error-prone. For instance, ribosomal RNA genes are known to engage in horizontal transfer [24,66,68] and to yield histories that are inconsistent with those inferred using other core genes [17,18,29,30]. Furthermore, ribosomal RNA genes often cannot be used when studying closely related species due to excessive sequence similarity. Similarly, concatenation based approaches, such as the widely used multilocus sequence analysis (MLSA) technique [23], essentially ignore horizontal gene transfer and aggregate the phylogenetic signal from several gene families with potentially distinct evolutionary histories [22,44]. Indeed, the tree resulting from the concatenation might represent neither the organismal phylogeny nor any of the genes included in the concatenation [36].

To overcome these limitations, several genome-scale methods have also been proposed for microbial phylogeny inference. These include methods such as Phylo SI that are based on gene order information [54,55], supertree-based methods such as SPR supertrees [64] and MRP [8,68] that allow for the use of multiple orthologous gene families, and methods based on average nucleotide identity (ANI) of genomes [26,28,33]. Such genome-scale methods are inherently preferable to methods that base phylogeny reconstruction on only a single gene or a small set of concatenated genes [44]. However, while these above methods all represent useful approaches for microbial phylogenomics, they are either targeted at analyzing closely related strains or species (gene order and ANI based methods), or are limited to using single-copy gene families or orthologous groups and do not model key evolutionary events affecting microbial gene family evolution (supertree based methods). Recently, truly genome-scale approaches for microbial phylogenomics, capable of using thousands of complete (multi-copy) gene families, have also been developed. The two most prominent such methods are ASTRAL-Pro 2 [67] and SpeciesRax [46], both of which take as input a collection of unrooted gene family trees, where each gene family tree may contain zero, one, or multiple genes from any species/strain under consideration. ASTRAL-Pro 2 is based on quartets and seeks a species tree that maximizes a quartet based score [67]. While ASTRAL-Pro 2 does not directly model any specific evolutionary processes, such as HGT or gene duplication, responsible for gene tree discordance, it can handle complete (multi-copy) gene families and previous research suggests that it's quartet based approach should be robust to HGT [16]. SpeciesRax uses an explicit Duplication-Transfer-Loss model of gene family evolution in microbes and seeks a species tree that maximizes the reconciliation likelihood of observing the input gene trees under that model [46].

In this work, we propose a new approach for microbial phylogenomics and systematically compare its performance with ASTRAL-Pro 2 and SpeciesRax using simulated and real datasets. The new approach, called *PhyloGTP*, is based on *gene tree parsimony* (GTP), a well-known technique for phylogenomic inference. GTP provides a framework for inferring species trees from a collection of gene trees impacted by complex evolutionary processes. Specifically, GTP seeks a species tree that most parsimoniously reconciles all the input gene trees under an appropriately chosen model of gene-tree/species-tree reconciliation. Facilitated by effective software implementations [13,62], GTP is widely used for phylogenomic studies of multicellular eukaryotes (e.g., [12,27,40,42,43]), where the most appropriate reconciliation model is often the duplication-loss (DL) model [25]. To apply GTP to microbes, one must account for HGT by using the more complex Duplication-Transfer-Loss (DTL) reconciliation model. Despite its promise, GTP has not yet been implemented with DTL reconciliation and has therefore not yet been applied to microbial genomes. PhyloGTP addresses this gap, allowing for the first systematic assessment of GTP's potential for microbial phylogenomics. We note that PhyloGTP is conceptually similar to SpeciesRax since both methods are based on explicit DTL models of microbial gene family evolution, and both methods seek species trees that best reconcile the input gene trees under their DTL models. However, there are two key differences between PhyloGTP and SpeciesRax: First, PhyloGTP uses a standard, widely-used parsimony-based DTL model [3,60] while SpeciesRax uses a different, probabilistic DTL model [46]. And second, PhyloGTP and SpeciesRax use different heuristic search strategies to find their best species tree estimates, as we discuss later.

We use an extensive simulation study to evaluate the accuracies of PhyloGTP, ASTRAL-Pro 2, and SpeciesRax, focusing especially on the impact of number of input gene trees, DTL rates, and input gene tree error rates. We find that PhyloGTP and SpeciesRax are more accurate than ASTRAL-Pro-2 across nearly all tested conditions, that PhyloGTP often substantially outperforms SpeciesRax when the number of input gene trees is small or when DTL rates are high, and that SpeciesRax generally outperforms PhyloGTP on datasets with high gene tree error but low DTL rates. We also used PhyloGTP and SpeciesRax to analyze two real microbial datasets; a more complex 174-taxon Archaeal dataset exhibiting extreme divergence and compositional biases, and a less complex dataset of 44 Frankiales exhibiting low divergence. While both PhyloGTP and SpeciesRax perform well on these real datasets, they do result in a few clearly incorrect placements for the Archaeal dataset. This suggests that both PhyloGTP and SpeciesRax are potentially susceptible to biases present in complex datasets.

While our prototype implementation of PhyloGTP is considerably slower than SpeciesRax, our results establish GTP as a promising approach for microbial phylogenomics and show that PhyloGTP is capable of yielding more accurate microbial species trees for many datasets. At the same time, our results show that even reconciliation-based phylogenomic approaches like PhyloGTP and SpeciesRax may not produce accurate results for certain complex microbial datasets and that their results should be interpreted with caution. An open-source prototype implementation of PhyloGTP is available from https://github.com/samsonweiner/PhyloGTP.

2   Basic Definitions and Preliminaries

Let T be a leaf-labeled tree with node, edge, and leaf sets denoted by $V(T)$, $E(T)$, and $Le(T)$. If T is rooted, we denote it's root by $rt(T)$. For any node $v \in V(T)$, where T is a rooted tree, the (maximal) subtree rooted at v is denoted T_v. Unless otherwise specified, all trees are binary and unrooted.

We use the term *species tree* for the tree depicting evolutionary relationships for the taxa (e.g., species, strains, etc.) under consideration. Given a gene family from the taxa under consideration, a *gene tree* is a tree that depicts the evolutionary relationships of the genes in the gene family. We assume that each leaf in a gene tree is labeled with the taxon from which that leaf (i.e., gene sequence) was taken. Note that a gene tree may have zero, one, or multiple genes from the same taxon.

Throughout this work, we assume that the taxon set under consideration is denoted by Ω and that the species tree, denoted S, depicts the evolutionary relationships for taxa in Ω, i.e., $Le(S) = \Omega$. We use $\mathcal{G}$ to denote a collection of gene trees $\{G_1, ..., G_k\}$, where each G_i, $1 \le i \le k$, describes the evolutionary history of a different gene family present in the taxon set Ω. We also implicitly assume that $Le(S) = \cup_{i=1}^{k} Le(G_i)$.

DTL Reconciliation. The DTL reconciliation model allows for the reconciliation of a given rooted gene tree with a given rooted species tree by postulating gene duplication, HGT, and gene loss events. DTL reconciliation is often performed in a maximum parsimony framework, in which each event type has an associated (user-defined) cost and the objective is to find a reconciliation of minimum total cost [3,15,19,58–60]. In the current work, we specifically use the DTL reconciliation model first developed in [3,60], for which optimal (most parsimonious) DTL reconciliations can be computed in $O(mn)$ time, where m and n denote the number of leaves in the gene tree and species tree being reconciled, respectively. Importantly, an unrooted gene tree can be reconciled with a rooted species tree within the same $O(mn)$ time complexity [3].

In the following, we denote the event costs for gene duplications, HGTs, and gene losses by P_d, P_t, and P_l, respectively. Given a gene tree $G \in \mathcal{G}$, species tree S, and event costs P_d, P_t, and P_l, we denote by $\mathcal{R}_{P_d,P_t,P_l}(G, S)$ the reconciliation cost of an optimal DTL reconciliation of G and S under the event costs P_d, P_t, and P_l.

Definition 1 (Total DTL Reconciliation Cost). *Given a species tree S, a collection of gene trees $\mathcal{G} = \{G_1, ..., G_k\}$, and event costs P_d, P_t, and P_l, the total DTL reconciliation cost of $\mathcal{G}$ with S is the sum of the DTL reconciliation costs of each $G \in \mathcal{G}$ with S, i.e., $\sum_{i=1}^{k} \mathcal{R}_{P_d,P_t,P_l}(G_i, S)$.*

GTP-Based Problem Formulation. To compute accurate genome-scale microbial phylogenies, we use a gene tree parsimony formulation based on DTL reconciliation. Specifically, given as input a collection of hundreds or thousands of gene trees, we seek a species tree that minimizes the total DTL reconciliation cost against the collection of input gene trees. More formally,

Problem 1 (Most Parsimonious Species Tree (MPST)). *Given a collection of gene trees $\mathcal{G}$ and event costs P_d, P_t, and P_l, find a species tree S that minimizes the total DTL reconciliation cost with $\mathcal{G}$.*

The MPST problem can be shown to be NP-hard, W[2]-hard, and inapproximable to within log factor through a reduction from the NP-hard gene duplication problem [6,38]. The gene duplication problem is a special case of MPST problem defined in this manuscript and seeks a species tree minimizing just the total number of gene duplications. Details of the reduction are straightforward and omitted for brevity. Given the NP-hardness of the MPST problem, PhyloGTP uses a local search heuristic to solve the problem, as described in the next section.

3   Description of PhyloGTP

The local search heuristic implemented in PhyloGTP is similar to those use for many other NP-hard phylogeny inference problems, including those used for other popular variants of gene tree parsimony [13,39,49,62]. The local search heuristic starts with an initial candidate rooted species tree and iteratively improves it using local search. Specifically, in each local search iteration, the heuristic finds a minimum reconciliation cost tree in the "local neighborhood" of the current species tree. The best tree found in that local neighborhood then becomes the starting point for the next local search iteration. The heuristic terminates when a lower cost tree cannot be found in the local neighborhood of the current species tree. Next, we describe how PhyloGTP computes the initial candidate species tree and how it defines the local neighborhood for each subsequent local search iteration.

Construction of Initial Candidate Species Tree. If an estimated user-defined initial species tree is unavailable, PhyloGTP uses a stepwise taxon-addition algorithm to compute a reasonable initial species tree for the local search. The stepwise taxon-addition algorithm works by starting from a two-taxon rooted species tree and iteratively placing taxa, one at a time, onto the species tree topology along the branch that minimizes the total DTL reconciliation cost. In our implementation, the taxa are added in order of decreasing coverage, where the coverage of taxon s is the number of gene trees that include a gene from s. At each iteration, each gene tree is pruned to reflect only the taxa present in the current (incomplete) species tree. Once all taxa have been added, the resulting rooted species tree is used as the starting species tree for the subsequent local search. We found that using this stepwise taxon-addition algorithm results in an average reduction of 93% in the number of local search iterations until convergence when compared to using a random species tree topology as the initial starting tree (detailed results not shown).

Description of Local Search Iterations. PhyloGTP implements a constrained (rooted) subtree prune and regraft (SPR) [9] based local search using the initial tree as a starting point. SPR is the most commonly used tree edit operation for phylogenetic local search and induces a local neighborhood of $\Theta(n^2)$ trees, where n is the number of leaves in the species tree [56]. Rather than always evaluating all trees in the full SPR neighborhood at each iteration, PhyloGTP first considers only the restricted set of trees obtained by regrafting a single pruned subtree S_v, rooted at a some node $v \in V(S)/rt(S)$, onto each possible edge in the current species tree S. It finds the lowest cost tree S' within that restricted neighborhood and, if S' has lower cost than S, then S is replaced by S' and PhyloGTP proceeds to the next local search iteration. If no improvement was found in the restricted neighborhood using S_v, then a new node $u \neq v \in V(S)$ is chosen and the restricted local search step is repeated using the pruned subtree S_u. Thus, PhyloGTP is initially constrained to a small subset of the full SPR search space, but will incrementally expand the set of trees under consideration until an improvement is found, or until the full SPR neighborhood is explored. In the latter case, if no improvement is found then the search is determined to have converged. Note that the order in which subtrees are considered for pruning is randomized at the beginning of each local search iteration. In addition, if there are multiple species trees with minimum reconciliation cost within a restricted neighborhood, then the new species tree S' is selected uniformly at random among them.

Observe that we use a search strategy based on restricted SPR local neighborhoods instead of exploring the full SPR local neighborhood at each local search iteration. This is motivated by the underlying computational complexity of the computation. If n denotes the number of taxa in the analysis and k the number of input gene trees then, assuming most of the k gene trees have $\Theta(n)$ leaves, the time complexity of naively evaluating all candidate species trees in a single SPR local neighborhood becomes $\Theta(n^2) \times \Theta(n^2) \times \Theta(k)$ which is $\Theta(k \cdot n^4)$. This does not scale well with increasing n. Furthermore, many local search iterations have to be performed during a single execution of the heuristic. By using a search strategy based on restricted SPR local neighborhoods, the number of candidate species trees evaluated during most local search iterations reduces to $\Theta(n)$, reducing the time complexity of most local search iterations to a more reasonable $\Theta(k \cdot n^3)$. Importantly, this approach retains the key advantage of using a full SPR-based search since the heuristic search only terminates if a better tree is not found in the full SPR local neighborhood. Previous work on a simpler GTP problem suggests that heuristics based on restricted SPR local neighborhoods perform as well as those based on using full SPR neighborhoods during each local search iteration [63].

DTL Event Costs Assignment. By default, PhyloGTP uses event costs of 2, 3, and 1 for gene duplications, HGTs, and gene losses, respectively (i.e., $P_d = 2$, $P_t = 3$, and $P_l = 1$). These are standard costs used in the DTL reconciliation literature and have been previously observed to work well in practice for microbial datasets [5,7,15]. All experimental results reported in this manuscript are based on these default event costs for PhyloGTP.

Parallelization. PhyloGTP implements parallelization to further improve its scalability and enable application to large-scale datasets. The parallelization strategy works by dynamically distributing the computation associated with obtaining the reconciliation costs of candidate species trees in the local search neighborhood across a user-defined number of cores. Thus, when using c cores, the running time of the heuristic is reduced by roughly a factor of c.

4   Results

We use both simulated and real biological datasets to carefully assess the reconstruction accuracy of PhyloGTP. We also compare the accuracy of PhyloGTP against two recently developed state-of-the-art methods: SpeciesRax [46] and ASTRAL-Pro 2 [67]. SpeciesRax first uses a novel distance-based method, miniNJ, which estimates leaf-leaf distances based on the input gene trees, to construct an initial species tree using Neighbor Joining, and then executes a lightweight local search heuristic to optimize the initial species tree based on a probabilistic DTL reconciliation model. ASTRAL-Pro 2 first constructs a constrained search space of candidate species trees based on greedily optimizing a quartet similarity score, and then uses dynamic programming to find the best tree within that constrained search space. Both SpeciesRax and ASTRAL-Pro 2 were run using default parameter settings as provided in their respective manuals.

4.1   Results on Simulated Data

Dataset Description. We used simulated datasets with known ground truth species trees to assess the impact of three key parameters on reconstruction accuracy: Number of input gene trees, rates of gene duplication, HGT, and gene loss (or DTL rates for short), and estimation error in the input gene trees.

Simulated datasets were created using a three-step pipeline: (1) simulation of a ground-truth species tree and corresponding true gene trees with varying DTL rates, (2) simulation of sequence alignments of different lengths for each gene tree, and (3) reconstruction of estimated gene trees from the sequence alignments. In the first step, we used SaGePhy [34] to first simulate ground-truth species trees, each with exactly 50 leaves (taxa) and a height (root to tip distance) of 1, under a probabilistic birth-death framework. We then used these species trees to simulate multiple gene trees under the probabilistic duplication-transfer-loss model implemented in SaGePhy. This resulted in 9 different datasets of simulated true gene trees, each corresponding to a different number of input gene trees (10, 100, or 1000), and a different DTL rate (low, medium, or high; see Table 1). Each dataset comprised of 10 replicates. The chosen DTL rates are based on the relative rates and frequencies of gene duplication and HGT events in real microbial datasets [7]. In each case, the gene loss rate is assigned to be equal to the gene duplication rate plus the additive HGT rate, so as to balance the number of gene gains with the number of gene losses (Table 1). Basic statistics on these simulated true gene trees, including average sizes and numbers of gene duplication and HGT events, are provided in Table 2.

In the second step, we used AliSim [37] to simulate DNA sequence alignments along each simulated gene tree under the General Time-Reversible (GTR) model (using default AliSim GTR model settings) with three different sequence lengths: 400, 100, and 50 bp. In the third and final step, maximum-likelihood gene trees were inferred using IQ-TREE 2 [45] from the simulated sequence alignments under the Jukes-Cantor (JC) model. We use the simpler JC model when estimating gene trees, instead of the GTR model used to generate the sequences, since this better captures the biases of applying substitution models to real sequences. Thus, from each dataset of true gene trees, we derive 3 additional datasets of estimated gene trees corresponding to the three sequence lengths. The purpose of the second and third steps above is to generate error-prone gene trees that reflect the reconstruction/estimation error present in real gene trees. We found that the estimated gene trees had average normalized Robinson-Foulds distances [53] (defined below) of 0.08, 0.22, and 0.35 for sequence lengths 400, 100, and 50 bp, respectively, to the corresponding true gene trees.

Table 1. Key parameters used in the simulation study. The table lists the main parameters and their values explored in the simulation study. All 36 ($= 3 \times 3 \times 4$) combinations of these three parameters were evaluated at 10 replicates each. DTL rates are specified in the form (d, t, l), where d is the gene duplication rate, t is the HGT rate (split evenly between additive and replacing HGTs), and l is the gene loss rate. The number of species was fixed at 50 for these datasets.

Parameter	Values
Number of gene trees	$10, 100, 1000$
DTL rates	low $= (0.3, 0.6, 0.6)$
	med $= (0.6, 0.12, 0.12)$
	high $= (0.12, 0.24, 0.24)$
Sequence length (nucleotides)	$400, 100, 50$, and true gene trees

Table 1 summarises the specific ranges of parameter values we explored for the number of gene trees, DTL rates, and sequence lengths. We evaluated all combinations of these parameter values, resulting in a total of 36 simulated datasets, with each dataset comprising of 10 replicates created using that specific assignment of parameter values. We also created some additional datasets with 10 and 100 taxa for the runtime analysis.

Evaluating Reconstruction Accuracy. To evaluate the species tree reconstruction accuracies of the different methods, we compare the species tree estimated by each method with the corresponding ground truth species tree. To perform this comparison we utilize the widely used (unrooted) normalized Robinson-Foulds distance (NRFD) [53] between the reconstructed and ground truth species trees. For any reconstructed species tree, the NRFD reports the fraction of non-trivial splits in that species tree that do not appear in the corresponding ground

Table 2. Basic statistics for simulated gene trees. Average number of leaves, duplications, and HGTs, and losses in the simulated low, medium, and high DTL gene trees. For each DTL rate, the number of losses is roughly equal to the number of duplications plus half the number of HGTs. The numbers shown are averaged over all 10 replicates of the 100 gene tree datasets.

DTL rate	Leaves	Duplications	HGTs
Low	53.618	3.408	6.586
Med	55.121	6.15	11.125
High	59.718	10.077	18.37

truth species tree. For ease of interpretation, we report results in terms of *percentage accuracy*, defined to be the percentage of non-trivial splits in the reconstructed species tree that also appear in the ground truth species tree. Thus, percent accuracy is simply $(1 - \text{NRFD}) \times 100$. Thus, for example, a percentage accuracy of 87% is equivalent to an NRFD of 0.13.

Accuracy on True (Error-Free) Gene Trees. We first evaluate the accuracy of the species tree reconstruction methods when given true (error-free) gene trees as input (effectively skipping steps 2 and 3 of the simulation pipeline). While error-free gene trees do not capture the complexities of real data, this analysis helps us understand how the different methods perform in a controlled, ideal setting. Figure 1 shows the results for low, medium, and high DTL rates with varying numbers of gene trees for 50-taxon datasets. Unsurprisingly, we find that both DTL rates and number of input gene trees are highly impactful parameters. The performance of all three methods worsens as DTL rates increase, and improves as the numbers of input gene trees increase. Both PhyloGTP and SpeciesRax substantially outperform ASTRAL-Pro 2, especially on the medium and high DTL datasets. In particular, we find that ASTRAL-Pro 2 is highly susceptible to high DTL rates, and that it also shows poor performance when the number of input gene tree is small. Interestingly, the accuracy of Astral-pro 2 improves rapidly as the number of gene trees increases, with the method performing equivalently to PhyloGTP and SpeciesRax on the low and medium DTL datasets when the input consists of 1000 gene trees. Between PhyloGTP and SpeciesRax, we find that PhyloGTP shows higher accuracy when the number of gene trees is small (100 or fewer), particularly when DTL rates are medium or high. For the remaining datasets, both PhyloGTP and SpeciesRax show nearly identical accuracies.

Accuracy on Estimated (Erroneous) Gene Trees. We next assess the accuracy of the reconstructed species trees when the input consists of estimated (erroneous) gene trees. Figure 2 shows the results of this analysis for all 27 combinations of number of input gene trees, DTL rates, and sequence lengths (or gene tree estimation error rates). As expected, the accuracy of all three methods is substantially affected by the quality of the estimated gene trees, with higher accuracies achieved using gene trees estimated from longer sequences.

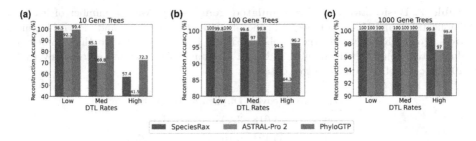

Fig. 1. Accuracy on true gene trees. Tree reconstruction accuracies are shown for PhyloGTP, SpeciesRax, and ASTRAL-Pro 2 when applied to error-free or 'true' gene trees. Results are shown for increasing numbers of input gene trees (10, 100, and 1000) and for low, medium, and high DTL rates. The number of taxa (i.e., number of leaves in the species tree) is fixed at 50. Higher percentages (y-axis) imply greater accuracy.

We also find that an increased number of input gene trees can partly make up for error in the input gene trees. For example, compared to using true input gene trees (Fig. 1), PhyloGTP shows a 5–21% reduction in accuracy with 10 estimated gene trees but only a 1–4% reduction with 1000 estimated gene trees, depending on sequence length. Similar trends are observed with SpeciesRax and ASTRAL-Pro2. Overall, we find that PhyloGTP and SpeciesRax still outperform ASTRAL-Pro-2 across most datasets and that ASTRAL-Pro 2 continues to be more susceptible to high DTL rates than the other methods. As before, the performance of ASTRAL-Pro 2 improves rapidly with increasing number of input gene trees, even sometimes outperforming SpeciesRax and PhyloGTP when DTL rates are low or medium. This suggests that ASTRAL-Pro 2 may be appropriate for microbial phylogenomics on datasets with lots of gene trees and relatively low prevalence of HGT. Comparing PhyloGTP with SpeciesRax, we find that both methods have similar performance overall, with PhyloGTP and SpeciesRax showing average percent accuracies of 88.36% and 86.87%, respectively, when averaged across all 27 datasets. However, PhyloGTP consistently outperforms SpeciesRax on datasets with high DTL rates (showing better accuracy, sometimes substantially better, in all by one high DTL dataset), as well as on datasets with 10 input gene trees. We also find that SpeciesRax tends to outperform PhyloGTP on datasets with high gene tree error and low DTL rates. This suggests that PhyloGTP may be especially useful for analyzing datasets with high levels of HGT or with a small number of gene trees.

Runtimes. We compare the runtimes of the three methods when varying the number of taxa (10, 50, and 100) over low, medium, and high DTL rates. In addition, we also evaluate the impact of the number of input gene trees (100 and 1000) using the 50-taxon dataset. These runtimes are shown in Table 3. All methods have parallel implementations and were allocated 12 cores on a 2.8 GHz × 4 Intel i7 processor with 16 GB of RAM. We find that ASTRAL-Pro 2 is, by far, the fastest method, requiring only about 5 s on the 50-taxon 1000 gene tree datasets and less than 10 s on the 100-taxon 100 gene tree datasets. SpeciesRax

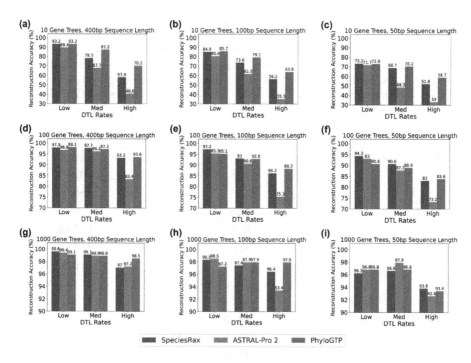

Fig. 2. Accuracy on estimated gene trees. Tree reconstruction accuracies are shown for PhyloGTP, SpeciesRax, and ASTRAL-Pro 2 when applied to estimated gene trees. Results are shown for all 27 combinations of number of input gene trees, sequence lengths (shorter sequence lengths imply greater gene tree estimation error), and DTL rates. The first, second, and third rows correspond to datasets with 10, 100, and 1000 gene families, respectively, and the first, second, and third columns correspond to 400, 100, and 50 base pair sequence lengths, respectively. The number of taxa (i.e., number of leaves in the species tree) is fixed at 50. Higher percentages imply greater accuracy.

is also extremely fast, requiring only about 60 s and 50 s, respectively, on those datasets. PhyloGTP is much slower than the other two methods, requiring about 3.5 h and 10.5 h on those same datasets. This is expected since this prototype implementation of PhyloGTP has a time complexity that is quartic (n^4) in the number of species. Unlike PhyloGTP, ASTRAL-Pro 2 does not rely on local search heuristics, instead using highly efficient algorithms for computing quartet similarity scores and for finding an optimal species tree within a constrained search space. SpeciesRax does implement a local search heuristic and uses DTL reconciliation, but it's heuristic is light-weight and searches over a smaller search space. SpeciesRax also uses a fast distance-based approach to compute a good initial species tree, which greatly reduces the number of local search steps needed. It may be possible to use some of these techniques to speed up PhyloGTP as well, without sacrificing accuracy.

Table 3. Impact of number of taxa and gene trees on running time. Runtimes in seconds are shown for the three methods for datasets with 10, 50, and 100 taxa and low medium, and high rates of DTL. For the 10- and 100-taxon datasets, the number of input gene trees is 100. For 50-taxon datasets, results are shown for both 100 and 1000 gene trees. The runtimes are based on simulated true input gene trees and are averaged over 10 replicate runs. Each method was allocated 12 cores on a 2.8 GHz × 4 Intel i7 processor with 16 GB of RAM.

Dataset size	DTL rate	SpeciesRax	ASTRAL-Pro 2	PhyloGTP
10 taxa, 100 gene trees	low	1.45	0.08	4.02
	med	1.36	0.08	4.45
	high	1.35	0.09	4.82
50 taxa, 100 gene trees	low	5.69	1.11	1,299.56
	med	6.25	1.14	1,374.92
	high	8.9	1.43	2,015.03
50 taxa, 1000 gene trees	low	50.34	5.38	10,011.79
	med	52.33	5.29	11,433.19
	high	59.95	5.55	13,137.93
100 taxa, 100 gene trees	low	22.05	3.61	19,871.48
	med	28.90	4.84	32,606.87
	high	47.19	7.15	38,259.04

4.2   Results on Biological Data

We assembled two previously used biological datasets of different size, composition, and complexity to assess the accuracy and consistency of species trees inferred by PhyloGTP as compared to SpeciesRax and traditional non-DTL cognizant methods such as MLSA and tANI [26] (Table 4). To examine the effect of extreme divergence and genome complexity variation on species tree inference, we used a dataset composed of 176 Archaea, which was drawn from [21]. The Archaea included in the dataset span 2–3 kingdoms (or superphylums), and radically different lifestyles (from extremophiles inhabiting Antarctic lakes to mammal gut constituents). Because the pan-genome of an entire domain would be immeasurably large and computationally infeasible to accurately infer, we have reduced the number of gene families in this dataset to 282 core genes, which are shared by all members. This also allows direct comparison of the PhyloGTP species tree to previously calculated phylogenies in [21] which used the same loci. It should be noted that the 282 gene families used in the PhyloGTP analysis have been expanded to include all homologs (paralogs, xenologs, etc.) found in each genome, while only orthologs were used in [21].

To examine the impact of low sequence divergence on PhyloGTP species tree inference, we used a dataset of 44 Frankiales genomes, drawn from [26]. These included taxa are all closely related members of the order Frankiales, and as such the entire pan-genome (8,862 gene families with at least 4 sequences) was used for inference in PhyloGTP and SpeciesRax. The order Frankiales are composed of nitrogen-fixing symbionts of pioneer flora [61], and although they demonstrate variation in GC content and genome size these factors were previously shown to not bias phylogenetic inference [26].

Table 4. Summary of the two biological datasets.

Dataset	Number of gene families	Potential biases	Previous methods used to infer species tree
176 Archaea (domain)	282	Extreme divergence, long branch attraction, compositional bias	tANI, MLSA, single gene
44 Frankiales (order)	8,862	Low divergence, contamination, genome size difference	tANI, MLSA

Archaeal Dataset. A myriad of controversies surround the phylogeny of Archaea. These controversies include the monophyly of the DPANN superphylum [2,10,21,47,52], the placement of extreme halophiles [1,21,47,57], and the root of the Archaea [51]. These differences in phylogenetic inference are driven by many factors including, but not limited to compositional bias, long branch attraction, extremely small genomes, numerous HGT events, and biased sampling of metagenome-assembled genomes. Thus, it is interesting to evaluate the performance of PhyloGTP in the face of these factors.

Using 282 unrooted input gene trees, both PhyloGTP and SpeciesRax inferred Archaeal species trees with small inaccuracies with respect to commonly accepted placements of groups in previous analyses. These inaccuracies should be interpreted in the context that for several Archaeal clades (mostly halophiles) there is no consistent, consensus position that is universally accepted amongst Archaeaologists. For example, the monophyly of the DPANN superphylum is considered by some to be an artifact (driven by long branch attraction or biased genome sampling) [2,21,69]. Both species trees (Fig. 3) recover a monophyletic DPANN superphylum and successfully resolve the TACK clade. PhyloGTP successfully recovers a monophyletic Euryarchaea kingdom (Fig 3a), whereas SpeciesRax has misplaced the Methanomada and Thermococcales (both euryarchaeotes) onto the branches leading to the TACK group. One major point of disagreement between the two methods is the placement of the Haloarchaea. SpeciesRax correctly recovers the Haloarchaea within the euryarchaeota (Fig. 3b), while this group has moved inside the DPANN superphylum, to be the sister group of the Nanohaloarchaea, in the PhyloGTP phylogeny (Fig. 3a). In addition, the position of the Methanonatronarchaeia (another halophile) in both

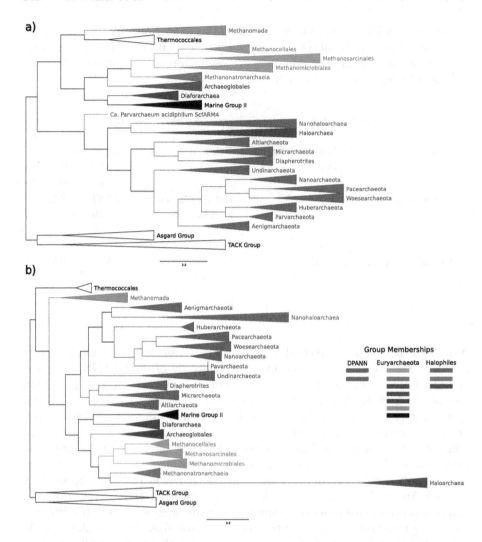

Fig. 3. Archaeal species tree reconstructions. Individual taxa on both trees have been collapsed into clades and are colored corresponding to higher level classifications (clades with the same color are part of the same class or phylum). The legend shows previously reported Kingdom memberships of these collapsed clades, and also the halophiles which may group together as a result of compositional bias. Part a) Unrooted Archaeal tree inferred by PhyloGTP; to be read as a cladogram since PhyloGTP does not infer branch lengths. Part b) Unrooted Archaeal tree inferred by SpeciesRax.

trees is recovered as later branching euryarchaeota (Fig. 3), in contrast to previous studies which report them as basal to the Methanotecta + Archaeoglobales superclass [1,21]. Incorrect placements of the Nanohaloarchaea, Haloarchaea and Methanonatronarchaeia are often attributed to compositional bias [21]. These halophiles prefer acidic amino acid residues (such as aspartate and glutamate), on account of their survival strategies in hypersaline environments, and these acidified proteomes attract the placement of these groups together in phylogenetic reconstructions.

Overall, these results demonstrate that PhyloGTP can produce a mostly accurate Archaeal tree, even in the face of the many biases present in the dataset (Table 4). At the same time, these results also show that PhyloGTP and SpeciesRax are both susceptible to the presence of problematic groups (such as the extreme halophiles) and other biases in complex datasets, potentially limiting their accuracy in some cases.

Frankiales Dataset. In the case of the Frankiales, reconstructions with PhyloGTP and SpeciesRax yield identical relationships between the major clades (Fig. 4). This suggest that both programs have comparable efficacy when the dataset analyzed is less complex and less divergent. Since this analysis used the entire pan-genome of the Frankiales, a possible concern is that small gene families (such as those that are only found in 4–8 genomes) may negatively impact these reconciliation based methods. To assess the impact of small gene families on species tree reconstruction, a subset of 1,702 genes families present in at least 20 genomes and in the smallest Frankia genome (*Frankia* sp. DG2) was used for inference using PhyloGTP and SpeciesRax. The trees produced from this subset recovered the same topologies for major clades as those in the full complement, indicating that the smaller gene families are not a problem for either method.

In comparison to previous trees inferred on the same genomes using previous non-DTL methods, such as those shown in [26], there are a few rearrangements of early branching clades in the backbone of the Frankiales. In phylogenies inferred using tANI and MLSA sequence methods, Group 1 (Fig. 4) is basal to the rest of the Frankiales. In the PhyloGTP and SpeciesRax trees, Group 3 is basal to the other Frankiales, with Group 1 as a later branching basal group. In addition to the movement of these clades, *Frankia* sp. NRRLB16219 and *Frankia* sp. CgIS1 have swapped positions, where *Frankia* sp. CgIS1 has moved from Group 2 to Group 5. These rearrangements may be attributed to the additional genomic data used to reconstruct the PhyloGTP and SpeciesRax trees. Only 24 loci were used in [26], and the inclusion of thousands of additional gene families have painted a slightly different picture of evolution throughout the Frankiales. This suggests that truly genome-scale methods like PhyloGTP could lead to more accurate phylogenomic inference on real datasets compared to other methods.

Comparison of Total DTL Reconciliation Costs. We also compared the total DTL reconciliation costs of the PhyloGTP and SpeciesRax species trees for these biological datasets. We find that, for both datasets, PhyloGTP species trees show considerably lower reconciliation costs. Specifically, for the Archaeal dataset, PhyloGTP and SpeciesRax species trees have total DTL reconciliation costs of 58,291 and 59,140, respectively. For the Frankiales dataset, these reconciliation costs are 148,898 and 156,376, respectively. These numbers show that, unlike PhyloGTP, speciesRax does not necessarily minimize the total DTL reconciliation cost. This is likely due to the different objective function used by SpeciesRax.

144   S. Weiner et al.

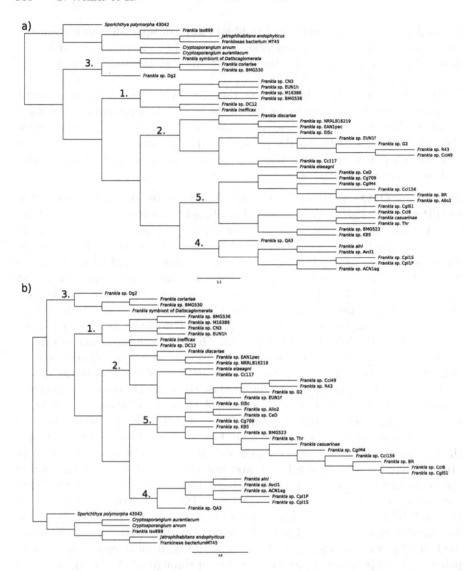

Fig. 4. Cladograms of the Frankiales. Clades on both trees are categorized and sorted based on the group designations described in [26]. Note that both trees show identical relationships among the labeled clades, but not necessarily within those clades. Part a) Frankiales cladogram inferred by PhyloGTP. Part b) Cladogram inferred by SpeciesRax.

Details of Dataset Assembly. Annotated genomes of 176 Archaea used in [21] were collected. The 282 core gene loci described in [21] were used as amino acid query sequences to search every collected genome, using blastp [11] with default parameters (-evalue was changed to 1e-10). All significant sequence for every loci across all genomes were collected (provided they met a length threshold of 50% in reference to the average gene family sequence size to filter partial sequences). Each gene family was then aligned using mafft-linsi [32] with default parameters and used as the basis for gene tree inference in IQ-Tree 2 [45], where the best substitution model for each gene family was determined using Bayesian Inference Criterion [31].

Annotated proteomes of the 44 Frankiales used in [26] were collected. Protein sequences were clustered into gene families and using the OrthoFinder2.4 pipeline [20] with default parameters (the search algorithm was changed to blast). Briefly, all-vs-all blastp (evalue of 1e−3) was used to find the best hits between input species. The set of query-matches were then clustered into gene families using the MCL algorithm, and the subsequent gene families were aligned using mafft-linsi with default parameters. Resulting alignments were used to create gene trees using FastTree [50] using the JTT model and default parameters.

5   Discussion and Conclusion

In this work, we introduced PhyloGTP, a new method for microbial species tree inference using GTP. PhyloGTP searches for the most parsimonious species tree under the DTL reconciliation model, making it the first GTP-based method suitable for microbial phylogenomics. Our simulation study shows that PhyloGTP can substantially outperform SpeciesRax when the number of input gene trees is small or when DTL rates are high. However, PhyloGTP is not consistently better than SpeciesRax and SpeciesRax tends to outperform PhyloGTP on datasets with high gene tree error but low DTL rates. We also find that both PhyloGTP and SpeciesRax almost always outperform ASTRAL-Pro 2, a highly scalable but HGT-naive method. Our results on the two biological datasets suggest that PhyloGTP works very well on real datasets overall, but also that both PhyloGTP and SpeciesRax can sometimes be misled by problematic taxa and compositional and other biases present in complex datasets.

While our experiments with PhyloGTP have yielded promising results, the prototype implementation of PhyloGTP is far slower, and hence far less scalable, than SpeciesRax or ASTRAL-Pro 2. However, we expect future work on improved algorithms and heuristics for GTP under DTL reconciliation to result in software implementations that are much faster and more accurate than the current PhyloGTP prototype. There are several possible directions for future research. First, PhyloGTP will likely benefit from differential weighting of the input gene trees. For example, the current implementation of PhyloGTP does not take into account the level of inference uncertainty in the input gene trees. Such measures of uncertainty are often readily available, such as bootstrap support values, and they could be used to distinguish between more reliable and less reliable gene trees. A simple multiplicative weight between 0 and 1 could then be assigned to each gene tree, reflecting confidence in that gene tree. It may also make sense to normalize reconciliation costs based on gene tree size and to down-weight gene trees exhibiting very high reconciliation costs. Exploring and carefully evaluating such weighting schemes is a promising direction for future research.

Second, PhyloGTP could be substantially sped up using alternative tree search strategies or improved algorithms for reconciliation cost computations. For example, it may be possible to constrain the search space of candidate species trees without sacrificing accuracy, or design algorithms to quickly approximate the total DTL reconciliation cost of candidate species trees to guide the local search heuristic. Third, given our findings on the Archaeal dataset, it would be useful to characterize the performance of PhyloGTP and SpeciesRax more carefully using more nuanced simulated datasets exhibiting some of the complications and biases observed in complex biological datasets. Finally, it may also be possible to devise asymptotically faster algorithms to compute the lowest DTL reconciliation cost tree within an entire SPR local neighborhood, as has been previously accomplished for simpler reconciliation models [4].

Acknowledgments. This work was supported in part by a University of Connecticut Research Excellence Program award to JPG and MSB.

Disclosure of Interests. The authors have no competing interests to declare that are relevant to the content of this article.

References

1. Aouad, M., Borrel, G., Brochier-Armanet, C., Gribaldo, S.: Evolutionary placement of methanonatronarchaeia. Nat. Microbiol. **4**(4), 558–559 (2019)
2. Aouad, M., Taib, N., Oudart, A., Lecocq, M., Gouy, M., Brochier-Armanet, C.: Extreme halophilic archaea derive from two distinct methanogen class ii lineages. Mol. Phylogenet. Evol. **127**, 46–54 (2018)
3. Bansal, M.S., Alm, E.J., Kellis, M.: Efficient algorithms for the reconciliation problem with gene duplication, horizontal transfer and loss. Bioinformatics **28**(12), 283–291 (2012)
4. Bansal, M.S., Eulenstein, O.: Algorithms for genome-scale phylogenetics using gene tree parsimony. IEEE/ACM Trans. Comput. Biol. Bioinf. **10**(4), 939–956 (2013)
5. Bansal, M.S., Kellis, M., Kordi, M., Kundu, S.: RANGER-DTL 2.0: rigorous reconstruction of gene-family evolution by duplication, transfer and loss. Bioinformatics **34**(18), 3214–3216 (2018)
6. Bansal, M.S., Shamir, R.: A note on the fixed parameter tractability of the gene-duplication problem. IEEE/ACM Trans. Comput. Biology Bioinform. **8**(3), 848–850 (2011)
7. Bansal, M.S., Wu, Y.-C., Alm, E.J., Kellis, M.: Improved gene tree error correction in the presence of horizontal gene transfer. Bioinformatics **31**(8), 1211–1218 (2015)
8. Beiko, R.G., Harlow, T.J., Ragan, M.A.: Highways of gene sharing in prokaryotes. Proc. Natl. Acad. Sci. U.S.A. **102**(40), 14332–14337 (2005)
9. Bordewich, M., Semple, C.: On the computational complexity of the rooted subtree prune and regraft distance. Ann. Comb. **8**, 409–423 (2004)
10. Brochier-Armanet, C., Forterre, P., Gribaldo, S.: Phylogeny and evolution of the archaea: one hundred genomes later. Curr. Opin. Microbiol. **14**(3), 274–281 (2011)
11. Camacho, C., et al.: BLAST+: architecture and applications. BMC Bioinform. **10**, 1–9 (2009)

12. Cerón-Romero, M.A., Fonseca, M.M., de Oliveira Martins, L., Posada, D., Katz, L.A.: Phylogenomic analyses of 2,786 genes in 158 lineages support a root of the eukaryotic tree of life between opisthokonts and all other lineages. Genome Biol. Evol. **14**(8), evac119 (2022)
13. Chaudhary, R., Bansal, M.S., Wehe, A., Fernandez-Baca, D., Eulenstein, O.: iGTP: a software package for large-scale gene tree parsimony analysis. BMC Bioinform. **11**(1), 574 (2010)
14. Ciccarelli, F.D., Doerks, T., von Mering, C., Creevey, C.J., Snel, B., Bork, P.: Toward automatic reconstruction of a highly resolved tree of life. Science **311**(5765), 1283–1287 (2006)
15. David, L.A., Alm, E.J.: Rapid evolutionary innovation during an Archaean genetic expansion. Nature **469**, 93–96 (2011)
16. Davidson, R., Vachaspati, P., Mirarab, S., Warnow, T.: Phylogenomic species tree estimation in the presence of incomplete lineage sorting and horizontal gene transfer. BMC Genom. **16**(Suppl 10), S1 (2015)
17. Doolittle, W.F.: Phylogenetic classification and the universal tree. Science **284**(5423), 2124–2128 (1999)
18. Doolittle, W.F., Boucher, Y., Nesbo, C.L., Douady, C.J., Andersson, J.O., Roger, A.J.: How big is the iceberg of which organellar genes in nuclear genomes are but the tip? Philosophical Transactions of the Royal Society of London. Ser. B: Biol. Sci. **358**(1429), 39–58 (2003)
19. Doyon, J.-P., Scornavacca, C., Gorbunov, K.Y., Szöllősi, G.J., Ranwez, V., Berry, V.: An efficient algorithm for gene/species trees parsimonious reconciliation with losses, duplications and transfers. In: Tannier, E. (ed.) RECOMB-CG 2010. LNCS, vol. 6398, pp. 93–108. Springer, Heidelberg (2010). https://doi.org/10.1007/978-3-642-16181-0_9
20. Emms, D.M., Kelly, S.: Orthofinder: phylogenetic orthology inference for comparative genomics. Genome Biol. **20**, 1–14 (2019)
21. Feng, Y., et al.: The evolutionary origins of extreme halophilic archaeal lineages. Genome Biol. Evol. **13**(8), evab166 (2021)
22. Gadagkar, S.R., Rosenberg, M.S., Kumar, S.: Inferring species phylogenies from multiple genes: concatenated sequence tree versus consensus gene tree. J. Exp. Zool. B Mol. Dev. Evol. **304B**(1), 64–74 (2005)
23. Glaeser, S.P., Kämpfer, P.: Multilocus sequence analysis (MLSA) in prokaryotic taxonomy. Syst. Appl. Microbiol. **38**(4), 237–245 (2015). Taxonomy in the age of genomics
24. Gogarten, J.P., Doolittle, W.F., Lawrence, J.G.: Prokaryotic evolution in light of gene transfer. Mol. Biol. Evol. **19**(12), 2226–2238 (2002)
25. Goodman, M., Czelusniak, J., Moore, G.W., Romero-Herrera, A.E., Matsuda, G.: Fitting the gene lineage into its species lineage. A parsimony strategy illustrated by cladograms constructed from globin sequences. Syst. Zool. **28**, 132–163 (1979)
26. Gosselin, S., Fullmer, M.S., Feng, Y., Gogarten, J.P.: Improving phylogenies based on average nucleotide identity, incorporating saturation correction and nonparametric bootstrap support. Syst. Biol. **71**(2), 396–409 (2022)
27. Green, R.E., Braun, E.L., Armstrong, J., Earl, D., et al.: Three crocodilian genomes reveal ancestral patterns of evolution among archosaurs. Science **346**(6215), 1254449 (2014)
28. Henz, S.R., Huson, D.H., Auch, A.F., Nieselt-Struwe, K., Schuster, S.C.: Whole-genome prokaryotic phylogeny. Bioinformatics **21**(10), 2329–2335 (2004)
29. Hilario, E., Gogarten, J.P.: Horizontal transfer of ATPase genes – the tree of life becomes a net of life. Biosystems **31**(2–3), 111–119 (1993)

30. Hirt, R.P., Logsdon, J.M., Healy, B., Dorey, M.W., Doolittle, W.F., Embley, T.M.: Microsporidia are related to fungi: evidence from the largest subunit of RNA polymerase ii and other proteins. Proc. Natl. Acad. Sci. **96**(2), 580–585 (1999)

31. Kalyaanamoorthy, S., Minh, B.Q., Wong, T.K., Von Haeseler, A., Jermiin, L.S.: ModelFinder: fast model selection for accurate phylogenetic estimates. Nat. Methods **14**(6), 587–589 (2017)

32. Katoh, K., Standley, D.M.: MAFFT multiple sequence alignment software version 7: improvements in performance and usability. Mol. Biol. Evol. **30**(4), 772–780 (2013)

33. Konstantinidis, K.T., Tiedje, J.M.: Genomic insights that advance the species definition for prokaryotes. Proc. Natl. Acad. Sci. **102**(7), 2567–2572 (2005)

34. Kundu, S., Bansal, M.S.: SaGePhy: an improved phylogenetic simulation framework for gene and subgene evolution. Bioinformatics **35**, 3496–3498 (2019)

35. Lang, J.M., Darling, A.E., Eisen, J.A.: Phylogeny of bacterial and archaeal genomes using conserved genes: supertrees and supermatrices. PLoS ONE **8**(4), e62510 (2013)

36. Lewis, P.O., et al.: Estimating Bayesian phylogenetic information content. Syst. Biol. **65**(6), 1009–1023 (2016)

37. Ly-Trong, N., Naser-Khdour, S., Lanfear, R., Minh, B.Q.: AliSim: a fast and versatile phylogenetic sequence simulator for the genomic era. Mol. Biol. Evol. **39**(5), msac092 (2022)

38. Ma, B., Li, M., Zhang, L.: From gene trees to species trees. SIAM J. Comput. **30**(3), 729–752 (2000)

39. Maddison, W.P., Maddison, D.: Mesquite: a modular system for evolutionary analysis. version 2.6 (2009). http://mesquiteproject.org

40. Marcet-Houben, M., Gabaldon, T.: Horizontal acquisition of toxic alkaloid synthesis in a clade of plant associated fungi. Fungal Genet. Biol. **86**, 71–80 (2016)

41. Markowitz, V.M., et al.: IMG 4 version of the integrated microbial genomes comparative analysis system. Nucleic Acids Res. **42**(D1), D560–D567 (2014)

42. Marlétaz, F., de la Calle-Mustienes, E., Acemel, R., et al.: The little skate genome and the evolutionary emergence of wing-like fins. Nature **616**, 495–503 (2023)

43. McCarthy, C.G.P., Fitzpatrick, D.A.: Phylogenomic reconstruction of the oomycete phylogeny derived from 37 genomes. mSphere **2**(2) (2017)

44. McInerney, J.O., Cotton, J.A., Pisani, D.: The prokaryotic tree of life: past, present and future? Trends Ecol. Evol. **23**(5), 276–281 (2008)

45. Minh, B.Q., et al.: IQ-TREE 2: new models and efficient methods for phylogenetic inference in the genomic era. Mol. Biol. Evol. **37**(5), 1530–1534 (2020)

46. Morel, B., Schade, P., Lutteropp, S., Williams, T.A., Szöllősi, G.J., Stamatakis, A.: SpeciesRax: a tool for maximum likelihood species tree inference from gene family trees under duplication, transfer, and loss. Mol. Biol. Evol. **39**(2), msab365 (2022)

47. Narasingarao, P., et al.: De novo metagenomic assembly reveals abundant novel major lineage of archaea in hypersaline microbial communities. ISME J. **6**(1), 81–93 (2012)

48. Olsen, G.J., Woese, C.R., Overbeek, R.: The winds of (evolutionary) change: breathing new life into microbiology. J. Bacteriol. **176**(1), 1–6 (1994)

49. Page, R.D.M.: GeneTree: comparing gene and species phylogenies using reconciled trees. Bioinform. (Oxf. Engl.) **14**(9), 819–820 (1998)

50. Price, M.N., Dehal, P.S., Arkin, A.P.: Fasttree 2-approximately maximum-likelihood trees for large alignments. PLoS ONE **5**(3), e9490 (2010)

51. Raymann, K., Brochier-Armanet, C., Gribaldo, S.: The two-domain tree of life is linked to a new root for the archaea. Proc. Natl. Acad. Sci. **112**(21), 6670–6675 (2015)

52. Raymann, K., Forterre, P., Brochier-Armanet, C., Gribaldo, S.: Global phylogenomic analysis disentangles the complex evolutionary history of DNA replication in archaea. Genome Biol. Evol. **6**(1), 192–212 (2014)

53. Robinson, D., Foulds, L.: Comparison of phylogenetic trees. Math. Biosci. **53**(1), 131–147 (1981)

54. Sevillya, G., Doerr, D., Lerner, Y., Stoye, J., Steel, M., Snir, S.: Horizontal gene transfer phylogenetics: a random walk approach. Mol. Biol. Evol. **37**(5), 1470–1479 (2019)

55. Shifman, A., Ninyo, N., Gophna, U., Snir, S.: Phylo SI: a new genome-wide approach for prokaryotic phylogeny. Nucleic Acids Res. **42**(4), 2391–2404 (2014)

56. Song, Y.S.: On the combinatorics of rooted binary phylogenetic trees. Ann. Comb. **7**(3), 365–379 (2003)

57. Sorokin, D.Y., et al.: Reply to 'evolutionary placement of methanonatronarchaeia'. Nat. Microbiol. **4**(4), 560–561 (2019)

58. Stolzer, M., Lai, H., Xu, M., Sathaye, D., Vernot, B., Durand, D.: Inferring duplications, losses, transfers and incomplete lineage sorting with nonbinary species trees. Bioinformatics **28**(18), 409–415 (2012)

59. Tofigh, A.: Using trees to capture reticulate evolution : lateral gene transfers and cancer progression. Ph.D. thesis, KTH Royal Institute of Technology (2009)

60. Tofigh, A., Hallett, M.T., Lagergren, J.: Simultaneous identification of duplications and lateral gene transfers. IEEE/ACM Trans. Comput. Biology Bioinform. **8**(2), 517–535 (2011)

61. Trujillo, M., et al.: Bergey's Manual of Systematics of Archaea and Bacteria. Wiley Online Library (2021)

62. Wehe, A., Bansal, M.S., Burleigh, J.G., Eulenstein, O.: DupTree: a program for large-scale phylogenetic analyses using gene tree parsimony. Bioinformatics **24**(13), 1540–1541 (2008)

63. Wehe, A., Burleigh, J.: Scaling the gene duplication problem towards the tree of life. In: 2nd International Conference on Bioinformatics and Computational Biology 2010, BICoB 2010, pp. 133–138 (2010)

64. Whidden, C., Zeh, N., Beiko, R.G.: Supertrees based on the subtree prune-and-regraft distance. Syst. Biol. **63**, 566–581 (2014)

65. Woese, C.R.: Bacterial evolution. Microbiol. Rev. **51**(2), 221–271 (1987)

66. Yap, W.H., Zhang, Z., Wang, Y.: Distinct types of rRNA operons exist in the genome of the actinomycete thermomonospora chromogena and evidence for horizontal transfer of an entire rRNA operon. J. Bacteriol. **181**(17), 5201–5209 (1999)

67. Zhang, C., Mirarab, S.: ASTRAL-Pro 2: ultrafast species tree reconstruction from multi-copy gene family trees. Bioinformatics **38**(21), 4949–4950 (2022)

68. Zhaxybayeva, O., Doolittle, W.F., Papke, R.T., Gogarten, J.P.: Intertwined evolutionary histories of marine synechococcus and prochlorococcus marinus. Genome Biol. Evol. **1**, 325–339 (2009)

69. Zhaxybayeva, O., Stepanauskas, R., Mohan, N.R., Papke, R.T.: Cell sorting analysis of geographically separated hypersaline environments. Extremophiles **17**, 265–275 (2013)

Genome Rearrangements

Maximum Alternating Balanced Cycle Decomposition and Applications in Sorting by Intergenic Operations Problems

Klairton Lima Brito[1], Alexsandro Oliveira Alexandrino[1],
Gabriel Siqueira[1(✉)], Andre Rodrigues Oliveira[2], Ulisses Dias[3],
and Zanoni Dias[1]

[1] Instituto de Computação, Universidade Estadual de Campinas (Unicamp),
Campinas, Brazil
{klairton,alexsandro,gabriel.siqueira,zanoni}@ic.unicamp.br
[2] Faculdade de Computação e Informática, Universidade Presbiteriana Mackenzie,
São Paulo, Brazil
[3] Faculdade de Tecnologia, Universidade Estadual de Campinas (Unicamp),
Limeira, Brazil
andrero@ic.unicamp.br, ulisses@ft.unicamp.br

Abstract. In the literature on genome rearrangement, several approaches use different structures to improve the results of genome rearrangement problems. In particular, graphs are structures widely used for the purpose of representing information retrieved from genomes. The breakpoint graph is a useful tool in this area because it allows representing in the same structure the gene orders of two genomes being compared. The maximum cycle decomposition of this graph brings immediate gain for deriving lower bounds for various genome rearrangement problems. This paper introduces a generalization of the Maximum Alternating Cycle Decomposition problem (MAX-ACD), called the Maximum Alternating Balanced Cycle Decomposition problem (MAX-ABCD). The MAX-ACD problem is closely related to the Sorting by Reversals problem and is a relevant topic of investigation in mathematics. The MAX-ABCD problem has applications in the Sorting by Intergenic Reversals problem, which is a problem that takes into account both the gene order and the information present in the intergenic regions. We present an algorithm with a constant approximation factor for the MAX-ABCD problem. Furthermore, we design an improved algorithm for the Sorting by Intergenic Operations of Reversal and Indel problem that guarantees an approximation factor of $\frac{3}{2}$ considering a scenario where the orientation of the genes is known. For the scenario where the orientation of the genes is unknown and based on an algorithm for the MAX-ABCD problem, we develop approximation algorithms for the Sorting by Intergenic Reversals and Sorting by Intergenic Operations of Reversal and Indel problems with an approximation factor of $2k$, where $k = \frac{31}{21} + \epsilon$.

Keywords: Cycle Decomposition · Approximation Algorithm · Reversal

C. Scornavacca and M. Hernández-Rosales (Eds.): RECOMB-CG 2024, LNBI 14616, pp. 153–172, 2024.
https://doi.org/10.1007/978-3-031-58072-7_8

1 Introduction

Genome rearrangement problems have been investigated since the 1920s. Nowadays, there are different ways to represent a genome. However, using a string of characters to represent genetic information is one of the most used methods. In particular, if each gene has only one copy the string corresponds to a permutation. In this case, the i-th gene of a genome is mapped into an element π_i of a permutation π. If the orientation of the genes is known, a sign "+" or "−" is associated with each element π_i indicating the gene orientation. Otherwise, the sign is omitted. The *signed case* and *unsigned case* are terms used to indicate whether the representation has information regarding the orientation of the genes.

When we are using the representation of genomes with permutations, we can estimate the evolutionary distance by calculating the minimum number of rearrangement events (mutations that may affect large stretches of DNA) capable of transforming a source genome (represented by a permutation π) into the target genome (represented by a permutation ι, which is normally set to the identity permutation (1 2 3 ...)). The reversal, which inverts a segment of the genome and, in the signed case, also flips the sign of the affected genes, is one of the most studied events in the literature that focus on mathematical models for genome rearrangement problems [6,12,16]. The Sorting by Reversals problem (SBR) seeks a minimum-length sequence of reversals that transforms one genome into another. The signed case of the SBR problem admits an exact polynomial-time algorithm [16]. However, Caprara [12] showed that the unsigned case of the problem is NP-hard, and today the best-known algorithm has an approximation factor of 1.375 [6].

One of the structures used to derive bounds for the SBR problem is the Breakpoint Graph [4]. Bafna and Pevzner proved that the number of breakpoints (b) minus the number of the cycles from a maximum alternating cycle decomposition of the breakpoint graph (c) is a lower bound for the SBR problem [4]. A breakpoint is a pair of consecutive elements in the source genome that are not consecutive in the target genome. The number of breakpoints is easily calculated, given the source and target genomes. Finding a decomposition of a breakpoint graph into alternating cycles that maximizes the number of cycles leads us to the Maximum Alternating Cycle Decomposition problem (MAX-ACD), which is NP-hard [12] in the unsigned case.

Besides, Caprara [11] presented a probabilistic analysis and showed that the number of reversals required to transform π (source genome representation) into ι (target genome representation) is $b - c$ with a probability of $1 - \Theta(\frac{1}{n^5})$ for a random signed permutation with n elements. His analysis characterizes the existence of the structures called hurdles and fortresses, because the presence of these structures in π ensures that more than $b - c$ reversals are needed to transform π into ι (such characterisation was later simplified by Swenson et al. [24]). This result motivates searching for algorithms to approximate the value of c [13,14,19]. The best-known approximation factor for that problem, in the unsigned case, is $\frac{17}{12} + \epsilon$ [14]. The results for the MAX-ACD problem also have applications in other genome rearrangement problems [14,22,23].

A genome has a lot of information, but research in the genome rearrangement field traditionally uses only the order and orientation of genes. Studies have pointed out that incorporating other genetic features into the models can make the results more realistic [8,10]. Intergenic regions are DNA sequences in between each pair of consecutive genes and at the ends of a linear genome. The number of nucleotides present in it gives the size of each intergenic region. Considering this new structure, studies have emerged representing a genome using two elements: i) a permutation that stores the information of gene order and orientation (when available) and ii) a list of non-negative integers storing the information regarding the size of each intergenic region. The genome rearrangement problems using this new genome representation remain with the same goal. However, the rearrangement events have the potential to affect both genes and intergenic regions.

Considering the exclusive use of the reversal event, we have the Sorting by Intergenic Reversals problem (SBIR), which is NP-hard in both the signed and unsigned cases [9,20]. Including the indel event, we have the Sorting by Intergenic Operations of Reversal and Indel problem (SBIRI). The unsigned case of the SBIRI is NP-hard, while the complexity of the signed case remains unknown. The best-known algorithms for the SBIR and SBIRI problems have an approximation factor of 2 for the signed case [20], and 4 for the unsigned case [9].

Considering that indels may affect both genes and intergenic regions, the SBIRI has a 2.5-approximation algorithm for the signed case, but the problem complexity remains unknown [1,2]. For the unsigned case, the SBIRI is NP-hard and has a 4-approximation algorithm [3].

We introduce a variation of the MAX-ACD problem to deal with intergenic regions, called the Maximum Alternating Balanced Cycle Decomposition problem (MAX-ABCD), and present an algorithm with a constant approximation factor. Furthermore, we design an improved algorithm for the signed case of the SBIRI problem that guarantees an approximation factor of $\frac{3}{2}$. For the unsigned case, we develop approximation algorithms for the SBIR and SBIRI problems with factors of $2k$ and $\frac{3k}{2}$, respectively, where $k = \frac{31}{21} + \epsilon$.

This paper is organized as follows. Section 2 introduces concepts and definitions used to obtain the results. Section 3 presents the theoretical results. Lastly, Sect. 4 concludes the manuscript.

2   Definitions

In this section, we introduce definitions, concepts, and graph structures used in the next section.

There are different ways to represent a genome, and depending on its characteristics, some approaches may have more advantages than others. This work assumes that each gene has no duplication and that the compared genomes share the same set of genes. Besides, we consider two types of representations. The first uses information restricted to the gene order, while the second also takes into account the information regarding the intergenic regions. There is an intergenic region between each pair of consecutive genes and at the extremities of a genome. Each intergenic region has a specific number of nucleotides, named *size*. Next, we describe the structures used in these two representations.

Given a genome $\mathcal{G} = (\mathcal{G}_1, \mathcal{G}_2, \ldots, \mathcal{G}_n)$, with n genes, we use a representation with a permutation $\pi = (\pi_1 \ \pi_2 \ \cdots \ \pi_n)$, such that π_i, with $1 \leq i \leq n$, is an integer number that represents the gene $\mathcal{G}_i$. If the gene orientation of $\mathcal{G}$ is known, each element π_i receives a "+" or "−" sign indicating its orientation. Otherwise, the sign is omitted. We call this form the *classic genome representation*.

Given a genome $\mathcal{G} = (\mathcal{R}_1, \mathcal{G}_1, \mathcal{R}_2, \mathcal{G}_2, \ldots, \mathcal{R}_n, \mathcal{G}_n, \mathcal{R}_{n+1},)$, with n genes and $n+1$ intergenic regions, we use a representation with two elements: (i) a permutation $\pi = (\pi_1 \ \pi_2 \ \cdots \ , \pi_n)$, such that π_i, with $1 \leq i \leq n$, is an integer number that represents the gene $\mathcal{G}_i$; (ii) a list of non-negative integers $\breve{\pi} = (\breve{\pi}_1, \breve{\pi}_2, \ldots, \breve{\pi}_{n+1})$, such that $\breve{\pi}_i$, with $1 \leq i \leq n+1$, represents the size of the intergenic region $\mathcal{R}_i$. If the gene orientation of $\mathcal{G}$ is known, each element π_i receives a "+" or "−" sign indicating its orientation; the sign is omitted otherwise. We call this form the *intergenic genome representation*.

The extended form of a classic genome $\mathcal{G} = \pi$ is obtained by adding two new elements $\pi_0 = 0$ and $\pi_{n+1} = n + 1$. Similarly, given an intergenic genome $\mathcal{G} = (\pi, \breve{\pi})$, we obtain its extended form by adding two new elements $\pi_0 = 0$ and $\pi_{n+1} = n + 1$. We hereafter assume that both genomes are in their extended form. In the following, we show how the reversal and the indel events affect an intergenic genome $\mathcal{G} = (\pi, \breve{\pi})$.

Definition 1. *Given an intergenic genome $\mathcal{G} = (\pi, \breve{\pi})$, a reversal $\rho_{(x,y)}^{(i,j)}$, with $1 \leq i \leq j \leq n$, $0 \leq x \leq \breve{\pi}_i$, and $0 \leq y \leq \breve{\pi}_{j+1}$, splits the intergenic regions $\breve{\pi}_i$ and $\breve{\pi}_{j+1}$, respectively, into (x, x') and (y, y'), such that $x' = \breve{\pi}_i - x$ and $y' = \breve{\pi}_{j+1} - y$. The segment $(x', \pi_i, \breve{\pi}_{i+1}, \ldots, \breve{\pi}_j, \pi_j, y)$ is inverted and, if the orientation of the genes is known, the signs of the elements from π_i to π_j are flipped. Lastly, the segments are reassembled. The reversal $\rho_{(x,y)}^{(i,j)}$ applied on $\mathcal{G} = (\pi, \breve{\pi})$, denoted by $(\pi, \breve{\pi}) \cdot \rho_{(x,y)}^{(i,j)}$, results in a new genome $G' = (\pi', \breve{\pi}')$, such that:*

$$\pi' = \quad (\pi_0 \ \pi_1 \ldots \pi_{i-1} \ -\pi_j \ -\pi_{j-1} \ldots -\pi_{i+1} \ -\pi_i \ \pi_{j+1} \ \pi_{j+2} \ldots \pi_n \ \pi_{n+1}),$$
$$\breve{\pi}' = \quad (\breve{\pi}_1, \breve{\pi}_2, \ldots, \breve{\pi}_{i-1}, \breve{\pi}'_i, \breve{\pi}_j, \ldots, \breve{\pi}_{i+1}, \breve{\pi}'_{j+1}, \breve{\pi}_{j+2}, \ldots, \breve{\pi}_n, \breve{\pi}_{n+1}),$$

if the orientation of the genes is known and,

$$\pi' = \quad (\pi_0 \ \pi_1 \ldots \pi_{i-1} \ \pi_j \ \pi_{j-1} \ldots \pi_{i+1} \ \pi_i \ \pi_{j+1} \ \pi_{j+2} \ldots \pi_n \ \pi_{n+1}),$$
$$\breve{\pi}' = \quad (\breve{\pi}_1, \breve{\pi}_2, \ldots, \breve{\pi}_{i-1}, \breve{\pi}'_i, \breve{\pi}_j, \ldots, \breve{\pi}_{i+1}, \breve{\pi}'_{j+1}, \breve{\pi}_{j+2}, \ldots, \breve{\pi}_n, \breve{\pi}_{n+1}),$$

otherwise. In both cases $\breve{\pi}'_i = x + y$ and $\breve{\pi}'_{j+1} = x' + y'$.

Definition 2. *Given an intergenic genome $\mathcal{G} = (\pi, \breve{\pi})$, an indel $\delta_{(x)}^{(i)}$, with $1 \leq i \leq n+1$ and $x \geq -\breve{\pi}_i$, changes the size of the intergenic region $\breve{\pi}_i$ to $\breve{\pi}_i + x$.*

The indel $\delta^{(i)}_{(x)}$ *applied on* $\mathcal{G} = (\pi, \breve{\pi})$, *denoted by* $(\pi, \breve{\pi}) \cdot \delta^{(i)}_{(x)}$, *results in a new genome* $G' = (\pi', \breve{\pi}')$, *such that:*

$$\pi' = (\pi_0 \, \pi_1 \ldots \pi_{i-1} \, \pi_i \, \pi_{i+1} \ldots \pi_{j-1} \, \pi_j \, \pi_{j+1} \, \pi_{j+2} \ldots \pi_n \, \pi_{n+1}),$$
$$\breve{\pi}' = (\breve{\pi}_1, \breve{\pi}_2, \ldots, \breve{\pi}_{i-1}, \underline{\breve{\pi}'_i}, \breve{\pi}_{i+1}, \ldots, \breve{\pi}_j, \breve{\pi}_{j+1}, \breve{\pi}_{j+2}, \ldots, \breve{\pi}_n, \breve{\pi}_{n+1}),$$

with $\breve{\pi}'_i = \breve{\pi}_i + x.$

Note that the indel event is a single notation to represent the non-conservative events of insertion and deletion. Observe that when the parameter x assumes a negative and positive value the indel event, respectively, deletes and inserts nucleotides.

Given an intergenic genome $\mathcal{G} = (\pi, \breve{\pi})$ and a sequence $S = (\beta_1, \beta_2, \ldots, \beta_n)$ of genome rearrangement events, $(\pi, \breve{\pi}) \cdot S = (\pi, \breve{\pi}) \cdot \beta_1 \cdot \beta_2 \cdot \cdots \cdot \beta_n$ denotes the sequence S being applied on $\mathcal{G}$, which results in a new genome.

Given a permutation π, the inverse permutation of π, denoted by π^{-1}, is the permutation that contains the information regarding the position of each element from π. For instance, the inverse permutation of $\pi = (0\ 2\ 4\ 1\ 3\ 5)$ is $\pi^{-1} = (0\ 3\ 1\ 4\ 2\ 5)$. Note that π_i^{-1} indicates the position of element i in π.

A *classic instance* $\mathcal{I} = (\pi, \iota)$ is composed of a pair of classic genomes, the source genome π and target genome ι. If the orientation of genes in both genomes is known, then the classic instance $\mathcal{I}$ is called *signed*. Otherwise, $\mathcal{I}$ is called *unsigned*.

Definition 3. *Given an unsigned classic instance* $\mathcal{I} = (\pi, \iota)$, *a pair of elements* (π_i, π_{i+1}), *such that* $0 \le i \le n$, *is a* **breakpoint** *if* $|\pi_{i+1} - \pi_i| \ne 1$.

Definition 4. *Given an unsigned classic instance* $\mathcal{I} = (\pi, \iota)$, *a pair* (a, b), *such that* $0 \le a \le n+1$ *and* $0 \le b \le n+1$, *is an* **adjacency** *if* $|a - b| = 1$ *and* $|\pi_a^{-1} - \pi_b^{-1}| = 1$.

An *intergenic instance* $\mathcal{I} = ((\pi, \breve{\pi}), (\iota, \breve{\iota}))$ is composed of a pair of intergenic genomes, the source genome $(\pi, \breve{\pi})$ and target genome $(\iota, \breve{\iota})$. Similarly to the classic instance, if the orientation of genes in both genomes is known, then the intergenic instance $\mathcal{I}$ is called *signed*. Otherwise, $\mathcal{I}$ is called *unsigned*.

Given an intergenic instance $\mathcal{I} = ((\pi, \breve{\pi}), (\iota, \breve{\iota}))$, the intergenic reversal distance $d_R(\mathcal{I})$ is the minimum number of reversals necessary to turn $(\pi, \breve{\pi})$ into $(\iota, \breve{\iota})$, and the intergenic reversal and indel distance $d_{RI}(\mathcal{I})$ is the minimum number of reversals and indels necessary to turn $(\pi, \breve{\pi})$ into $(\iota, \breve{\iota})$.

Next, we generalize the breakpoint and adjacency definition to consider the size of the intergenic regions in an unsigned intergenic instance.

Definition 5. *Given an unsigned intergenic instance* $\mathcal{I} = ((\pi, \breve{\pi}), (\iota, \breve{\iota}))$, *a pair of elements* (π_i, π_{i+1}), *such that* $0 \le i \le n$, *is an* **intergenic breakpoint** *if one of the following cases occur:* (i) $|\pi_{i+1} - \pi_i| \ne 1$; (ii) $|\pi_{i+1} - \pi_i| = 1$ *and* $\breve{\pi}_{i+1} \ne \breve{\iota}_x$, *such that* $x = \max(\pi_i, \pi_{i+1})$.

Given an unsigned intergenic instance $\mathcal{I}$, $ib(\mathcal{I})$ denotes the number of intergenic breakpoints in $\mathcal{I}$.

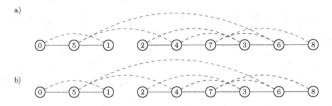

Fig. 1. The breakpoint graph $G_B(\mathcal{I})$ for $\mathcal{I} = (\pi = (0\,5\,1\,2\,4\,7\,3\,6\,8), \iota = (0\,1\,2\,3\,4\,5\,6\,7\,8))$. The reality edges are represented using a solid line and placed horizontally, while desire edges use a dashed line forming arcs over the vertices. Figures (a) and (b) show two different decompositions of $G_B(\mathcal{I})$ into edge-disjoint alternating cycles, represented by red and blue colors. (Color figure online)

Definition 6. *Given an unsigned intergenic instance* $\mathcal{I} = ((\pi, \breve{\pi}), (\iota, \breve{\iota}))$, *a pair* (a, b), *such that* $0 \le a \le n + 1$ *and* $0 \le b \le n + 1$, *is an* **intergenic adjacency** *if the following conditions are fulfilled: (i)* $|a - b| = 1$; *(ii)* $|\pi_a^{-1} - \pi_b^{-1}| = 1$; *(iii)* (π_x, π_{x+1}) *is not an intergenic breakpoint, with* $x = \min(\pi_a^{-1}, \pi_b^{-1})$.

In the following, we present the breakpoint graph structure and its extended form created to deal with intergenic instances, called weighted breakpoint graph.

2.1 Breakpoint Graph

Given a classic unsigned instance $\mathcal{I} = (\pi, \iota)$, the graph $G_B(\mathcal{I}) = (V, E, w)$ is an edge-colored graph composed by the set of vertices $V = \{\pi_0, \pi_1, \ldots, \pi_n, \pi_{n+1}\}$ and the set of edges $E = E_b \cup E_g$, which is divided into reality edges (E_b) and desire edges (E_g). For all $0 \le i \le n$, there exists a reality edge (π_i, π_{i+1}) if the pair of elements (π_i, π_{i+1}) is a breakpoint in $\mathcal{I}$. For all $0 \le i \le n$, there exists a desire edge $(i, i + 1)$ if the pair $(i, i + 1)$ is not an adjacency in $\mathcal{I}$.

Note that there is no cycle in G_B with edges of one color exclusively. A cycle C such that every pair of consecutive edges has distinct colors is called an alternating cycle. An alternating cycle C with k reality edges is called a k-cycle. Note that, by construction, there are no 1-cycles in G_B since when the pair of elements (π_i, π_{i+1}) is not a breakpoint it implies that the pair (π_i, π_{i+1}) is an adjacency. Figure 1 shows two possibilities of decomposition of the breakpoint graph $G_B(\mathcal{I})$ into edge-disjoint alternating cycles using the classic unsigned instance $\mathcal{I} = (\pi = (0\,5\,1\,2\,4\,7\,3\,6\,8), \iota = (0\,1\,2\,3\,4\,5\,6\,7\,8))$.

2.2 Weighted Breakpoint Graph on Unsigned Intergenic Instances

The weighted breakpoint graph is an extension of the breakpoint graph created to deal with the additional information from an intergenic instance. The weighted breakpoint graph differs by considering the intergenic breakpoint and intergenic adjacency definitions to construct the set of edges. Besides, each edge of the graph has a weight associated with the size of a specific intergenic region from

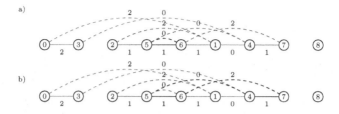

Fig. 2. The weighted breakpoint graph $G_{IB}(I)$ for $I = ((\pi = (0\,3\,2\,5\,6\,1\,4\,7\,8), \breve{\pi} = (2,5,1,1,1,0,1,3)), (\iota = (0\,1\,2\,3\,4\,5\,6\,7\,8), \breve{\iota} = (2,2,5,0,0,0,2,3)))$. The reality edges are represented using a solid line and placed horizontally, while desire edges use a dashed line forming arcs over the vertices. Figures (a) and (b) show two decompositions of $G_{IB}(I)$ into edge-disjoint alternating cycles, represented by black, red, and blue colors. (Color figure online)

the instance. In the following, we formally define the weighted breakpoint graph structure.

Given an unsigned intergenic instance $I = ((\pi, \breve{\pi}), (\iota, \breve{\iota}))$, the weighted breakpoint graph $G_{IB}(I) = (V, E, w)$ is composed by the set of vertices $V = \{\pi_0, \pi_1, \ldots, \pi_n, \pi_{n+1}\}$ and the set of edges $E = E_b \cup E_g$, which is divided into reality edges (E_b) and desire edges (E_g). The weighted function $w : E \to \mathbb{N}_0$ relates the size of the intergenic regions from the source and target genomes with weights of the edges in E. For all $0 \le i \le n$, there exists a reality edge (π_i, π_{i+1}) with $w((\pi_i, \pi_{i+1})) = \breve{\pi}_{i+1}$ if the pair (π_i, π_{i+1}) is an intergenic breakpoint in I. For all $0 \le i \le n$, there exists a desire edge $(i, i+1)$ with $w((i, i+1)) = \breve{\iota}_{i+1}$ if the pair $(i, i+1)$ is not an intergenic adjacency in I.

The alternating cycle and the k-cycle definitions are exactly as described in Sect. 2.1. An alternating cycle C from G_{IB} is *balanced* if the sum of the weights of its reality edges minus the sum of the weights of its desire edges is equal to zero. Otherwise, C is *unbalanced*. Note that there may be 1-cycles in G_{IB}, since the intergenic region between two adjacent elements in π and ι may have different sizes. However, all 1-cycles in G_{IB} are unbalanced by construction.

Figure 2 shows two decompositions of the weighted breakpoint graph $G_B(I)$ into edge-disjoint alternating cycles for $I = ((\pi = (0\,3\,2\,5\,6\,1\,4\,7\,8), \breve{\pi} = (2,5,1,1,1,0,1,3)), (\iota = (0\,1\,2\,3\,4\,5\,6\,7\,8), \breve{\iota} = (2,2,5,0,0,0,2,3)))$. Note that the decomposition in Fig. 2(a) has one balanced cycle, while the decomposition in Fig. 2(b) has three balanced cycles.

Given an unsigned intergenic instance $I = ((\pi, \breve{\pi}), (\iota, \breve{\iota}))$, we use $c_b(G_{IB}(I))$ to denote the maximum number of balanced cycles obtained from all possible decompositions of $G_{IB}(I)$ into edge-disjoint alternating cycles.

2.3 Weighted Breakpoint Graph on Signed Intergenic Instances

The weighted breakpoint graph on signed intergenic instances uses the information regarding the orientation of the genes to create a structure where the decomposition into edge-disjoint alternating cycles is unique.

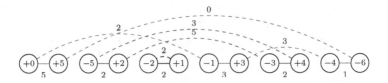

Fig. 3. The weighted breakpoint graph $G^+_{IB}(\mathcal{I})$ of the signed intergenic instance $\mathcal{I}$ $= ((\pi = (+0 -5 -2 -1 -3 -4 +6), \breve{\pi} = (5,2,2,3,2,1)), (\iota = (+0 +1 +2 +3 +4 +5 +6),$ $\breve{\iota} = (2,2,5,3,3,0)))$. The reality edges are represented using a solid line and placed horizontally, while desire edges use a dashed line forming arcs over the vertices.

Given a signed intergenic instance $\mathcal{I} = ((\pi, \breve{\pi}), (\iota, \breve{\iota}))$, the graph $G^+_{IB}(\mathcal{I}) =$ (V, E, w) is composed by the set of vertices $V = \{+\pi_0, -\pi_1, +\pi_1, -\pi_2, +\pi_2,$ $\dots, -\pi_n, +\pi_n, -\pi_{n+1}\}$ and the set of edges $E = E_b \cup E_g$, which is divided into reality edges (E_b) and desire edges (E_g). The weighted function $w : E \to \mathbb{N}_0$ relates the size of the intergenic regions from the source and target genomes with weights of the edges in E. For all $0 \le i \le n$, there is a reality edge $(+\pi_i, -\pi_{i+1})$ with $w((+\pi_i, -\pi_{i+1})) = \breve{\pi}_{i+1}$. For all $0 \le i \le n$, there exists a desire edge $(+i, -(i+1))$ with $w((+i, -(i+1))) = \breve{\iota}_{i+1}$.

The alternating cycle and a k-cycle definitions are exactly as described in Sect. 2.1. Besides, the balanced and unbalanced cycle definitions are the same as described in Sect. 2.2. Note that each vertex in G^+_{IB} has two incident edges (one reality and one desire). For this reason, the decomposition of $G^+_{IB}(\mathcal{I})$ into edge-disjoint alternating cycles is unique. There are different ways to draw the weighted breakpoint graph. However, we use a representation called *standard* to obtain a unique representation of the cycles in the graph. The standard representation places the vertices in a horizontal line from the left to the right following the order: $+0, -\pi_1, +\pi_1, -\pi_2, +\pi_2, \dots, -(n + 1)$. Next, the reality edges are placed horizontally while the desire edges form an arc over the vertices. Figure 3 shows a weighted breakpoint graph created using the signed intergenic instance $\mathcal{I} = ((\pi = (+0 -5 -2 -1 -3 -4 +6), \breve{\pi} = (5,2,2,3,2,1)),$ $(\iota = (+0 +1 +2 +3 +4 +5 +6), \breve{\iota} = (2,2,5,3,3,0)))$.

An unbalanced cycle C in G^+_{IB} is *positive* if the sum of the weight in its desire edges minus the sum of the weight in its reality edges is greater than zero. In the opposite way, an unbalanced cycle C in G^+_{IB} is *negative* if the sum of the weight in its desire edges minus the sum of the weight in its reality edges is less than zero. A k-cycle is called *non-trivial* if $k \ge 2$ and *trivial* otherwise. Each cycle in G^+_{IB} is identified by the list of its reality edges, such that the first reality edge is the rightmost considering the standard representation, and is traversed from right to left. A reality edge that is traversed from right to left is called *convergent*, and it is called *divergent* otherwise. A non-trivial cycle such that all reality edges are convergent is called *convergent*, and *divergent* otherwise. For instance, Fig. 3 shows a weighted breakpoint graph with three cycles $C_1 =$ $((-6, -4), (+3, -1), (+0, +5))$, $C_2 = ((+4, -3), (+2, -5))$, and $C_3 = ((+1, -2))$. Note that C_3 is a trivial balanced cycle, while C_1 and C_2 are negative divergent and positive convergent cycles, respectively.

Given a non-trivial cycle C from $G_{\mathcal{IB}}^+$, we say that a desire edge of C is an *open gate* if it does not intersect with any other desire edge from C. For instance, in Fig. 3, the desire edges $(-3, +2)$ and $(-5, +4)$ from the cycle C_2 are open gates. An open gate is closed if a desire edge from another cycle intersects with it. In Fig. 3, the desire edges $(-4, +3)$ and $(-1, +0)$ close the open gates $(-3, +2)$ and $(-5, +4)$. It is important to note that given any signed intergenic instance $\mathcal{I}$, all the open gates from $G_{\mathcal{IB}}^+(\mathcal{I})$ are closed [4].

Given a signed intergenic instance $\mathcal{I}$, $c_b(G_{\mathcal{IB}}^+(\mathcal{I}))$ denotes the number of balanced cycles in $G_{\mathcal{IB}}^+(\mathcal{I})$ while $c_b^t(G_{\mathcal{IB}}^+(\mathcal{I}))$ and $c_b^{nt}(G_{\mathcal{IB}}^+(\mathcal{I}))$ denote the number of trivial balanced and non-trivial balanced cycles in $G_{\mathcal{IB}}^+(\mathcal{I})$, respectively. Note that $c_b(G_{\mathcal{IB}}^+(\mathcal{I})) = c_b^t(G_{\mathcal{IB}}^+(\mathcal{I})) + c_b^{nt}(G_{\mathcal{IB}}^+(\mathcal{I}))$.

3   Theoretical Results

In this section, we formally present the Maximum Alternating Balanced Cycle Decomposition (MAX-ABCD) problem. Next, we describe the decision version of the MAX-ACD problem.

MAX-ACD (decision version)
Input: A breakpoint graph $G_\mathcal{B}$ of an unsigned classic instance $\mathcal{I} = (\pi, \iota)$ and a natural number s'.
Question: Is it possible to obtain an alternating-cycle decomposition of $G_\mathcal{B}$ with s' cycles?

Caprara [12] showed that the Maximum Alternating Cycle Decomposition (MAX-ACD) problem is NP-hard. In the following, we describe the decision version of the MAX-ABCD problem.

MAX-ABCD (decision version)
Input: A weighted breakpoint graph $G_{\mathcal{IB}}$ of an unsigned intergenic instance $\mathcal{I} = ((\pi, \breve{\pi}), (\iota, \breve{\iota}))$ and a natural number s''.
Question: Is it possible to obtain an alternating-cycle decomposition of $G_{\mathcal{IB}}$ with s'' balanced cycles?

Note that the MAX-ABCD is a generalization of the MAX-ACD problem, which leads us to the following lemma.

Lemma 1. *The* MAX-ABCD *problem is NP-hard.*

Proof. Given a classic unsigned instance $\mathcal{I} = (\pi, \iota)$ of the MAX-ACD problem and a natural number s', we create an intergenic unsigned instance $\mathcal{I}' = ((\pi', \breve{\pi}'), (\iota', \breve{\iota}'))$ for the MAX-ABCD problem and define a natural number s'' as follow: (i) $\pi' = \pi$; (ii) $\iota' = \iota$; (iii) $\breve{\pi}' = \breve{\iota}' = (0, 0, \ldots, 0)$; (iv) $s'' = s'$. Note that all the edges (reality or desire) in $G_{\mathcal{IB}}(\mathcal{I}')$ have weight zero, so it follows that any cycle in $G_{\mathcal{IB}}(\mathcal{I}')$ is balanced. Note that maximizing the number of balanced cycles in $G_{\mathcal{IB}}(\mathcal{I}')$ implies maximizing the number of cycles in $G_\mathcal{B}(\mathcal{I})$. □

3.1   An Approximation Algorithm for the MAX-ABCD Problem

In this section, we adapted the approach presented by Caprara and Rizzi [13] and Lin and Jiang [19] to design an approximation algorithm for the MAX-ACD problem. Following their approach we used know solutions to some sub-problems [5,17]. This sub-problems are the origin of the parameter ϵ in the approximation, and the original papers have more details on how the complexity of the algorithm is affected by this parameter.

For the results in this section, we consider an arbitrary optimal decomposition of $G_{\mathcal{IB}}(\mathcal{I})$ into edge-disjoint alternating cycles. The goal is to show that the algorithm finds a solution with a constant factor of $c_b(G_{\mathcal{IB}}(\mathcal{I}))$, which is the cost of the optimal decomposition. For $k \geq 2$, c_{bk} denotes the maximum number of balanced k-cycles obtained from the fixed decomposition. Note that $c_b(G_{\mathcal{IB}}(\mathcal{I})) = \sum_{k \geq 2} c_{bk}$.

The main idea is to obtain the best solution considering only cycles with two reality edges and considering cycles with up to three reality edges. It is important to note that, for any $k \geq 2$, finding the largest collection of edge-disjoint cycles with at most k reality edges in G_B is NP-hard [7]. Consequently, this result is also valid considering the $G_{\mathcal{IB}}$ graph.

Using the $G_{\mathcal{IB}}$ graph, create another graph G^* as follows: (i) for each 2-cycle in $G_{\mathcal{IB}}$ add a vertex in G^*; (ii) add an edge between two vertices in G^* if and only if the corresponding 2-cycles in $G_{\mathcal{IB}}$ share at least one edge.

Lemma 2. *For any graph G^* created from any graph $G_{\mathcal{IB}}$, we have that $\Delta(G^*) \leq 6$.*

Proof. Caprara and Rizzi [13] showed that G^* created from the G_b graph has a maximum degree of six. The proof is based on the maximum number of edges shared by two different 2-cycles and the maximum number of 2-cycle that an edge can belong to. Since the cycles structures in G_b and $G_{\mathcal{IB}}$ are the same, the result holds considering G^* created from $G_{\mathcal{IB}}$.   □

Given a graph $G = (V, E)$, an independent set S is a set of vertices of G, such that each edge in G has at most one endpoint in S. In other words, for every pair of vertices in S, there is no edge connecting them in G. The Maximal Independent Set (MIS) problem is an NP-hard problem [18], in which the goal is to find the maximum independent set in an input graph $G = (V, E)$. The MIS problem does not admit an algorithm with a constant approximation factor [15]. However, when the maximum degree of a graph is upper bounded by B, with $B > 2$, the problem admits algorithms with a constant approximation factor [21]. In that case, B is called the graph degree.

Lemma 3 (Berman and Fürer [5]). *Let $B > 2$ be a graph degree and $\epsilon > 0$. The MIS problem in bounded degree graphs has an approximation algorithm with a factor of $\frac{5}{B+3} - \epsilon$ if B is even and $\frac{5}{B+3.25} - \epsilon$ if B is odd.*

Finding the largest collection of edge-disjoint balanced 2-cycles in a weighted breakpoint graph $G_{\mathcal{IB}}$ is equivalent to the MIS problem in the corresponding graph G^*, which leads us to the following corollary.

Corollary 1. *Finding the largest collection of edge-disjoint balanced 2-cycles in G_{IB} can be approximated in polynomial time with a factor of $\frac{5}{9} - \epsilon$ for any positive ϵ.*

It is important to mention that the theoretical approximation factor of $\frac{5}{9} - \epsilon$ is achieved considering the worst scenario, which can happen only in the case when G^* has a degree of six. Let c_{b2}^* denote the maximum number of edge-disjoint balanced cycles with two reality edges in a given weighted breakpoint graph G_{IB}. Note that $c_{b2}^* \geq c_{b2}$.

Given a set U and a collection $C = \{C_1, C_2, \ldots, C_m\}$ of subsets of U, the Set Packing (SP) problem seeks the largest sub-collection of pairwise disjoint subsets. The SP problem belongs to the NP-hard class [18]. If every subset in C has a size at most k, we obtain the k-Set Packing (K-SP) problem. For $k > 2$ the K-SP problem admits an algorithm with a constant approximation factor, which leads us to the following lemma.

Lemma 4. (Hurkens and Schrijver [17]). *There is an approximation algorithm for the K-SP problem with a factor of $\frac{2}{k} - \epsilon$ for any positive ϵ.*

Given a weighted breakpoint graph $G_{IB}(\mathcal{I}) = (V, E, w)$, we create an instance $\mathcal{I}_{KSP}$ of the K-SP problem as follows: (i) $U = E$; (ii) for each balanced alternating cycle with at most three reality edges add the cycle edges as a subset in C. Observe that each subset in C has a size of at most six since we are dealing with balanced alternating cycles with at most three reality edges (and also three desire edges).

Note that finding the largest collection of edge-disjoint balanced alternating cycles with at most three reality edges in weighted breakpoint graph G_{IB} is equivalent to the K-SP problem in the corresponding instance $\mathcal{I}_{KSP}$, which leads us to the following corollary.

Corollary 2. *Finding the largest collection of edge-disjoint balanced alternating cycles with at most three reality edges in G_{IB} can be approximated in polynomial time with a factor of $\frac{1}{3} - \epsilon$ for any positive ϵ.*

Let c_{b3}^* denote the maximum number of edge-disjoint balanced cycles with up to three reality edges in a given weighted breakpoint graph G_{IB}. Note that $c_{b3}^* \geq c_{b2} + c_{b3}$. Let z_2 and z_3 denote the number of edge-disjoint balanced cycles obtained from corollaries 1 and 2, respectively. Observe that $z_2 \geq (\frac{5}{9} - \epsilon)c_{b2}^* \geq (\frac{5}{9} - \epsilon)c_{b2}$ and $z_3 \geq (\frac{1}{3} - \epsilon)c_{b3}^* \geq (\frac{1}{3} - \epsilon)(c_{b2} + c_{b3})$. Now we derive the following theorem.

Theorem 1. *Given an intergenic instance $\mathcal{I} = ((\pi, \breve{\pi}), (\iota, \breve{\iota}))$, the parameter $ib(\mathcal{I}) - c_b(G_{IB}(\mathcal{I}))$ can be approximated within $\frac{31}{21} + \epsilon$, for any positive ϵ, in polynomial time.*

Proof. We will use a proof similar to that presented by Lin and Jiang [19] considering the balanced cycles in the weighted breakpoint graph. Let c_b^* denote the largest collection of balanced cycles between z_2 and z_3. Adopting $\epsilon = \frac{5}{3}\eta$ in Corollary 1 and $\epsilon = \eta$ in Corollary 2, we obtain that:

$$c_b^* \geq \max\left\{\left(\frac{5}{9} - \frac{5}{3}\eta\right)c_{b2}, \left(\frac{1}{3} - \eta\right)\left(c_{b2} + c_{b3}\right)\right\}.$$

Since each intergenic breakpoint in $\mathcal{I}$ implies in one reality edge in $G_{IB}(\mathcal{I})$, we have that $c_b(G_{IB}(\mathcal{I})) \leq c_{b2} + c_{b3} + \frac{ib(\mathcal{I}) - 2c_{b2} - 3c_{b3}}{4}$. Note that $G_{IB}(\mathcal{I})$ has $ib(\mathcal{I})$ reality edges and removing the reality edges used by the balanced cycles with two and three reality edges in an optimal decomposition remains $ib(\mathcal{I}) - 2c_{b2} - 3c_{b3}$ reality edges. In the best scenario, a collection of edge-disjoint balanced cycles with four reality edges each can be obtained. Therefore, we have that:

$$\frac{ib(\mathcal{I}) - c_b^*}{ib(\mathcal{I}) - c_b(G_{IB}(\mathcal{I}))} \leq \frac{ib(\mathcal{I}) - c_b^*}{ib(\mathcal{I}) - c_{b2} - c_{b3} - \frac{ib(\mathcal{I}) - 2c_{b2} - 3c_{b3}}{4}}.$$

Now, we will consider the maximum and the minimum value that $ib(\mathcal{I})$ can assume and the ratio achieved. Observe that $2c_{b2} + 3c_{b3} \leq ib(\mathcal{I}) \leq \infty$. Setting $ib(\mathcal{I})$ to be ∞ the ratio $\frac{4}{3}$ is achieved. Considering the value of $ib(\mathcal{I})$ as $2c_{b2} + 3c_{b3}$, we obtain the ratio of $\frac{2c_{b2} + 3c_{b3} - c_b^*}{c_{b2} + 2c_{b3}}$. Now, we analyze the two following cases:

– Case 1: $(\frac{5}{9} - \frac{5}{3}\eta)c_{b2} \geq (\frac{1}{3} - \eta)(c_{b2} + c_{b3})$. In this case, we have that $c_{b3} \leq \frac{2 - 6\eta}{\frac{1 - 3\eta}{3}}c_{b2} = \frac{2}{3}c_{b2}$. Thus,

$$\frac{2c_{b2} + 3c_{b3} - c_b^*}{c_{b2} + 2c_{b3}}$$

$$= \frac{\left(\frac{13}{9} + \frac{5}{3}\eta\right)c_{b2} + 3c_{b3}}{c_{b2} + 2c_{b3}}$$

$$= \frac{\frac{13c_{b2} + 15\eta c_{b2} + 27c_{b3}}{9}}{\frac{c_{b2} + 2c_{b3}}{1}}$$

$$= \frac{13c_{b2} + 15\eta c_{b2} + 27c_{b3}}{9c_{b2} + 18c_{b3}}$$

$$\leq \frac{13c_{b2} + 15\eta c_{b2} + 18c_{b2}}{9c_{b2} + 12c_{b2}}$$

$$= \frac{31c_{b2} + 15\eta c_{b2}}{21c_{b2}}$$

$$= \frac{31}{21} + \frac{5}{7}\eta.$$

– Case 2: $(\frac{5}{9} - \frac{5}{3}\eta)c_{b2} \leq (\frac{1}{3} - \eta)(c_{b2} + c_{b3})$. In this case, we have that $c_{b2} \leq \frac{1 - 3\eta}{\frac{2 - 6\eta}{9}}c_{b3} = \frac{3}{2}c_{b3}$. Thus,

$$\frac{2c_{b2} + 3c_{b3} - c_b^*}{c_{b2} + 2c_{b3}}$$

$$= \frac{\left(\frac{5}{3} + \eta\right)c_{b2} + \left(\frac{8}{3} + \eta\right)c_{b3}}{c_{b2} + 2c_{b3}}$$

$$= \frac{\frac{5c_{b2} + 3\eta c_{b2} + 8c_{b3} + 3\eta c_{b3}}{3}}{\frac{c_{b2} + 2c_{b3}}{1}}$$

$$= \frac{5c_{b2} + 3\eta c_{b2} + 8c_{b3} + 3\eta c_{b3}}{3c_{b2} + 6c_{b3}}$$

$$\leq \frac{\frac{15c_{b3}}{2} + \frac{9c_{b3}}{2}\eta + 8c_{b3} + 3\eta c_{b3}}{\frac{9c_{b3}}{2} + 6c_{b3}}$$

$$= \frac{31c_{b3} + 15\eta c_{b3}}{21c_{b3}}$$

$$= \frac{31}{21} + \frac{5}{7}\eta.$$

Adopting $\eta = \frac{7}{5}\epsilon$, we have that $\frac{ib(\mathcal{I}) - c_b^*}{ib(\mathcal{I}) - c_b(G_{IB}(\mathcal{I}))} \leq \frac{31}{21} + \epsilon$, and the theorem follows.　□

3.2 Improved Approximation Algorithm for the $S_B IRI$ Problem on Signed Intergenic Instances

In this section, we present a $\frac{3}{2}$-approximation algorithm for the SBIRI problem on signed intergenic instances based on the weighted breakpoint graph structure. In the following, we present lemmas that are used by the algorithm steps.

Lemma 5 (Lemma 7.2 from Oliveira et al. [20]). *Given a signed intergenic instance $\mathcal{I}$, $d_{\mathrm{RI}}(\mathcal{I}) \geq n + 1 - c_b(G_{\mathcal{IB}}^+(\mathcal{I}))$.*

Lemma 6. *Let C be a trivial negative cycle in $G_{\mathcal{IB}}^+$, then there is an indel that increases the number of balanced cycles by one.*

Proof. Note that an indel changes the weight of only one reality edge in $G_{\mathcal{IB}}^+$. Besides, a trivial cycle has only two edges (one reality and one desire). In this case, the indel decreases the weight of the reality edge to be equal to the weight present in the desire edge of the cycle, and the lemma follows. □

Lemma 7. *Let C be a positive cycle in $G_{\mathcal{IB}}^+$, then there is an indel that increases the number of balanced cycles by one.*

Proof. In this case, just apply an indel on the first reality edge of the cycle to increase its weight in order to turn C into a balanced cycle. □

Lemma 8 (Lemma 5.1 from Oliveira et al. [20]). *Let C be a non-trivial divergent cycle in $G_{\mathcal{IB}}^+$ that is either balanced or negative, then there is a reversal that increases the number of balanced cycles by one.*

Next, we present two important remarks that are fundamental to proving the next lemma.

Remark 1. Given a signed intergenic instance $\mathcal{I} = ((\pi, \breve{\pi}), (\iota, \breve{\iota}))$, if all the non-trivial cycles in $G_{\mathcal{IB}}^+(\mathcal{I})$ are convergent, then each element π_i of π has a positive sign.

Remark 2. Given a signed intergenic instance $\mathcal{I} = ((\pi, \breve{\pi}), (\iota, \breve{\iota}))$, if any element π_i of π has a negative sign, then $G_{\mathcal{IB}}^+(\mathcal{I})$ has at least one divergent cycle.

Now, we show how to deal with a weighted breakpoint graph in which each non-trivial cycle is either balanced convergent or negative convergent.

Lemma 9. *Given a signed intergenic instance $\mathcal{I} = ((\pi, \breve{\pi}), (\iota, \breve{\iota}))$ and let $G_{\mathcal{IB}}^+(\mathcal{I})$ be a graph in which each non-trivial cycle is either balanced convergent or negative convergent. Then there is a sequence of three reversals that increases the number of balanced cycles by two.*

Proof. Oliveira et al. [20] showed that given a negative or balanced convergent cycle C from $G_{\mathcal{IB}}^+$ in which all negative and balanced cycles are convergent, then it is possible to increase the number of balanced cycles using two reversals. Since each non-trivial cycle in $G_{\mathcal{IB}}^+$ is either balanced convergent or negative convergent, the sequence of two reversals described by Oliveira *et al.* [20] can be applied. Besides, we will show that exists a third reversal that increases the number of balanced cycles by one.

Before applying any reversal note that, by Remark 1, each element π_i of π has a positive sign. The first reversal is used to create a divergent cycle C in $G_{\mathcal{IB}}^+(\mathcal{I})$ without splitting the cycles. This can be done using only one reversal (see Lemma 5.2 from Oliveira *et al.* [20]).

After creating a divergent cycle, a reversal from Lemma 8 can be applied on cycle C to increase the number of balanced cycles in one unit. Observe that the second reversal splits C generating a new balanced cycle, then each non-trivial cycle in $G_{\mathcal{IB}}^+(\mathcal{I})$ remains balanced or negative.

Now, we show that there is at least one divergent cycle in $G_{\mathcal{IB}}^+(\mathcal{I})$ and Lemma 8 can be applied again. Note that the first reversal creates a segment in the permutation π, such that the elements have a negative sign. After the second reversal, at least one element from π remains with a negative sign. Otherwise, the second reversal would have undone the action of the first, which implied that after applying these two reversals the resulting permutation would be equal to the original permutation and, therefore, the number of cycles would not increase after these reversals. Since at least one element π_i of π has a negative sign, by Remark 2, we know that $G_{\mathcal{IB}}^+(\mathcal{I})$ has at least one divergent (balanced or negative) cycle and Lemma 8 can be applied. Thus, a sequence of three reversals increases the number of balanced cycles in two units and the lemma follows. □

Now consider Algorithm 1, which is divided into four steps. Step I: turn trivial negative cycles into balanced. Step II: turn positive cycles into balanced. Step III: deal with divergent cycles. Step IV: deal with convergent cycles.

Algorithm 1 The input is a signed intergenic instance $\mathcal{I} = ((\pi, \breve{\pi}), (\iota, \breve{\iota}))$ and the output is a sequence S of reversal and indel operations such that $(\pi, \breve{\pi}) \cdot S = (\iota, \breve{\iota})$.

$S \leftarrow [\,]$
while $(\pi, \breve{\pi}) \neq (\iota, \breve{\iota})$ **do**
    if there is a trivial negative cycle in $G_{\mathcal{IB}}^+(\mathcal{I})$ **then**
        $S' \leftarrow [\delta_1]$ ▷ Step I: Lemma 6.
    else if there is a positive cycle in $G_{\mathcal{IB}}^+(\mathcal{I})$ **then**
        $S' \leftarrow [\delta_1]$ ▷ Step II: Lemma 7.
    else if there is a divergent cycle in $G_{\mathcal{IB}}^+(\mathcal{I})$ **then**
        $S' \leftarrow [\rho_1]$ ▷ Step III: Lemma 8.
    else
        $S' \leftarrow [\rho_1, \rho_2, \rho_3]$ ▷ Step IV: Lemma 9.
    end if
    $\mathcal{I} \leftarrow ((\pi, \breve{\pi}) \cdot S', (\iota, \breve{\iota}))$
    $S \leftarrow S + S'$
end while
return S

Lemma 10. *Given a signed intergenic instance* $\mathcal{I} = ((\pi, \breve{\pi}), (\iota, \breve{\iota}))$, *Algorithm 1 turns* $(\pi, \breve{\pi})$ *into* $(\iota, \breve{\iota})$ *using at most* $\frac{3(n+1-c_b(G_{\mathcal{IB}}^+(\mathcal{I})))}{2}$ *reversals and indels.*

Proof. Observe that if Algorithm 1 reaches Step III, then steps I and II ensure that any trivial cycle is balanced and any non-trivial cycle is either negative or balanced. Note that both steps I and II increase the number of balanced cycles in one unit each. If exists any divergent cycle, then Step III also increases the number of balanced cycles in one unit. Otherwise, each non-trivial cycle is either balanced convergent or negative convergent and Step IV increases the number of balanced cycles in two units. Since each iteration applies one of the four steps and the number of balanced cycles increases by at least one unit, then eventually $(\pi, \breve{\pi})$ will be turned into $(\iota, \breve{\iota})$ and the algorithm stops. Note that to turn $(\pi, \breve{\pi})$ into $(\iota, \breve{\iota})$ it is necessary to increase the number of balanced cycles in $n + 1 - c_b(G_{\mathcal{IB}}^+(\mathcal{I}))$ units. The worst case of Algorithm 1 occurs in Step IV, which has a cost of $\frac{3}{2}$ operations per balanced cycle. Thus, no more than $\frac{3(n+1-c_b(G_{\mathcal{IB}}^+(\mathcal{I})))}{2}$ reversals and indels are used to turn $(\pi, \breve{\pi})$ into $(\iota, \breve{\iota})$, and the lemma follows. □

Note that in Algorithm 1 steps I, II, and III require a linear time to be performed, while Step IV requires a quadratic time. Since the number of iterations is $\mathcal{O}(n)$, then the time complexity of Algorithm 1 is $\mathcal{O}(n^3)$.

Lemma 11. *Given a signed intergenic instance* $\mathcal{I}$*, Algorithm 1 is a* $\frac{3}{2}$*-approximation for the* SBIRI *problem.*

Proof. Directly by Lemmas 5 and 10. □

3.3   Improved Approximation Algorithms for SbIR and SbIRI Problems on Unsigned Intergenic Instances

In this section, we present for the SBIR and SBIRI problems, on unsigned intergenic instances, approximation algorithms with factor of $2k$. The value of k in both algorithms is $\frac{31}{21} + \epsilon < 1.48 + \epsilon$. In the following, we present a lower bound for both problems and lemmas that are used by the algorithms to ensure the approximation ratio.

Given an unsigned intergenic instance $\mathcal{I} = ((\pi, \breve{\pi}), (\iota, \breve{\iota}))$ and a genome rearrangement event σ, $\Delta c_b(I, \sigma) = c_b(G_{\mathcal{IB}}(\mathcal{I}')) - c_b(G_{\mathcal{IB}}(\mathcal{I}))$, such that $\mathcal{I}' = ((\pi, \breve{\pi}) \cdot \sigma, (\iota, \breve{\iota}))$, denotes the variation in the number of balanced cycles after the application of σ on the source genome from $\mathcal{I}$. Similarly, $\Delta ib(I, \sigma) = ib(\mathcal{I}') - ib(\mathcal{I})$, such that $\mathcal{I}' = ((\pi, \breve{\pi}) \cdot \sigma, (\iota, \breve{\iota}))$, denotes the variation in the number of intergenic breakpoints after the application of σ on the source genome from $\mathcal{I}$. Next, we present lemmas used to derive a lower bound for the SBIR and SBIRI problems.

The proof of Lemmas 12 and 13 is similar to the presented in Theorem 1 from Bafna and Pevzner [4] but considering an intergenic context.

Lemma 12. *For every unsigned intergenic instance* $\mathcal{I}$ *and reversal* ρ*, we have that* $\Delta c_b(I, \rho) - \Delta ib(I, \rho) \le 1$.

Proof. Note that a reversal ρ removes the reality edges (π_{i-1}, π_i) and (π_j, π_{j+1}) from $G_{\mathcal{IB}}(\mathcal{I})$ and may add the reality edges (π_{i-1}, π_j) and (π_i, π_{j+1}). Besides, observe that each reality edge in $G_{IB}(\mathcal{I})$ represents an intergenic breakpoint. Suppose that the number of reality edges after applying the reversal ρ does not change, it implies that $\Delta ib(I, \rho) = 0$. However, in the best scenario, the number of balanced cycles may increase at most by one unit either by merging two unbalanced cycles generating a new balanced cycle or splitting a cycle into two new cycles. In this case, we have that $\Delta c_b(I, \rho) - \Delta ib(I, \rho) \leq 1$. Now, suppose that the number of reality edges after applying the reversal ρ decreases in one unit, it implies that $\Delta ib(I, \rho) = -1$. Note that to remove one reality edge from G_{IB} the reversal ρ affected a k-cycle generating a $k - 1$-cycle and one intergenic adjacency, but the number of balanced cycles remains the same ($\Delta c_b(I, \rho) = 0$), which lead us to $\Delta c_b(I, \rho) - \Delta ib(I, \rho) = 1$. Lastly, suppose that the number of reality edges after applying the reversal ρ decreases in two units, it implies that $\Delta ib(I, \rho) = -2$. Note that to remove two reality edges from G_{IB} the reversal ρ must affect a balanced 2-cycle generating two intergenic adjacencies. Consequently, we have that $\Delta c_b(I, \rho) = -1$ and $\Delta c_b(I, \rho) - \Delta ib(I, \rho) = 1$. Considering the three possibilities, we have that $\Delta c_b(I, \rho) - \Delta ib(I, \rho) \leq 1$, and the lemma follows.   □

Lemma 13. *For every unsigned intergenic instance $\mathcal{I}$ and indel δ, we have that $\Delta c_b(I, \delta) - \Delta ib(I, \delta) \leq 1$.*

Proof. Note that an indel δ only changes the weight associated with the reality edge (π_{i-1}, π_i) in $G_{IB}(\mathcal{I})$. We have three possibilities: (i) the indel affects the weight of one reality edge of a k-cycle, such that $k > 1$, turning it into a balanced cycle. In this case, we have that $\Delta c_b(I, \delta) = 1$, $\Delta ib(I, \delta) = 0$, and $\Delta c_b(I, \delta) - \Delta ib(I, \delta) = 1$; (ii) the indel affects the weight of one reality edge of a k-cycle, such that $k > 1$, turning it into an unbalanced cycle or keeping it unbalanced. In this case, we have that $\Delta c_b(I, \delta) < 1$, $\Delta ib(I, \delta) = 0$, and $\Delta c_b(I, \delta) - \Delta ib(I, \delta) < 1$; (iii) the reality edge affected by the indel belongs to a 1-cycle (recall that every 1-cycle in $G_{IB}(\mathcal{I})$ is unbalanced) and generates a new intergenic adjacency. In this case, we have that $\Delta c_b(I, \delta) = 0$, $\Delta ib(I, \delta) = -1$, and $\Delta c_b(I, \delta) - \Delta ib(I, \delta) = 1$. Considering the three possibilities, we have that $\Delta c_b(I, \delta) - \Delta ib(I, \delta) \leq 1$, and the lemma follows.   □

Lemma 14. *Given an unsigned intergenic instance $\mathcal{I}$, we have that $d_R(\mathcal{I}) \geq ib(\mathcal{I}) - c_b(G_{IB}(\mathcal{I}))$ and $d_{RI}(\mathcal{I}) \geq ib(\mathcal{I}) - c_b(G_{IB}(\mathcal{I}))$.*

Proof. The proof is similar to that presented in Theorem 2 from Bafna and Pevzner [4] considering that $\Delta c_b(I, \sigma) - \Delta ib(I, \sigma) \leq 1$ for any $\sigma \in \{\rho, \delta\}$ (Lemmas 12 and 13).   □

Given an unsigned intergenic instance $\mathcal{I}$ and a decomposition D of $G_{IB}(\mathcal{I})$ into edge-disjoint alternating cycles, $c_b^D(G_{IB}(\mathcal{I}))$ denotes the number of balanced cycles obtained in $G_{IB}(\mathcal{I})$ from the D decomposition.

Next, we present important characteristics considering a signed intergenic instance induced from a decomposition of G_{IB} into edge-disjoint alternating cycles.

Lemma 15. *Let $\mathcal{I}$ be an unsigned intergenic instance and D a decomposition of $G_{\mathcal{IB}}(\mathcal{I})$ into edge-disjoint alternating cycles, then we have that:*

$$ib(\mathcal{I}) - c_b^D(G_{\mathcal{IB}}(\mathcal{I})) = n + 1 - c_b(G_{\mathcal{IB}}^+(\mathcal{I}')),$$

such that $\mathcal{I}'$ is a signed intergenic instance induced by D.

Proof. Note that $c_b(G_{\mathcal{IB}}^+(\mathcal{I}')) = c_b^t(G_{\mathcal{IB}}^+(\mathcal{I}')) + c_b^{nt}(G_{\mathcal{IB}}^+(\mathcal{I}'))$. Besides, the maximum number that $ib(\mathcal{I})$ can reach is $n + 1$, and each trivial balanced cycle in $G_{\mathcal{IB}}^+(\mathcal{I}')$ represents an intergenic adjacency in $\mathcal{I}$. Thus, we have that $ib(\mathcal{I}) = n + 1 - c_b^t(G_{\mathcal{IB}}^+(\mathcal{I}'))$. Since $\mathcal{I}'$ was induced by the decomposition D of $G_{\mathcal{IB}}(\mathcal{I})$, then for each non-trivial balanced cycle in $G_{\mathcal{IB}}^+(\mathcal{I}')$ we have a corresponding balanced cycle with the same number of reality edges in $G_{\mathcal{IB}}(\mathcal{I})$. Therefore, $c_b(G_{\mathcal{IB}}^+(\mathcal{I}')) = c_b^{nt}(G_{\mathcal{IB}}^+(\mathcal{I}'))$ and the lemma follows.   □

Remark 3. Given an unsigned intergenic instance $\mathcal{I} = ((\pi, \breve{\pi}), (\iota, \breve{\iota}))$ and a D decomposition of $G_{\mathcal{IB}}(\mathcal{I})$ into edge-disjoint alternating cycles. Let $\mathcal{I}' = ((\pi', \breve{\pi}'), (\iota', \breve{\iota}'))$ be a signed intergenic instance induced by D, then any sequence of operations that transforms $(\pi', \breve{\pi}')$ into $(\iota', \breve{\iota}')$ also transforms $(\pi, \breve{\pi})$ into $(\iota, \breve{\iota})$.

Note that the results for the SBIR and SBIRI problems on signed intergenic instances are important to derive results on unsigned intergenic instances, which leads us to the following lemma.

Lemma 16 (Algorithms 1 and 2 from Oliveira *et al.* [20]). *Given a signed intergenic instance $\mathcal{I} = ((\pi, \breve{\pi}), (\iota, \breve{\iota}))$, there is an algorithm that turns $(\pi, \breve{\pi})$ into $(\iota, \breve{\iota})$ using at most $2(n + 1 - c_b(G_{\mathcal{IB}}^+(\mathcal{I})))$ reversals, if $\mathcal{I}$ is balanced, and at most $2(n + 1 - c_b(G_{\mathcal{IB}}^+(\mathcal{I})))$ reversals and indels, if $\mathcal{I}$ is unbalanced.*

Let ALG_D an algorithm for the MAX-ABCD problem that follows the steps described in Theorem 1 and ensures an approximation ratio of $k = \frac{31}{21} + \epsilon$ for the parameter $ib(\mathcal{I}) - c_b(G_{\mathcal{IB}}(\mathcal{I}))$. Now consider the Algorithm 2 for the SBIR problem on unsigned intergenic instances.

Algorithm 2 The input is an unsigned intergenic instance $\mathcal{I} = ((\pi, \breve{\pi}), (\iota, \breve{\iota}))$ and the output is a sequence S of reversal operations such that $(\pi, \breve{\pi}) \cdot S = (\iota, \breve{\iota})$.

Let D be a decomposition obtained from $ALG_D(\mathcal{I})$
Let $\mathcal{I}'$ be a signed intergenic instance induced by D
Let S be a sequence of reversals provided by Algorithm 1 from Oliveira *et al.* [20] for the instance $\mathcal{I}'$
return S

Lemma 17. *Given an unsigned intergenic instance $\mathcal{I} = ((\pi, \breve{\pi}), (\iota, \breve{\iota}))$, Algorithm 2 turns $(\pi, \breve{\pi})$ into $(\iota, \breve{\iota})$ using at most $2k(ib(\mathcal{I}) - c_b(G_{\mathcal{IB}}(\mathcal{I})))$ reversals, such that $k = \frac{31}{21} + \epsilon$.*

Proof. First of all note that, by Theorem 1, we have that $ib(\mathcal{I}) - c_b^D(G_{\mathcal{IB}}(\mathcal{I})) \le k(ib(\mathcal{I}) - c_b(G_{\mathcal{IB}}(\mathcal{I})))$. Since $\mathcal{I}'$ was induced by D, then the sequence S of reversals provided by the algorithm from Oliveira *et al.* [20] also transforms $(\pi, \breve{\pi})$ into $(\iota, \breve{\iota})$ (Remark 3). By Lemma 16, the sequence S has no more than $2(n + 1 - c_b(G_{\mathcal{IB}}^+(\mathcal{I}')))$ reversals. By Lemma 15, we have that $ib(\mathcal{I}) - c_b^D(G_{\mathcal{IB}}(\mathcal{I})) = n + 1 - c_b(G_{\mathcal{IB}}^+(\mathcal{I}'))$, then the sequence S has no more than $2(ib(\mathcal{I}) - c_b^D(G_{\mathcal{IB}}(\mathcal{I}))) \leq 2k(ib(\mathcal{I}) - c_b(G_{\mathcal{IB}}(\mathcal{I})))$ reversals, and the lemma follows.   □

Theorem 2. *Given an unsigned intergenic instance $\mathcal{I}$, Algorithm 2 is a $2k$-approximation for the* SBIR *problem, such that $k = \frac{31}{21} + \epsilon$.*

Proof. Straightforward from Lemmas 14 and 17.   □

Now consider a new algorithm (Algorithm 2), which is similar to Algorithm 2 but using the algorithm from Oliveira *et al.* [20] for the SBIRI problem.

Lemma 18. *Given an unsigned intergenic instance $\mathcal{I} = ((\pi, \breve{\pi}), (\iota, \breve{\iota}))$, Algorithm 2 turns $(\pi, \breve{\pi})$ into $(\iota, \breve{\iota})$ using at most $2k(ib(\mathcal{I}) - c_b(G_{\mathcal{IB}}(\mathcal{I})))$ reversals and indels, such that $k = \frac{31}{21} + \epsilon$.*

Theorem 3. *Given an unsigned intergenic instance $\mathcal{I}$, Algorithm 2 is a $2k$-approximation for the* SBIRI *problem, such that $k = \frac{31}{21} + \epsilon$.*

4 Conclusion

In this work, we introduced a generalization of the Maximum Alternating Cycle Decomposition problem, named the Maximum Alternating Balanced Cycle Decomposition problem. We also designed an enhanced algorithm for the Sorting by Intergenic Operations of Reversal and Indel problem on signed intergenic instances, which guarantees an approximation factor of $\frac{3}{2}$. Combined with an algorithm for the Maximum Alternating Balanced Cycle Decomposition problem, we showed how to obtain better approximation algorithms for the Sorting by Intergenic Reversals and the Sorting by Intergenic Operations of Reversal and Indel problems.

It is important to mention that the use of unsigned genomes is less applicable to a real scenario than the use of signed genomes, but it is still relevant from the theoretical point of view, because the complexity of the problem arises from that scenario. Besides, this problem can still be used when we are not considering gene orientation and can give us insights on the signed variant of the problems.

In future works, practical experiments can be performed to verify the practical applicability of the proposed algorithms and it is possible to investigate how to achieve better results for intergenic models in unsigned scenarios using the Maximum Alternating Balanced Cycle Decomposition problem. Another direction for future work is to consider problems where indel operations can insert or delete genes.

Acknowledgments. This work was supported by the Coordenação de Aperfeiçoamento de Pessoal de Nível Superior - Brasil (CAPES) - Finance Code 001 and the São Paulo Research Foundation, FAPESP (grants 2013/08293-7 and 2021/13824-8).

References

1. Alexandrino, A.O., Brito, K.L., Oliveira, A.R., Dias, U., Dias, Z.: Reversal distance on genomes with different gene content and intergenic regions information. In: Martín-Vide, C., Vega-Rodríguez, M.A., Wheeler, T. (eds.) AlCoB 2021. LNCS, vol. 12715, pp. 121–133. Springer, Cham (2021). https://doi.org/10.1007/978-3-030-74432-8_9

2. Alexandrino, A.O., Brito, K.L., Oliveira, A.R., Dias, U., Dias, Z.: Reversal and indel distance with intergenic region information. IEEE/ACM Transactions on Computational Biology and Bioinformatics, pp. 1–13 (2022)

3. Alexandrino, A.O., Oliveira, A.R., Dias, U., Dias, Z.: Incorporating intergenic regions into reversal and transposition distances with indels. J. Bioinform. Comput. Biol. **19**(06), 2140011 (2021)

4. Bafna, V., Pevzner, P.A.: Genome rearrangements and sorting by reversals. SIAM J. Comput. **25**(2), 272–289 (1996)

5. Berman, P., Fürer, M.: Approximating maximum independent set in bounded degree graphs. In: SODA'94: Proceedings of the Fifth Annual ACM-SIAM Symposium on Discrete Algorithms, pp. 365–371. Society for Industrial and Applied Mathematics (1994)

6. Berman, P., Hannenhalli, S., Karpinski, M.: 1.375-approximation algorithm for sorting by reversals. In: Möhring, R., Raman, R. (eds.) ESA 2002. LNCS, vol. 2461, pp. 200–210. Springer, Heidelberg (2002). https://doi.org/10.1007/3-540-45749-6_21

7. Berman, P., Karpinski, M.: On some tighter inapproximability results (extended abstract). In: Wiedermann, J., van Emde Boas, P., Nielsen, M. (eds.) ICALP 1999. LNCS, vol. 1644, pp. 200–209. Springer, Heidelberg (1999). https://doi.org/10.1007/3-540-48523-6_17

8. Biller, P., Knibbe, C., Beslon, G., Tannier, E.: Comparative genomics on artificial life. In: Beckmann, A., Bienvenu, L., Jonoska, N. (eds.) CiE 2016. LNCS, vol. 9709, pp. 35–44. Springer, Cham (2016). https://doi.org/10.1007/978-3-319-40189-8_4

9. Brito, K.L., Jean, G., Fertin, G., Oliveira, A.R., Dias, U., Dias, Z.: Sorting by genome rearrangements on both gene order and intergenic sizes. J. Comput. Biol. **27**(2), 156–174 (2020)

10. Bulteau, L., Fertin, G., Komusiewicz, C.: (Prefix) Reversal distance for (signed) strings with few blocks or small alphabets. J. Discret. Algorithms **37**, 44–55 (2016)

11. Caprara, A.: On the tightness of the alternating-cycle lower bound for sorting by reversals. J. Comb. Optim. **3**(2), 149–182 (1999)

12. Caprara, A.: Sorting permutations by reversals and Eulerian cycle decompositions. SIAM J. Discret. Math. **12**(1), 91–110 (1999)

13. Caprara, A., Rizzi, R.: Improved approximation for breakpoint graph decomposition and sorting by reversals. J. Comb. Optim. **6**(2), 157–182 (2002)

14. Chen, X.: On sorting unsigned permutations by double-cut-and-joins. J. Comb. Optim. **25**(3), 339–351 (2013)

15. Garey, M.R., Johnson, D.S.: Computers and Intractability; A Guide to the Theory of NP-Completeness. W. H. Freeman & Co., New York, NY, USA (1979)

16. Hannenhalli, S., Pevzner, P.A.: Transforming cabbage into turnip: polynomial algorithm for sorting signed permutations by reversals. J. ACM **46**(1), 1–27 (1999)

17. Hurkens, C.A.J., Schrijver, A.: On the size of systems of sets every t of which have an SDR, with an application to the worst-case ratio of heuristics for packing problems. SIAM J. Discret. Math. **2**(1), 68–72 (1989)

18. Karp, R.M.: Reducibility among combinatorial problems. In: Jünger, M., et al. (eds.) 50 Years of Integer Programming 1958-2008, pp. 219–241. Springer, Heidelberg (2010). https://doi.org/10.1007/978-3-540-68279-0_8

19. Lin, G., Jiang, T.: A further improved approximation algorithm for breakpoint graph decomposition. J. Comb. Optim. **8**(2), 183–194 (2004)

20. Oliveira, A.R., et al.: Sorting signed permutations by intergenic reversals. IEEE/ACM Trans. Comput. Biol. Bioinform. **18**(6), 2870–2876 (2021)

21. Papadimitriou, C.H., Yannakakis, M.: Optimization, approximation, and complexity classes. J. Comput. Syst. Sci. **43**, 425–440 (1991)

22. Pinheiro, P.O., Alexandrino, A.O., Oliveira, A.R., de Souza, C.C., Dias, Z.: Heuristics for breakpoint graph decomposition with applications in genome rearrangement problems. In: BSB 2020. LNCS, vol. 12558, pp. 129–140. Springer, Cham (2020). https://doi.org/10.1007/978-3-030-65775-8_12

23. Rahman, A., Shatabda, S., Hasan, M.: An approximation algorithm for sorting by reversals and transpositions. J. Discret. Algorithms **6**(3), 449–457 (2008)

24. Swenson, K.M., Lin, Yu., Rajan, V., Moret, B.M.E.: Hurdles hardly have to be heeded. In: Nelson, C.E., Vialette, S. (eds.) RECOMB-CG 2008. LNCS, vol. 5267, pp. 241–251. Springer, Heidelberg (2008). https://doi.org/10.1007/978-3-540-87989-3_18

On the Distribution of Synteny Blocks Under a Neutral Model of Genome Dynamics

Sagi Snir[1(✉)], Yuri Wolf[2], Shelly Brezner[1], Eugene Koonin[2], and Mike Steel[3]

[1] Department of Evolutionary and Environmental Biology, University of Haifa, Haifa, Israel
ssagi@research.haifa.ac.il
[2] National Center for Biotechnology Information, National Institutes of Health, Bethesda, USA
[3] Biomathematics Research Centre, University of Canterbury, Christchurch, New Zealand

Abstract. Prokaryotes are a rich source of versatile molecular functional systems that typically consist of multiple, interacting proteins. The study of such systems leads to fundamental biological discoveries, for example, understanding of the origins of innate and adaptive immunity in animals and also provides for the development of various biotechnology applications. The discovery of functional systems by microbial genome mining is facilitated by the fact that functionally coupled genes in bacterial and archaeal genomes often cluster in operons that are conserved across long evolutionary spans. However, accurate differentiation of operons from spurious gene clusters by genome comparison is a non-trivial task that depends on an underlying model of neutral genomes evolution. Here, we investigate the predictions of a gene clustering based on a recently developed stochastic model of genome rearrangement arising from horizontal gene transfer between evolving species along a phylogenetic tree. We focus on synteny blocks, that is, strings of genes conserved across genomes and derive analytic expressions for the expected number of synteny blocks of a given size (or of maximal size) in terms of the temporal separation between the genomes and the rates of evolutionary events. Our setting is similar to the heavily studied stick breaking problem family, but its discrete structure and the stochastic nature of the underlying process suggest a simple, independent model. We demonstrate the predictive power of this model both in simulations and on real data from the ATGC data base.

Keywords: Bacterial evolution · lateral gene transfer · stochastic processes · synteny blocks · the stick breaking model

1 Introduction

Genome evolution involves a broad range of processes, from point mutations to gene duplication and deletion to local and large scale genome rearrange-

C. Scornavacca and M. Hernández-Rosales (Eds.): RECOMB-CG 2024, LNBI 14616, pp. 173–188, 2024.
https://doi.org/10.1007/978-3-031-58072-7_9

ments [11,27]. In the early days of comparative genomics, it has been demonstrated and subsequently amply validated that gene order in prokaryotes evolves much faster than the gene sequences, at least, those of evolutionarily conserved genes [19,31,37]. In particular, the genomes of bacteria and archaea are highly dynamic [5,6,40]. Indeed, deletions, inversions and translocations of genomic segments as well as insertion of new sequences acquired via horizontal gene transfer (HGT) rapidly disrupt genomic synteny, that is, gene order conservation across evolutionary lineages [16,20,24,32]. This process of synteny loss is countered by the conservation of operons, arrays of genes (typically, two to five genes, but in a few cases, longer ones) that encode functionally linked proteins [12,17,34]). Under the pressure of selection that favors operonic organization, operons persist even across long evolutionary distances, forming *synteny blocks*. These synteny blocks are often transferred horizontally as a single unit, increasing their spread across the diversity of prokaryotes [15,38].

Nevertheless, despite intensive comparative genomic studies, both theoretical and empirical (see [2,3,22,23] among many), it remains generally unclear to what degree synteny blocks in bacterial and archaeal genomes, are common and relevant functionally, apart from and beyond operonic organization. This is partly due to the lack of a robust theoretical approach for modeling these genome dynamic events, to embed them in a likelihood framework under which a neutral model of genome synteny evolution could be constructed. In [28], the first stochastic model was proposed *The Jump Model*, that included a single operation, in which a gene "jumps" to a random location, a "slot" between two other genes in the genome. Under the Jump model, a dissimilarity measure called *synteny index* (SI) [1,29], measuring differences in gene order and content between genomes, was shown to be *consistent* and *additive*, meaning that, with an infinite amount of information, SI reconstructs correct edge lengths, allowing to accurately reconstruct tree [10]. As gene order and content are established markers for phylogenetic reconstruction [4,9,25,26,30,33,36,39], a model that encompasses these events in a time related context, lays the ground work for a likelihood-based analysis of genome evolution.

Furthermore, the Jump model provides for analysis beyond gene Content and order, namely, the length distribution of conserved gene strings *synteny blocks* (SBs) across evolutionary lineages under a model of neutral genome evolution. Modeling neutral is essential to decipher between random and functional SBs that are subject to evolutionary constraints.

In this work we make a first step towards this goal by developing the theoretical basis of the neutral model of genome evolution. We start with simple observations regarding SBs and then the Jump model. The key ingredient of our approach is the analysis of the *synteny block length distribution* (SBD) which characterizes the evolutionary divergence between genomes evolving under the neutral Jump model. We draw direct parallels between this model and the classical problem of stick breaking [8,35] and show that this model accurately reconstitutes the SBD for the idealized case of infinite genome size and asymptotically low jump rate. We further devised an approach to adjust the model for finite genome size and high jump rates, achieving a good fit to simulation results and a high predictive power.

2   Preliminaries

We start with the Jump Model introduced in [28] and investigated further in [10] as our basic model of genome rearrangement under horizontal gene transfer (HGT) event between species in a phylogenetic tree T. Given an ancestral sequence of genes at the root ρ of T (say $g_1 g_2 \ldots g_n$), consider the path from the root of the tree to two extant (present-day) species (say x, y) in which the same genes are present, but perhaps in different orders. To take a specific example, suppose that $n = 5$, and x has genome $g_1 g_2 g_3 g_5 g_4$ and y has genome $g_1 g_4 g_2 g_3 g_5$. There are many scenarios that could explain this: one is that a copy of g_5 from the path leading to y was transferred by an HGT event into the genome on the path from ρ to x and landed between g_3 and g_4 and, in addition, a copy of g_4 from the path from ρ to x was transferred into the genome on the path from ρ to y and landed between g_1 and g_2. If we relabel x as $h_1 h_2 h_3 h_4 h_5$ then we can view y as a genome arising from two Jump operations on this sequence: h_4 moves to lie after h_5 (to give a new sequence $h'_1 h'_2 h'_3 h'_4 h'_5$) and then h'_4 jumps to land between h'_1 an h'_2. This example (for a specific scenario) illustrates that we can fix the sequence at x and then consider y as being obtained from x by a sequence of gene Jump operations.

More precisely, given a sequence of n genes, $g_1 g_2 \cdots g_n$, a *Jump* operation moves one of the genes (say g_i) to a different location. Thus each gene has $n - 1$ possible positions it can move to. A simple stochastic model to model gene jump operations assumes that:

- when each Jump operation occurs, each gene is equally likely to move(jump), and the location of where it lands is uniformly distributed amongst the $n - 1$ possible target positions, and
- gene jumps proceed independently of each other.

We refer to this random process as the *Jump Model*. If, in addition, the rate of Jump operations takes a time-independent constant value λ, then the number of jump operations that occur over a time period t follows a Poisson distribution with expected value $\lambda \cdot t$, where t measures the total time separating x and y (thus if x, y are extant, and ρ is t_0 time units in the past, then $t = 2t_0$). We refer to this as the *Poisson Jump Model*. As the process is normally described following an edge (or a *branch*) we often denote λt as the *edge length* or equivalently as the *path length* for a path. We remark that along the paper we may assume $t = 1$ and then the rate λ corresponds to the entire period of the process.

Under this model, a key result from [10,28] is that a certain measure (the synteny index) can be transformed to estimate the evolutionary distance (time) separating the two genomes. This approach considered fixed neighborhoods of each gene. In this work, we investigate a further aspect of jump operations (and the general and Poisson Jump Model), namely the occurrence and frequency of 'synteny blocks' of conserved subsequences shared by two genomes that have undergone a certain (deterministic or random) number of jump operations.

2.1 Synteny Blocks

Consider a genome of length $n \geq 2$. Rather than writing the genome as $(g_1, \ldots, g_n)$ we will henceforth write it instead as an ordered sequence $S_0 = 123 \cdots n$. Let S_k ($k \geq 0$) be a sequence obtained from S_0 by applying k independent jump operations under the Jump Model. Thus S_k is a random variable taking values in the set of all total orderings (i.e. permutations) of the set $[n] = \{1, 2, \ldots, n\}$. A *synteny block* of length $i > 1$ in S_k shared by S_0 is a consecutive subsequence of S_k of length $i > 1$ that also occurs consecutively (and in the same order) in S_0. Such a synteny block is said to be *maximal* if it is not contained in a larger synteny block.

For example, for $n = 7$ and $S_0 = 1234567$ a single jump that moves 4 to lie between 5 and 6 produces $S_1 = 1235467$ which has three synteny blocks of length 2 (12, 23 and 67) shared with S_0 and one synteny block of length 3 (namely, 123) shared with S_0. There are two maximal synteny blocks (123 and 67). If 2 is then moved in S_1 to appear after 7 we obtain $S_2 = 1354672$, and so S_2 has just one synteny block of length 2 that is shared with S_0, namely, 67.

For each $i \geq 1$, let $N_k^{(i)}$ (respectively $M_k^{(i)}$) be the number of synteny blocks (respectively maximal synteny blocks) of length i in S_k.

Lemma 1.

(i) $N_k^{(i)} \leq n - i + 1$ *with equality if and only if* $S_k = S_0$.
(ii) $M_k^{(i)} = N_k^{(i)} - 2N_k^{(i+1)} + N_k^{(i+2)}$, *with* $N_k^{(n+1)} = N_k^{(n+2)} = 0$.

Proof: For Part (i), observe that if $S_k = S_0$ we have $N_k^{(i)} = n - i + 1$. Moreover, there are only $n - i + 1$ consecutive subsequences of length i of S_n so the number shared with S_0 is as most this number, and if they are all shared, then $S_0 = S_k$.

For Part (ii), we have:

$$N_k^{(i)} = \sum_{j=i}^{n}(j - i + 1) \cdot M_k^{(j)}$$

since each maximal synteny block of size j contains $j - i + 1$ synteny blocks. Moreover,

$$\sum_{j=1}^{n} j \cdot M_k^{(j)} = n,$$

since the maximal synteny blocks partition the sequence S_0 (i.e. every element in $[n]$ occurs in exactly one maximal synteny block), and if a maximal block has size j then j elements of $[n]$ are covered by that block. Solving this linear system for $M_k^{(j)}$ gives:

$$M_k^{(j)} = N_k^{(j)} - 2\,N_k^{(j+1)} + N_k^{(j+2)}, \tag{1}$$

subject to the convention that $N_k^{(n+1)} = N_k^{(n+2)} = 0$.

□

We now define the distribution of the lengths of shared *maximal* synteny blocks induced by two genomes.

Definition 1. *Given two permutations S and S'. let $m_i(S, S')$ denote the number of maximal synteny blocks of length i shared by S and S', and let $m(S, S') = \sum_{i \geq 1} m_i(S, S')$. Then the induced SB distribution (SBD), denoted as $SBD(S, S')$ is defined:*

$$\frac{m_i(S, S')}{m(S, S')}, \tag{2}$$

for $1 \leq i \leq n$. We also denote this induced SBD as the observed SBD between S and S'.

3   Expected Number of Synteny Blocks of Length i as $k \to \infty$

As k becomes large S_k has the distribution of a random permutation, as the signal, or the order, is rapidly lost. Let N^i be the number of synteny blocks shared between S_0 and a random permutation. Thus, $\mathbb{E}[N^{(i)}] = \lim_{k \to \infty} \mathbb{E}[N_k^{(i)}]$ which is given as follows.

Proposition 1.

(i)

$$\mathbb{E}[N^{(i)}] = \frac{(n-i+1)^2(n-i)!}{n!} = \begin{cases} n, & \text{if } i = 1; \\ 1 - \frac{1}{n}, & \text{if } i = 2; \\ \frac{n-i+1}{n(n-1)....(n-i+2)} = \Theta(n^{-(i-2)}), & \text{if } i > 2. \end{cases}$$

(ii) $\lim_{k \to \infty} \mathbb{E}[N_k^{(i)}] = \mathbb{E}[N^{(i)}]$, *which is given by Part (i).*

Proof: We can write

$$N^{(i)} = \sum_{j=1}^{n-i+1} \mathbb{I}_j, \tag{3}$$

where $\mathbb{I}_j$ is the 0/1 indicator random variable that takes the value 1 if $j, j + 1, \cdots j+i-1$ appear consecutively (and in this order) in the random permutation of $[n]$, and $\mathbb{I}_j = 0$ otherwise. The number of permutations σ of $[n]$ that have this property is the number of places in σ where this consecutive sequence could start (namely $n-i+1$) times the number of choices for placement of the $n-i$ elements that occur before and after this permutation (namely $(n - i)!$). Dividing by $n!$ (the total number of permutations) we obtain:

$$\mathbb{P}(I_j = 1) = \frac{(n - i + 1)(n - i)!}{n!}.$$

Since the expression on the right is the same for all values of j we obtain from Eq. (3):

$$\mathbb{E}[N^{(i)}] = \mathbb{E}\left[\sum_{j=1}^{n-i+1} \mathbb{I}_j\right] = \sum_{j=1}^{n-i+1} \mathbb{P}(\mathbb{I}_j = 1) = \frac{(n-i+1)^2(n-i)!}{n!},$$

which for simplifies to the expressions given above for the cases $i = 2$ and $i > 2$.

For Part (ii), observe that as $k \to \infty$ the random variable S_k converges in distribution to the uniform distribution on all permutations of $[n]$, and so, as $k \to \infty$, the random variable I_k converges in distribution to the number of synteny blocks of length i that are shared between a random permutation on $[n]$ and S_0.   □

4   The Poisson Jump Model

Under the Poisson Jump Model, as originally stated and studied in [10, 28], jump operations occur at a constant rate λ per gene, and so the total number of jump operations occur at rate $\lambda_n = \lambda \cdot n$. In this paper we allow greater generality by requiring only that $\lambda_n \le \lambda \cdot n$. Let K_t denote the number of Jump operations that occur (for a given genome of length n) in a given period of time of duration t. Then K_t has a Poisson distribution with mean $\lambda_n t$. In other words,

$$\mathbb{P}(K_t = k) = e^{-\lambda_n t}\frac{(\lambda_n t)^k}{k!}, \tag{4}$$

for all $k \ge 0$. The Jump Model induces a probability distribution over the lengths of SBs, that is, the probability to observe an ℓ-long SB. We denote it as the *synteny blocks distribution* (SBD) and we note that each SBD is characteristic of a rate λ and hence should be written as SBD(λ). In Fig. 1 there are three characteristic SBD's obtained empirically by simulating the Jump model for three different values of λ.

Let $S^{(t)}$ be the genome obtained by applying the Jump Model for time t to the genome S_0 (thus $S^{(t)} = S_{K_t}$), let $\mathcal{N}_t^{(i)}$ denote the number of synteny blocks of S_0 of length i that are shared by S_0 and $S^{(t)}$, and for $i > 1$ let

$$\mathcal{L}_t^{(i)} = (n-i+1) - \mathcal{N}_t^{(i)}$$

denote the number of blocks of S_0 of length i that are **not** shared by S_0 and $S^{(t)}$.

If we now let $\mu_t^{(i)} = \mathbb{E}[\mathcal{N}_t^{(i)}]$, then

$$\mu_t^{(i)} = \sum_{k=0}^{\infty} \mathbb{E}[N_k^{(i)}] \cdot e^{-\lambda_n t}\frac{(\lambda_n t)^k}{k!}. \tag{5}$$

with $\mu_0^{(i)} = n - i + 1$, and (for fixed n) $\lim_{t\to\infty} \mu_t^{(i)} = \frac{(n-i+1)^2(n-i)!}{n!}$, by Proposition 1.

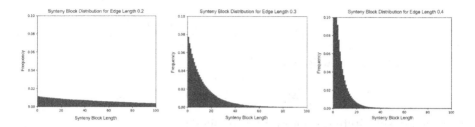

Fig. 1. SB Length Distribution: Simulated SB length frequency as a function of jump rate (edge length) for various rates.

Proposition 2. *Let $i > 1$. Then:*

$$\mathbb{E}[\mathcal{N}_t^{(i)}] = \mu_t^{(i)} \sim (n - i + 1) \exp\left(-\frac{(2i - 1)\lambda_n t}{n}\right),\tag{6}$$

and

$$\mathbb{E}[\mathcal{L}_t^{(i)}] \sim (n - i + 1)\left(1 - \exp\left(-\frac{(2i - 1)\lambda_n t}{n}\right)\right),\tag{7}$$

where $\sim$ refers to asymptotic equivalence as n grows.

Proof of Proposition 2: We have

$$N_k^{(i)} = \sum_{j=1}^{} \mathbb{I}_j^{(k)},$$

where $\mathbb{I}_j^{(k)}$ is the 0/1 indicator random variable that takes the value 1 if $j, j + 1, \ldots, j + i - 1$ appear consecutively (and in this order) in $S^{(t)}$ and $\mathbb{I}_j^{(k)} = 0$ otherwise. The sequence $j, j + 1, \ldots, j + i - 1$ appears consecutively after k jump operations whenever each of the jump operations satisfies the following two properties for each jump: (i) the gene that moves is not one of the i genes from $j, j+1, \ldots j+i-1$, and (ii) the location of where the jumped gene lands is not one of the $i-1$ positions between the i genes from $j, j+1, \ldots, j+i-1$. There may be other ways by which $j, j + 1, \ldots, j + i - 1$ appears consecutively after k jump operations but these have probability $o(1)$ as $n \to \infty$, as the probability of two adjacent jumped genes to remain adjacent in the same order is $1/n$ (given they both jumped which is also of low probability for large n and k constant). Thus, ignoring this vanishing difference, we obtain the following probability for the event that $\mathbb{I}_j^{(k)} = 1$ as follows:

$$\mathbb{P}(\mathbb{I}_j^{(k)} = 1) = \left[\frac{n - i}{n} \times \frac{n - (i - 1)}{n + 1}\right]^k.$$

and since this expression on the right is (again) independent of j we obtain:

$$\mathbb{E}[N_k^{(i)}] = (n - i + 1)\left[\frac{n - i}{n} \times \frac{n - (i - 1)}{n + 1}\right]^k.\tag{8}$$

Let

$$\nu_i = (i-1)/n, \text{ and } \alpha_i = \frac{n-i}{n} \times \frac{n-(i-1)}{n+1}.$$

Then Eq. (8) can be rewritten as:

$$\mathbb{E}[N_k^{(i)}] = n(1-\nu_i)\alpha_i^k.$$

We now substitute this into Eq. (5) to obtain:

$$\mu_t^{(i)} = \sum_{k=0}^{\infty} n(1-\nu_i)\alpha_i^k \cdot e^{-\lambda_n t}\frac{(\lambda_n t)^k}{k!}.$$

Rearranging gives:

$$\mu_t^{(i)} = n(1-\nu_i)e^{-\lambda_n t}\sum_{k=0}^{\infty}\frac{(\alpha_i \lambda_n t)^k}{k!}$$
$$= n(1-\nu_i)e^{-\lambda_n t}e^{\alpha_i \lambda_n t} = n(1-\nu_i)\exp(-\lambda_n t(1-\alpha_i)).$$

Now,

$$\mathbb{E}[\mathcal{L}_t^{(i)}] = n(1-\nu_i) - \mu_t^{(i)} = n(1-\nu_i)(1-\exp(-\lambda_n t(1-\alpha_i))),$$

we obtain the expression in Part (i).

$\square$

Example: Taking $n = 1000$ and $i = 11$, Proposition 2 predicts that the expected value of $\mathcal{L}_t^{(11)}/9990$ is approximately $1-\exp(-0.02\lambda_n t)$. Note also that if $\lambda_n/n \to 0$ then $\mathbb{E}[\mathcal{L}_t^{(i)}] \sim (2i-1)\lambda_n t$; in particular, if $\lambda_n = \lambda$ (constant) then $\mathbb{E}[\mathcal{L}_t^{(i)}]$ is asymptotically independent of n.

5   Partitioning into Maximal Synteny Blocks and the Distribution of Their Lengths

Given a genome (permutations of $[n]$), S, a *maximal synteny block* of S shared by $S_0 = (123\cdots n)$ is a synteny block of S that is shared by S_0 and which is not contained in any larger synteny block of S shared by S_0 (note in particular that a maximal synteny block need not be a largest synteny block). By definition, the maximal synteny blocks of S shared by S_0 are mutually disjoint. For example, if $n = 9$ and $S = (124578963)$ then the maximal synteny blocks of S shared by S_0 are $12, 45, 789$.

5.1   Maximal Synteny Blocks Under $\lambda_n = o(n)$

We now describe the behavior and distribution of the SBs under two regimes: when $\lambda_n = o(n)$, and for $\lambda_n = O(n)$. While the first regime requires one type of analysis due to basic assumptions, the second regime cannot rely on these assumptions and requires a different analysis. We start with $\lambda_n = o(n)$.

Recall that $m_i(S_0, S_k)$ holds the number of maximal synteny blocks of length i shared by S_0 and S_k, and $m(S_0, S_k)$ is the sum of these blocks over all lengths i.

Starting with S_0 apply k jump operations under the Jump Model, and let $l_1, l_2, \ldots, l_m$ (where $m = m(S_0, S_k)$) be the lengths of the resulting maximal synteny blocks (each of length > 1) shared by S_k and S_0. Consider the vector

$$\ell(S_0, S_k) = (l_1/n, l_2/n, \cdots, l_m/n)$$

where the components are ordered by increasing size. For example, consider $S_0 = (123456789)$ and suppose that gene 5 moves to lie between 7 and 8. Then $S_1 = (123467589)$ and so S_0 and S_1 share three synteny blocks (one of length 4 (i.e. 1234), and two of length 2 (i.e. 67 and 89)), so $\ell(S_1, S_0) = (2/9, 2/9, 4/9)$.

We now describe the limiting distribution of the random vector $\ell(S_0, S_k)$ as n becomes large, by considering the following 'stick breaking' model. Start with the continuous interval $[0, 1]$ (a 'stick'), and independently place $2k$ points uniformly at random along the stick. Now cut the stick at each of these points. Let $X_{2k} = (l'_1, \ldots, l'_{2k+1})$ be the resulting random variable corresponding to the lengths of the pieces ordered by size (thus, $l'_1 < l'_2 < \cdots < l'_{2k+1}$ with probability $1 - o(1)$). The expected length of l'_r for all $1 \leq r \leq 2k + 1$ is given by $\mathbb{E}[l'_r] = \frac{1}{2k+1} \sum_{j=2k+2-r}^{2k+1} \frac{1}{j}$ ([7], p. 167).

Proposition 3.

(i) *Under the Jump Model, the following hold;*
  (a) *For $k = O(n^\beta)$ with $\beta < \frac{1}{2}$ we have $\lim_{n \to \infty} \mathbb{P}(m(S_0, S_k) = 2k + 1) = 1$.*
  (b) *For k fixed, $\ell(S_0, S_k)$ converges in distribution to X_{2k} as $n \to \infty$.*
(ii) *Under the Poisson Jump Model the following hold:*
  (a)

$$\mathbb{E}[m_i(S_0, S^{(t)})] = \mu_t^{(i)} - 2\mu_t^{(i+1)} + \mu_t^{(i+2)},$$

  where $\mu_t^{(i)}$ is given by Eq. (5).
  (b) *If $\lambda_n = O(n^\beta)$ for $\beta < \frac{1}{2}$ then:*

$$\lim_{n \to \infty} \mathbb{P}(m(S_0, S^{(t)}) = j) = \begin{cases} e^{-\lambda_n t} \dfrac{(\lambda_n t)^{(j-1)/2}}{((j-1)/2)!}, & \text{if } j \geq 1 \text{ is odd;} \\ 0, & \text{if } j \text{ is even.} \end{cases}$$

Proof: For Part (i-a), observe that there are $O(n)$ positions where a gene can jump from and $O(n)$ positions where it can land. For the claim to hold, we want all these positions to be distinct. Thus if $k = O(n^\beta)$ for $\beta < \frac{1}{2}$, then under the jump model, by the birthday paradox [18] and k is asymptotically smaller than the boundary for a collision $\sqrt{n}$, the probability of the event E_n that all these $O(2n)$ positions are distinct is $1 - o(1)$ (where in this proof $o(1)$ refers to a term that tend to zero as $n \to \infty$). Conditional on E_n, the number of maximal synteny blocks is $2k+1$, which establishes Part (a). Part (i-b) now follows directly from Part (i-a) combined with the observation that $l_1 < l_2 < \cdots < l_{2k+1}$ with probability $1 - o(1)$ (i.e. asymptotically in n, there are no ties).

Part (ii-a) follows from Lemma 1 and linearity of expectation, and Part (ii-b) follows from Part (i-a), noting that if $\lambda_n = O(n^\beta)$ for $\beta < \frac{1}{2}$ then $k_n \le n^{\beta'}$ with probability $1 - o(1)$, where $\beta < \beta' < \frac{1}{2}$.

5.2   Maximal Synteny Blocks Under $\lambda_n = O(n)$

In this part we relax the assumption of asymptotically low rate of jumps λ, specifically $\lambda = c$ for c a constant smaller than one, implying $\lambda_n = O(n)$. Here we cannot anymore no ties in $\ell(S_0, S_k)$, or even that a jump does not occur more than once. As before, we assume a single time unit t, i.e. the rate λ corresponds for the entire process and $\lambda t = \lambda$ and hence t is ignored from the analysis.

Recall that under the Poisson Jump Model we have a rate λ operating on every element (gene) in the genome S. Let $P = \mathbb{P}_\lambda(\text{ jump})$ and $Q = \mathbb{P}_\lambda(\text{no jump})$ be the probabilities of a gene jumps or remains intact respectively, and clearly we have:

Observation 1

- $P = e^{-\lambda}$
- $Q = 1 - P = 1 - e^{-\lambda}$

Now, using these two probabilities, we can express the expected number of SBs of any length along the genome.

Proposition 4. *For any j with $1 < j < n-k-1$, let $\mathbb{I}_j^{(k)}$ denote a 0/1 indicator random variable that indicates whether a k-long (maximal) SB starts at position j. Then under the Poisson Jump Model the following hold:*

$$\mathbb{P}(\mathbb{I}_j^{(k)} = 1) = \begin{cases} P + 4QP^2 + o(1), & for\ k = 1; \\ \\ 4Q^{2k-1}P^2 + o(1), & for\ k > 1. \end{cases} \tag{9}$$

Proof: We start with the case of $k > 1$. For a SB to start at position j, first there needs to be either a jump from position $j - 1$ or a jump into the slot between position $j - 1$ and position j. The same also applies to positions $j + k$ and $j + k + 1$. There are four options for such an event. Additionally, we need a "no-jump out" by all genes at positions j to $j + k$, and a "no-jump in" to the $k - 1$ slots between positions j to $j + k$, in total $2k - 1$ "no-jump" events. There are other scenarios resulting in a k-long SB, however their probability is proportional to $1/n$ and become negligible as n grows.

For the case of $k = 1$, in addition to the previous scenario, we also obtain a 1-long SB in the case of a jump into position j and hence the additional P term. This completes the proof.   ∎

6   Experimental Results

We now describe experimental results we obtained by applying the theory described above for the SB length distribution, under real life sizes. The power of Proposition 4 stated above, is that it accounts for the probability of multiple events at a position and corrects for that. This scenario becomes realistic for a constant λ as we see in the experimental part below. Indeed Eq. (9) provides very accurate estimations for the SBD under realistic values as shown in simulation, to the degree it can entirely replace as the empirical SBD.

6.1   Measuring Goodness of Fit

We now describe a series of experiments to test the power of our model.

Our first experiment was aimed at measuring how our theoretical measurement performs under a range of jump rates and under realistic finite genome sizes. We simulated the Jump process for various values of rate λ and under realistic sizes of genomes and contrasted the results obtained empirically, with the analytical ones attained by Proposition 4. The results for two value of lambda $\lambda = 0.001, 0.01$ are shown in Fig. 2. Figure 2 contains two columns where in

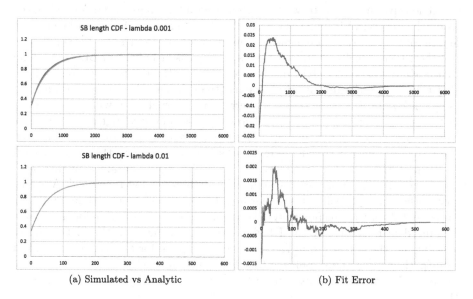

(a) Simulated vs Analytic   (b) Fit Error

Fig. 2. (a) **Goodness of Fit**: Comparison between analytically obtained expressions for the cumulative mass function, for two values of rate, $\lambda = 0.001, 0.01$. At each row a different value of λ is tested. The left graph at each row depicts the cumulative mass function (CMF) where the x axis is the SB length while the y-axis depicts the accumulated frequency for every SB length. At every graph in the left, the blue curve represents simulated values and the orange curve represents the analytic formula. The graphs on the right at every row, depicts the difference between the two curves as a function of SB length. Simulation was run on a genome size of 5000. (Color figure online)

the left column the empirical (blue curve) distribution generated by simulations is contrasted with the analytical distribution as computed by Proposition 4. The curves represent the cumulative distribution (CDF, i.e. the *accumulated* frequency of SB length, the y axis) as a function of SB length (x axis). At the right column, we see the difference between the two values. We see an almost perfect fit between the two curves, to the degree that the two curves are almost indistinguishable at small values of λ.

6.2   Measuring SBD Specificity

Our second experiment dealt with the predicting, or reconstruction, power of the SB-length distribution. Based on the previous experiment, we concluded that the analytic formula of Eq. (9) of Proposition 4 accurately reconstructs the PDF of every edge length.

The goal was to measure how accurate is the reconstruction of edge lengths from the induced SBD obtained by simulation. For this goal we devised the following experiment. For several generative models (or configurations, not to be confused with the Jump Model applied to these generative models), differing by either edge lengths or topology, we simulated the Poisson Jump Model down the generative model. Next, we generated several SBD's for several edge lengths using the analytical formula from Eq. (9) that creates a SBD for a given edge length (or rate λ). We subsequently compared the SBD induced by the simulation to other, theoretically produced SBD's. Comparison between SBD's was done by the Kullback-Leibler (KL) divergence [14]. The three topologies we used are shown in Fig. 3. All are directed graphs (trees) where the left two are a path (single leaf) and the right tree is a cherry - an internal node with two children as leaves. For every topology, several combinations of edge lengths were used. A genome at the root node (in-degree zero) is evolved by the Jump Model down the tree. The SBD between the endpoints is calculated and the closest model (λ-induced SBD) is chosen. Figure 2 depicts the results obtained in this part. For every topology from Fig. 3 several rate combinations were applied. At every row in the figure the same total length (rate λ) was used with 0.1 and 0.22 for upper and lower rows respectively. The graphs depicts a histogram for the closest λ-induced SBD chosen by the KL distance. As can be seen, the SBD is fairly specific and the correct total rate is almost uniquely for any topology or total rate (path length). This confirms that the SBD concept as a model statistic is quite expressive and can distinguish between various GD rates operating along the tree.

Results from Real Data Trees. Our last experiment was done using real data from a collection of closely related bacterial and archaeal genomes, the Alignable Tight Genomic Clusters (ATGC) database [13,21], which was constructed for the purpose of evolutionary genomics studies. The goal here is to test both our problem definition of a neutral model of SBs, and the tools we developed, on real prokaryotic data. The procedure was as follows. We applied our Jump model

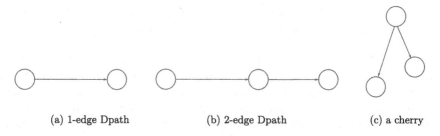

(a) 1-edge Dpath   (b) 2-edge Dpath   (c) a cherry

Fig. 3. Fit topologies: (**L**) A 1-edge directed path (**M**) A 2-edge directed path (**R**) A Cherry. At every topology several combination of edge lengths were assigned and a genome was evolved along the directed paths to the leaves. SBD was computed between the endpoints of the path and compared to a model SBD representing edge lengths.

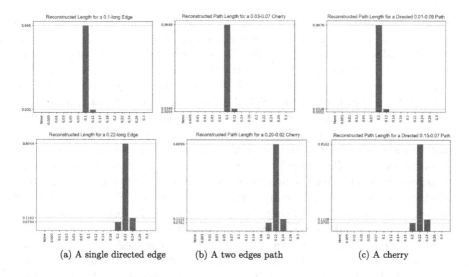

(a) A single directed edge   (b) A two edges path   (c) A cherry

Fig. 4. SB distribution specificity: Top: total path length (jump rate) 0.1, Bottom: total path length 0.22. At each row the left graph corresponds to a single directed edge, the middle graph to a cherry, and the right graph to a two-edge directed path. The simulated SBD between the two endpoints is compared to several model SBD by the Kullback-Leibler divergence distance. The closest model (rate) is selected. The graph shows a histogram of the selected rates.

down a (rooted) tree with Jump rates (λt) to obtain genomes at the leaves of the tree. Next, for every pair of leaves (u, v), we computed the obtained (observed) SBD, $\widehat{SBD}(u, v)$. Then we found the nearest model λ such that the theoretical SBD(λ) is the nearest to $\widehat{SBD}(u, v)$. Finally we compared it to the tree distance (path length) between the two leaves u, v (Fig. 4).

The experiment was run for two trees from the ATGC DB - the ATGC013 on 5 taxa from the from *Streptococcus* I/*parasanguinis*/*A12*, and the ATGC017 tree on 77 taxa from *Bacillus* strains. The trees with their edge length (Jump rates) are shown in Fig. 6 in the Appendix. For each of the trees we ran three runs of the experiment. The results in a form of a scatter plot, are shown in Fig. 7 (also in the Appendix, for lack of space). In the figure, the x-axis holds the real pairwise distance on the tree for every pair (u, v), while the y-axis holds the closest inferred distance (rate λ). Note that as we increase our nearest distance with increments, we cannot match it precisely always. Ideally we'd expect to see a 45-degree line. Indeed, we see a very near fit to this. We ran three experiments that are represented by the colors of the dots. We can see here that the observed SBD reconstructs almost perfectly the generative model, for every tree distance.

7   Conclusions

In this work we have defined and investigated the neutral model of the evolution of synteny blocks that is induced under a simple stochastic model of genome dynamics (GD). This model accounts for sequences of consecutive genes that are identical in both composition and order in two or more genomes. These gene sequences, which we denote synteny blocks (SBs), result from GD events that scramble the ancestral gene order. Specifically, we focused on the Jump Model in which genes transpose into different locations in a genome in a time-dependent stochastic process. We first provide a theoretical analysis of the problem and point to the equivalence to the classical stick-breaking problem. These results rely on the concept of SB length distribution (SBD) that uniquely characterizes the generative model. We provide theoretical results for the case of asymptotically large genomes are and asymptotically low jump rate. We also describe an approach that corrects for multiple events at a gene for the Poisson process with finite genome size and jump rate values and produces close agreement with the results of simulated gene order evolution. We tested our model under both simulated and real data-guided scenarios and demonstrated its ability to produce a unique SBD allowing to accurately reproduce the original model parameters. The results obtained here suggest that our model can identify both the characteristic rates of GD events between species and the deviations of the SB evolution from the neutral model. By fitting the best theoretical λ-induced model by the procedure illustrated in the experimental part, we can readily produce a p-value for every suspicious deviation from the theoretical distribution. Furthermore, the GD model used here (the Jump model) lays the basis for more advanced extensions such as analysis of gene insertions and deletions that we aim to pursue in the future.

Acknowledgments. We wish to thank the reviewers and in particular Reviewer 1 for his very meticulous examination and enlightening comments.

References

1. Adato, O., Ninyo, N., Gophna, U., Snir, S.: Detecting horizontal gene transfer between closely related taxa. PLoS Comput. Biol. **11**, e1004408 (2015)
2. Bafna, V., Pevzner, P.A.: Genome rearrangements and sorting by reversals. SIAM J. Comput. **25**(2), 272–289 (1996)
3. Bejerano, G., et al.: Ultraconserved elements in the human genome. Science **304**(5675), 1321–5 (2004)
4. Biller, P., Guéguen, L., Tannier, E.: Moments of genome evolution by double cut-and-join. BMC Bioinform. **16**(14), S7 (2015)
5. Doolittle, W.: Lateral genomics. Trends Cell Biol. **9**, M5–M8 (1999)
6. Gogarten, J., Doolittle, W., Lawrence, J.: Prokaryotic evolution in light of gene transfer. Mol. Biol. Evol. **19**, 2226–2238 (2002)
7. Grimmett, G., Stirzaker, D.: Probability and Random Processes, 3rd edn. Oxford University Press, Oxford (2001)
8. Holst, L.: On the lengths of the pieces of a stick broken at random. J. Appl. Probab. **17**(3), 623–634 (1980)
9. Huson, D.H., Steel, M.: Phylogenetic trees based on gene content. Bioinformatics **20**(13), 2044–2049 (2004)
10. Katriel, G., et al.: Gene transfer-based phylogenetics: analytical expressions and additivity via birth-death theory. Syst. Biol. **72**(6), 1403–1417 (2023)
11. Koonin, E.V., Wolf, Y.I.: Genomics of bacteria and archaea: the emerging dynamic view of the prokaryotic world. Nucleic Acids Res. **36**(21), 6688–6719 (2008)
12. Korbel, J.O., Jensen, L.J., von Mering, C., Bork, P.: Analysis of genomic context: prediction of functional associations from conserved bidirectionally transcribed gene pairs. Nat. Biotechnol. **22**(7), 911–917 (2004)
13. Kristensen, D.M., Wolf, Y.I., Koonin, E.V.: ATGC database and ATGC-COGs: an updated resource for micro- and macro-evolutionary studies of prokaryotic genomes and protein family annotation. Nucleic Acids Res. **45**(D1), D210–D218 (2017)
14. Kullback, S., Leibler, R.: On information and sufficiency. The Ann. Math. Stat. **22**(1), 79–86 (1951)
15. Lawrence, J.: Selfish operons: the evolutionary impact of gene clustering in prokaryotes and eukaryotes. Curr. Opin. Genet. Dev. **9**, 642–648 (1999)
16. Libeskind-Hadas, R., Wu, Y.-C., Bansal, M.S., Kellis, M.: Pareto-optimal phylogenetic tree reconciliation. Bioinformatics **30**(12), i87–i95 (2014)
17. Malke, H.: J. H. Miller and W. S. Reznikoff (editors), the operon (2nd edition). vii, 469 s., 128 abb., 36 tab. cold spring harbor 1980. cold spring harbor laboratory. Zeitschrift für allgemeine Mikrobiologie **21**(9), 697–697 (1981)
18. Mathis, F.H.: A generalized birthday problem. SIAM Rev. **33**(2), 265–270 (1991)
19. Mushegian, A., Koonin, E.: Gene order is not conserved in bacterial evolution. Trends Genet. **12**, 289–290 (1996)
20. Nadeau, J.H., Taylor, B.A.: Lengths of chromosomal segments conserved since divergence of man and mouse. Proc. Natl. Acad. Sci. **81**(3), 814–818 (1984)
21. Novichkov, P.S., Ratnere, I., Wolf, Y.I., Koonin, E.V., Dubchak, I.: ATGC: a database of orthologous genes from closely related prokaryotic genomes and a research platform for microevolution of prokaryotes. Nucleic Acids Res. **37**, D448-454 (2009)
22. Sankoff, D., Blanchette, M.: Multiple genome rearrangement and breakpoint phylogeny. J. Comput. Biol. **5**(3), 555–570 (1998)

23. Sankoff, D., El-Mabrouk, N.: Genome rearrangement. In: Jiang, T., Xu, Y., Zhang, M. (eds.) Current Topics in Computational Molecular Biology. CRC Press (2002)

24. Sankoff, D., Leduc, G., Antoine, N., Paquin, B., Lang, B.F., Cedergren, R.: Gene order comparisons for phylogenetic inference: evolution of the mitochondrial genome. Proc. Natl. Acad. Sci. **89**(14), 6575–6579 (1992)

25. Sankoff, D., Nadeau, J.H.: Conserved synteny as a measure of genomic distance. Discrete Appl. Math. **71**(1–3), 247–257 (1996)

26. Serdoz, S., et al.: Maximum likelihood estimates of pairwise rearrangement distances. J. Theor. Biol. **423**, 31–40 (2017)

27. Setubal, J.C., Almeida, N.F., Wattam, A.R.: Comparative genomics for prokaryotes. Methods Mol. Biol. **1704**, 55–78 (2018)

28. Sevillya, G., Doerr, D., Lerner, Y., Stoye, J., Steel, M., Snir, S.: Horizontal gene transfer phylogenetics: a random walk approach. Mol. Biol. Evol. **37**(5), 1470–1479 (2019)

29. Shifman, A., Ninyo, N., Gophna, U., Snir, S.: Phylo SI: a new genome-wide approach for prokaryotic phylogeny. Nucleic Acids Res. **42**(4), 2391–2404 (2013)

30. Sjöstrand, J., Tofigh, A., Daubin, V., Arvestad, L., Sennblad, B., Lagergren, J.: A Bayesian method for analyzing lateral gene transfer. Syst. Biol. **63**(3), 409–420 (2014)

31. Snel, B., Bork, P., Huynen, M.A.: Genomes in flux: the evolution of archaeal and proteobacterial gene content. Genome Res. **12**(1), 17–25 (2002)

32. Stolzer, M., Lai, H., Xu, M., Sathaye, D., Vernot, B., Durand, D.: Inferring duplications, losses, transfers and incomplete lineage sorting with nonbinary species trees. Bioinformatics **28**(18), i409–i415 (2012)

33. Szöllősi, G.J., Tannier, E., Lartillot, N., Daubin, V.: Lateral gene transfer from the dead. Syst. Biol. **62**(3), 386–397 (2013)

34. Teichmann, S.A., Babu, M.M.: Conservation of gene co-regulation in prokaryotes and eukaryotes. Trends Biotechnol. **20**(10), 407–410 (2002)

35. Verreault, W.: MacMahon partition analysis: a discrete approach to broken stick problems. J. Comb. Theory Ser. A **187**, 105571 (2022)

36. Wang, L.-S., Warnow, T.: Estimating true evolutionary distances between genomes. In: Proceedings of the Thirty-Third Annual ACM Symposium on Theory of Computing, pp. 637–646. ACM (2001)

37. Wolf, Y.I., Makarova, K.S., Lobkovsky, A.E., Koonin, E.V.: Two fundamentally different classes of microbial genes. Nat. Microbiol. **2**, 16208 (2016)

38. Wolf, Y.I., Rogozin, I.B., Kondrashov, A.S., Koonin, E.V.: Genome alignment, evolution of prokaryotic genome organization, and prediction of gene function using genomic context. Genome Res. **11**(3), 356–372 (2001)

39. Woodhams, M., Steane, D.A., Jones, R.C., Nicolle, D., Moulton, V., Holland, B.R.: Novel distances for Dollo data. Syst. Biol. **62**(1), 62–77 (2012)

40. Zhaxybayeva, O., Gogarten, J.P., Charlebois, R.L., Doolittle, W.F., Papke, R.T.: Phylogenetic analyses of cyanobacterial genomes: quantification of horizontal gene transfer events. Genome Res. **16**(9), 1099–1108 (2006)

Sampling Gene Adjacencies and Geodesic Points of Random Genomes

Poly H. da Silva[1,3], Arash Jamshidpey[2,3(✉)], and David Sankoff[4]

[1] Department of Statistics, Columbia University, New York, NY 10027, USA
[2] Department of Mathematics, Columbia University, New York, NY 10027, USA
[3] Irving Institute for Cancer Dynamics, Columbia University, New York, NY 10027, USA
{phd2120,aj2963}@columbia.edu
[4] Department of Mathematics and Statistics, University of Ottawa, Ottawa, ON K1N 6N5, Canada
sankoff@uottawa.ca

Abstract. The breakpoint distance employed in comparative genomics is not a geodesic distance, which makes it difficult to study genomes (i.e. permutations) that are intermediate between two given genomes G and G'. An intermediate genome, also called a *geodesic point*, is a genome whose sum of breakpoint distances to G and G' is equal to the breakpoint distance of G and G'. To construct an intermediate genome M, it is necessary to find sets of gene adjacencies I and J selected from G and G' whose union forms M. This means that the set of adjacencies of M is $I \cup J$. Any given set of adjacencies I selected from G may put some constraints on some adjacencies of G' so that they cannot be used in J to construct M or if they can, they must be used in specific ways. For instance, a gene adjacency of G' whose gene extremities are used in the *"middle"* of segments of I cannot be used to construct M. Based on these constraints, we classify the set of all adjacencies of G' with respect to I into four distinct groups. For two unichromosomal random genomes of the same gene-content, namely ξ_1 and ξ_2, as the number of genes tends to infinity, we study the limiting behaviour of the frequencies of adjacencies of each type in ξ_2 with respect to a random or deterministic set of adjacencies selected from ξ_1. We use the limiting results to provide necessary conditions for the size and the shape of the set of adjacencies selected from the first genome for the purpose of constructing an intermediate genome between ξ_1 and ξ_2. These results can help to shed light on how to construct *"accessible breakpoint medians"* far from the input genomes (corners).

Keywords: Random genomes · Breakpoint median · Geodesic points

1 Introduction

In comparative genomics, it is often important to compare the order of the same set of genes on the chromosomes of two or more different species. In the basic

C. Scornavacca and M. Hernández-Rosales (Eds.): RECOMB-CG 2024, LNBI 14616, pp. 189–210, 2024.
https://doi.org/10.1007/978-3-031-58072-7_10

case, this comes down to comparing two or more different permutations on the integers $1, \ldots, n$, where each number represents a gene. The simplest way is to count the number of *breakpoints* in two permutations (genomes). A pair of adjacent genes in a single-chromosome genome G is called a breakpoint of G with respect to G', another single-chromosome genome, if these genes are not adjacent in G'. The number of breakpoints of G with respect to G', denoted $d(G, G')$, is equal to the number of breakpoints of G' with respect to G. Formalized in [7], the breakpoint distance d counts the number of breakpoints in the set of gene adjacencies of two unichromosomal genomes with an identical set of genes.

For the study of genomic evolution, for every $\alpha \in (0, 1)$ such that $\alpha d(G, G')$ is an integer, it would be useful to be able to construct a genome M_α on a path between given genomes G and G', namely $d(G, M_\alpha) = \alpha d(G, G')$ and $d(M_\alpha, G') = (1 - \alpha) d(G, G')$. In this case, M_α is called a geodesic point in the space of genomes or permutations. The breakpoint distance, however is not geodesic [4]; it is not true that for every two permutations G and G' with $d(G, G') = r$, we can obtain a sequence of permutations $G = x_0, x_1, .., x_{r-1}, x_r = G'$ such that the $d(x_i, x_{i+1}) = 1$. This means that we cannot be sure, for any $\alpha \in (0, 1)$, where αr, is an integer, of finding a geodesic permutation M_α such that $d(G, M_\alpha) = \alpha r$ and $d(M_\alpha, G') = (1 - \alpha) r$. In particular, we may not be able to find a compromise geodesic genome, where $\alpha \approx \frac{1}{2}$. Even though it may not be a compromise between G and G', a geodesic genome M_α is always a "*median*", i.e. a genome minimizing $d(., G) + d(., G')$.

The notion of a median genome originally [7] derives not from the comparison of two genomes, but of three or more, as part of the construction of a gene-order phylogenetic tree [8]: given $k > 2$ genomes, $G_1, \ldots, G_k$, find a genome M that is as close as possible to all k of them, minimizing the total distance $\sum_{i=1}^{k} d(G_i, .)$. Of particular interest are those medians of $G_1, \ldots, G_k$ which are far from the input genomes (corners) and inherit its gene adjacencies from all of them, i.e. when there are n genes, there exist $\alpha_1, \cdots, \alpha_k > \varepsilon > 0$ s.t. $\alpha_1 + \cdots + \alpha_k = 1$ and $d(G_i, M) \approx \alpha_i n$ for $i = 1, \cdots, k$. More specifically, can we find medians of $G_1, \ldots, G_k$ which are also a compromise among them, i.e. $\alpha_i \approx 1/k$, for $i = 1, \cdots, k$?

Simulating this problem, however, leads to the conjecture that a median M is seldom a compromise among the k input genomes, but rather a genome that is very close to one of the given G_i [2], and this tendency becomes stronger as n increases.

[4] investigated this conjecture further, and proved a weaker conjecture also stated in [2], that after a convenient rescaling of the breakpoint distance, with high probability, the median value (i.e. the total distance to a median) of $G_1, \cdots, G_k$ is close to $(k - 1)(n - 1)$. This readily implies that, with high probability, each G_i is an asymptotic median of $G_1, \cdots, G_k$, and hence some medians are placed close to the random input genomes (corners). This however does not shed any light on the stronger conjecture of rareness of medians far from the corners. Hoping to explore this part of the conjecture further, [4] introduced the notion of accessible medians.

The definition of the accessible medians is strongly based on the geodesic points. Any accessible point is in fact identified as a result of determining geodesic points for some pairs of permutations. In particular, for two permutations, the notions of medians, accessible medians, and geodesic points, all coincide. Although, studying the density of geodesic points between two random permutations is interesting on its own, its connection to accessible points and medians makes it more important.

Any geodesic point M located between genomes G and G' must exclusively take its adjacencies from the union of gene adjacencies of G and G'. Furthermore, it must contain any gene adjacencies common between G and G' [4]. The latter condition is however less important when G and G' are large and picked independently at random from the space of genomes, as in this case the number of common adjacencies of two random genomes converges to a Poisson random variable with mean 2 as the number of genes tends to infinity. Assuming G and G' have a negligible number of common gene adjacencies, in order to find a geodesic point M sufficiently far from both G and G', there must be two sets of adjacencies I and J, neither very small nor very large, selected from G and G' such that their union forms M, i.e. $I \cup J$ is the set of adjacencies of M.

Motivated by this, given two genomes G and G' and a set of adjacencies I selected from G, a natural problem is to find all adjacencies of G' which are useful to construct a geodesic point between G and G' that contains I. This is equivalent to finding a set of gene adjacencies J selected from G' such that $I \cup J$ forms a genome. It is important to note that any choice of I from the gene adjacencies of G may place various constraints on the gene adjacencies of G', either rendering them unusable in J or imposing specific restrictions on how they can be used. In light of these constraints in the construction of geodesic points between G and G' that contains I, we categorize the set of all adjacencies of G' with respect to I into four distinct groups. These types are defined in the same spirit of the types of adjacencies used to construct a near median genome in [6].

For two random genomes ξ_1 and ξ_2 having the same gene contents, as the number of genes tends to infinity, we explore the limiting behaviour of the frequencies of adjacencies of each of these four types in ξ_2 with respect to a randomly or deterministically chosen set of adjacencies from ξ_1. We leverage these limiting results to establish necessary conditions for the size and shape of the set of adjacencies selected from the first genome for constructing an intermediary genome (geodesic point) between ξ_1 and ξ_2. These findings contribute insights into the construction of accessible breakpoint medians far from the corners.

This paper is laid out as follows. Section 2 provides a comprehensive overview of the definitions and concepts used in the rest of the paper. This includes representing genomes as permutations, breakpoint distance, geodesic points, partial geodesics between two genomes, and various notions associated with sets of adjacencies of a genome. In particular in Sect. 2.2, for given genomes G, G' and a set of adjacencies I selected from G, we introduce four different classes of adjacencies of G' with respect to I. Section 3 concerns the analysis of these types of adjacencies with respect to a deterministic or random set of adjacencies. In Sect. 3.1, we discuss the distributions of the size and shape of a set of gene adjacencies sampled at random from a given genome. In Sect. 3.2, we compute the expectation and variance of the number of adjacencies of each type in a random permutation of size n (a random genome with n genes), namely $\xi^{(n)}$, with respect to a given or random set of adjacencies of another permutation. As n tends to infinity, we then establish an L^2-convergence theorem for the normalized number (divided by n) of different types of adjacencies of $\xi^{(n)}$. This leads to a further discussion, in Sect. 3.3, on the possible size and shape of a set of adjacencies I for which there may exists a genome x containing a set of adjacencies J such that $I \cup J$ forms an intermediate genome.

2   Genomes Endowed with Breakpoint Distance

In the absence of duplication, linear unichromosomal genomes can be represented by permutations, where numbers indicate genes or markers in the corresponding genomes. Denote by S_n the set of all permutations on $[n] := \{1, ..., n\}$. For a permutation $\pi := \pi_1 \ ... \ \pi_n$, any unordered pair $\{\pi_i, \pi_{i+1}\} = \{\pi_{i+1}, \pi_i\}$, for $i = 1, ..., n - 1$, is called an adjacency of π. We denote by $\mathcal{A}_\pi$ the set of all adjacencies of π and by $\mathcal{A}_{x_1,...,x_k}$ the set of all common adjacencies of $x_1, ..., x_k \in S_n$. The *breakpoint* distance (bp distance) between x and y is defined by $d(x, y) = d^{(n)}(x, y) := n - 1 - |\mathcal{A}_{x,y}|$ which in fact equals half of the symmetric difference between sets $\mathcal{A}_x$ and $\mathcal{A}_y$.

A geodesic between permutations $x, y \in S_n$ is a chain of length $k = d(x, y)$ in S_n, namely $z_0 = x, z_1..., z_{k-1}, z_k = y$, such that $d(z_i, z_{i+1}) = 1$, for $i = 0, ..., k-1$. It is not hard to see that (S_n, d) is not a geodesic space, that is we can find at least one pair of permutations between which there is no geodesic chain. For instance, for $n = 4$, there is no geodesic between permutations 1 2 3 4 and 1 3 2 4. On the other hand, we find a geodesic path between 1 2 3 4 and 3 1 4 2 which is precisely given by $z_0 = 1\ 2\ 3\ 4, z_1 = 1\ 4\ 3\ 2, z_2 = 4\ 1\ 3\ 2, z_3 = 3\ 1\ 4\ 2$. A class of examples are given in [4]. For a non-geodesic metric or pseudometric space, the concept of a geodesic between two points can be extended to the concept of a *partial geodesic* or *geodesic patch* (p-geodesic), as introduced in [3,4]. More precisely, a p-geodesic between $x, y \in S_n$ is a maximal subset of S_n containing x and y which can be isometrically embedded into $\{0, 1, \cdots, d(x, y) - 1, d(x, y)\}$. In other words, a p-geodesic between x and y is a maximal chain $z_0 = x, z_1, ..., z_r = y$ in S_n such that $\sum_{i=0}^{r-1} d(z_i, z_{i+1}) = d(x, y)$. We notice that $r \leq d(x, y)$; $r = d(x, y)$ means that the p-geodesic is in fact a geodesic. Note that any geodesic is a p-geodesic, but not vice versa.

Any permutation on a p-geodesic between $x, y \in S_n$ is called a geodesic point (or geodesic permutation) of x and y. In other words, $z \in S_n$ is a geodesic point of x and y, if and only if we have $d(x, y) = d(x, z) + d(z, y)$. By a non-trivial p-geodesic between two permutations, we mean a p-geodesic with at least three geodesic points. We denote by $\overline{[x, y]} = \overline{[x, y]}_n := \{z \in S_n : d(x, y) = d(x, z) + d(z, y)\}$ the set of all geodesic points of x and y.

2.1 Segment Sets of Permutations

Common adjacencies of permutations can be regarded as a set of segments. A segment (of S_n) is a set of consecutive adjacencies of a permutation of length n. More explicitly, a segment of length $k \in [n-1]$ is a set of adjacencies $\{\{n_0, n_1\}, \{n_1, n_2\}, ..., \{n_{k-2}, n_{k-1}\}, \{n_{k-1}, n_k\}\}$, where $n_0, n_1, ..., n_k \in [n]$ are different natural numbers. It can also be denoted by $[n_0, n_1, ..., n_k]$ or equivalently by $[n_k, ..., n_1, n_0]$. In particular, any segment of length $n-1$ is the set of adjacencies of a permutation and vice versa. By convention, we assume that the empty set $\emptyset$ is a segment. We say a segment s is a subsegment of a segment s' if $s \subset s'$. For a given permutation $\pi = \pi_1 ... \pi_n \in S_n$, for $i \leq j$, the segment $[\pi_i, \pi_{i+1}, ..., \pi_j] = [\pi_j, ..., \pi_{i+1}, \pi_i]$ is denoted by $s_{ij} = s_{ij}^\pi$ and is called a segment of π. We denote by $|s|$ the length of a segment s. For a segment $s := [n_0, ..., n_k]$, n_0 and n_k are called end genes (points), and $n_1, ..., n_{k-1}$ are called internal genes (points) of the segment. Any gene which is not either an end gene or an internal gene of s is called an external gene with respect to s. We denote by $End(s)$, $Int(s)$, and $Ext(s)$, the set of end genes, internal genes, and external genes of s, respectively. Note that a segment is originally defined as a set of adjacencies and therefore all set operations can be applied on it. Two segments $s = [n_0, ..., n_k]$ and $s' = [m_0, ..., m_{k'}]$ are said to be strongly disjoint if $\{n_0, ..., n_k\} \cap \{m_0, ..., m_{k'}\} = \emptyset$. They are disjoint if $s \cap s' = \emptyset$, otherwise we say that they intersect.

Also, by a set of segments (segment set) of S_n, we mean the union of some pairwise strongly disjoint segments of S_n. In other words, a set of segments or a segment set I is a subset of $\mathcal{A}_\pi$ for some permutation π. In this case, we say I is a segment set of π or π contains I. It is clear that a segment set can be contained in more than one permutation, or in other words, it can be contained in the intersection of adjacencies of several permutations. By a segment (or component) of a segment set I we mean a maximal segment contained in I. To show a segment s is a segment of I, we write $s \hat{\in} I$. Although a segment set I containing segments $s_1, ..., s_k$ is in principle the union of adjacencies of s_i's, that is $I = \cup_{i=1}^k s_i$, to ease the notation, we sometime denote it by $\{s_1, ..., s_k\}$. Also we denote by $\|I\| := k$, the number of segments of I. Note that the notation $|\,.\,|$ is used for both cardinality of a set and absolute value of a real number. For example, as we already indicated for a segment s, $|s|$ is the number of adjacencies of s, and by the original definition of a segment set as a union of segments, $|I| = \sum_{i=1}^{\|I\|} |s_i|$ is the number of adjacencies of I.

Denote by $\mathscr{I}_{m,k}^{(n)}$ the set of all segment sets of S_n with m adjacencies and k segments, i.e. $\mathscr{I}_{m,k}^{(n)}$ is the set of all segment sets I with $|I| = m$ and $\|I\| = k$. Similarly, let $\mathscr{I}_m^{(n)}$ be the set of all segment sets of S_n with m adjacencies. Finally, denote by $\mathscr{I}^{(n)}$, the set of all segment sets of S_n. Note that $\mathscr{I}_{m,k}^{(n)}$ may be empty for some m, k, n. To have $\mathscr{I}_{m,k}^{(n)}$ non-empty, it is necessary to have $k \leq m \leq n - k$, where the last inequality holds since for a segment set $I \in \mathscr{I}_{m,k}^{(n)}$ and for any arbitrary permutation π containing I, there should be at least $k-1$

adjacencies of π that are not used in I in order to separate k segments of I, and therefore m should be bounded by $(n-1) - (k-1) = n - k$.

It is clear that the intersection of segments (and in general, the intersection of segment sets) is always a segment set. Two segment sets I and J (in particular two segments s and s', respectively) are said to be consistent, if their union is contained in $\mathcal{A}_\pi$, for some permutation π. In particular, any two segment sets of any $\pi \in S_n$ are consistent. For example, for $n = 10$, two segments $[3, 7, 10, 2, 5]$ and $[2, 5, 8, 1]$ are consistent and their union is the segment $[3, 7, 10, 2, 5, 8, 1]$, while two segments $[2, 6, 3, 8, 1]$ and $[8, 1, 4, 7, 6, 3, 5]$ are not consistent. When we speak of the union of two or more segment sets (respectively, two or more segments) we always assume that they are pairwise consistent. We say segment sets $I_1, ..., I_k$ complete each other if there exists a permutation π such that $\cup_{i=1}^{k} I_i = \mathcal{A}_\pi$. The complement of a segment set I contained in a permutation π, is $\bar{I}_\pi := \mathcal{A}_\pi \setminus I$. In other words, for a segment set $I = \{s_{i_1 j_1}^\pi, s_{i_2 j_2}^\pi, ..., s_{i_k j_k}^\pi\}$ contained in π, $\bar{I}_\pi = \{s_{j_0 i_1}^\pi, s_{j_1 i_2}^\pi, ..., s_{j_k i_{k+1}}^\pi\}$, with $j_0 = 1, i_{k+1} = n$. For $r = 1, ..., k + 1$, we denote by $\bar{I}_\pi^{(r)}$ the rth segment of $\bar{I}_\pi$ on π from left, that is $\bar{I}_\pi^{(r)} = s_{j_{r-1} i_r}^\pi$. When we write $\bar{I}_\pi$, we assume that I is contained in π. We can extend the notions of *end gene*, *internal gene*, and *external gene* to the case of segment sets as follows. A number $u \in \{1, ..., n\}$ is an end gene or point (respectively, an internal gene) of a non-empty segment set $I = \{s_1, ..., s_l\}$ if it is an end gene (respectively, an internal gene) of exactly one of the segments of I. It is an external gene of I if it is neither an end gene nor an internal gene of I, or equivalently if it is an external gene of all of segments of I. In other words, with a slight abuse of notation

$$End(I) = \bigcup_{s \hat{\in} I} End(s), \quad Int(I) = \bigcup_{s \hat{\in} I} Int(s), \quad Ext(I) = \bigcap_{s \hat{\in} I} Ext(s).$$

When I is the empty segment set, we define $End(I) = Int(I) = \emptyset$ and $Ext(I) = [n]$. For example, when $n = 10$, $I = \{[2, 3, 9, 4], [5, 6]\}$ is a segment set having $3, 9$ as its internal genes, $2, 4, 5, 6$ as its end genes, and $1, 7, 8, 10$ as its external genes. We say two segments $s, s' \hat{\in} I$ are neighbours with respect to π, if there exist $i < j$ such that $\pi_i, \pi_j \in End(s) \cup End(s')$ and for any k with $i < k < j$ (if there is any), $\pi_k \in Ext(I)$. We say a segment s connects two disjoint segments s_1 and s_2, if $s_1 \cup s \cup s_2$ is a segment.

2.2   Types of Adjacencies with Respect to a Segment Set

Given a segment set I from the identity permutation $id = id^{(n)} \in S_n$, we aim to count the number of all permutations $x \in S_n$ for which there exists a permutation $\pi \in \overline{[id, x]} \setminus \{id, x\} \neq \emptyset$ containing I such that $\bar{I}_\pi = \mathcal{A}_\pi \setminus I$ is a segment set in x. In order to identify all such permutations, it is useful to classify the set of all adjacencies of S_n, i.e. $\mathcal{A}^* := \{\{a, b\} \mid 1 \leq a \neq b \leq n\}$, with respect to I. Given I, an ideal classification should determine the possible ways in which one can use exactly $n - 1 - |I|$ adjacencies from $\mathcal{A}^*$ to construct a segment set J such that J and I complete each other. Having any such segment set J identified, one can construct all permutations x containing J. To formalize the idea, we say an adjacency $\{a, b\}$ is *2-free-end*, with respect to I, if a and b are both external genes of I. It is called *1-free-end*, w.r.t. I, if either a or b is an external gene of I, and the other is an end gene of I. It is a *trivial segment*, w.r.t. I, if a and b are both end genes of I. Finally, $\{a, b\}$ is *0-free-end*, w.r.t. I, if either a or b is an internal gene of I. An example is given in Fig. 1.

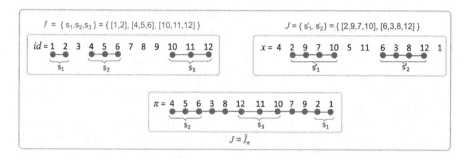

Fig. 1. An example of $\pi \in \overline{[id, x]}$ with $d(id, \pi) = 6$, $d(\pi, x) = 5$ and $\mathcal{A}_\pi = I \cup J$, i.e. $J = \bar{I}_\pi$. In this example, we have $End(I) = \{1, 2, 4, 6, 10, 12\}$, $Int(I) = \{5, 11\}$ and $Ext(I) = \{3, 7, 8, 9\}$. Also, the types of adjacencies of x with respect to the segment set I are: *2-free-end* $= \{\{7, 9\}, \{3, 8\}\}$, *1-free-end* $= \{\{2, 9\}, \{7, 10\}, \{3, 6\}, \{8, 12\}\}$, *trivial segment* $= \{\{2, 4\}, \{1, 12\}\}$ and *0-free-end* $= \{\{5, 10\}, \{5, 11\}, \{6, 11\}\}$.

Given a permutation x, the goal is to use some adjacencies of x to construct a segment set J, with $|J| = n - 1 - |I|$, such that J completes I. Although in most cases, it is impossible to construct such a segment set J from the adjacencies of a given permutation x, one can instead find a segment set J' with the maximum number of adjacencies from x, such that J' is consistent with I and $J' \cap I = \emptyset$. We notice that J' may not be unique. To this end, note that any 2-free-end adjacency of x must be used in J', while a 0-free-end adjacency of x may never be used. On the other hand, depending on x and I, a 1-free-end or a trivial segment adjacency may or may not be used in J'. However, it is not hard to see that at least half of 1-free-end and trivial segment adjacencies of x can always be used in J'.

Motivated by this, to construct a permutation π in $\overline{[id, x]}$ containing I such that $\bar{I}_\pi$ is contained in x, we cannot use any 0-free-end adjacency of x w.r.t. I, since both numbers (genes) in the extremities of this type of adjacency are already used in I as its internal genes. Therefore, to be able to construct π with this property, we must take 2-free-end adjacencies of x w.r.t I, to choose the segment set J contained in x as mentioned above. Both the other two types of adjacencies, i.e., 1-free-end and trivial segment adjacencies, can be used only as extremities of segments of J. More precisely, a 1-free-end adjacency of x w.r.t I may be used in extremities of segments of any size in J, while a trivial segment adjacency of x w.r.t. I may be used only as a segment of length 1 in J. The analysis of the number of adjacencies of each type is given in Sect. 3. The asymptotics for the normalized number of different types of adjacencies is given in Theorem 3.

3   Analysis of the Adjacency Types

In this section, we compute the expected value (Theorem 1) and variance (Theorem 2) of the number of all four types of adjacencies for a random permutation w.r.t. a given segment set in $\mathscr{I}_{m,k}$, w.r.t. a random segment set in $\mathscr{I}_{m,k}$, and w.r.t. a random segment set in $\mathscr{I}_m$. We then establish an L^2-convergence theorem (Theorem 3) for the normalized number of adjacencies of each type. The proofs of the results come in the Appendix. Following this, we study the possibility of constructing a permutation π in $[\overline{id}, x]$ containing segment set I from identity permutation id such that $\overline{I}_\pi$ is contained in x.

Letting S_n be equipped with composition as the multiplication operation, one can easily see that the breakpoint distance is a left-invariant distance on S_n, that is for any $x, y, z \in S_n$, we have $d(zx, zy) = d(x, y)$. This implies that for any $x \in S_n$ there exists a distance-preserving bijection $\zeta_x^{(n)}(y) = x^{-1}y$ form (S_n, d) to itself that sends x to id, i.e. $\zeta_x(x) = x^{-1} \cdot x = id$ and from the left-invariance property $d(x, y) = d(id, x^{-1}y) = d(\zeta_x(x), \zeta_x(y))$. On the other hand, for permutations ξ_1, ξ_2, ξ picked independently and uniformly at random from S_n, $\xi_2^{-1}\xi_1$ is the same as ξ in distribution. These facts imply that studying the geodesic point and partial geodesics between two random permutations ξ_1 and ξ_2 is equivalent to studying those between id and a random permutation ξ. Although the results of this paper are only stated for the types of adjacencies of a random genome ξ with respect to a given or random segment set selected from id, they are also true for the types of adjacencies of a random genome ξ_2 with respect to a given or random segment set selected from an independent random genome ξ_1.

3.1   Sampling a Random Segment

In the sequel, we assume that all random elements and variables are defined on a probability space $(\Omega, \mathbb{P}, \mathcal{F})$, and denote by $\mathbb{E}[\,.\,]$ and $Var(\,.\,)$, the expected value and variance of a random variables with respect to $\mathbb{P}$, respectively. We need the following lemma to obtain the distribution of the number of segments of a random set of adjacencies of size m. The proof is given in the Appendix.

Lemma 1. *Given a permutation $x \in S_n$, there exist*

$$\binom{m-1}{k-1}\binom{n-m}{k}$$

segment sets of x with $k > 0$ non-empty segments and $m \leq n - 1$ adjacencies.

Consider a permutation $x \in S_n$ and pick m adjacencies of x uniformly at random, without replacement. Lemma 1 indicates that the distribution of the number of segments of the resulting random segment set is hypergeometric $H(n-1, n-m, m)$, that is the probability that a random segment set of any permutation x has exactly k segments is

$$P_{n,m,k} := \frac{\binom{n-m}{k}\binom{m-1}{m-k}}{\binom{n-1}{m}}.$$

We can extend this to define more generalized random segment sets on $\mathscr{I}_m$ and $\mathscr{I}_{m,k}$. More precisely, consider a random permutation $\tilde{\xi} = \tilde{\xi}^{(n)}$ picked uniformly from S_n, and let $\mathcal{I}_m = \mathcal{I}_m^{(n)}$ be a segment set chosen uniformly at random from the set of all segment sets of $\tilde{\xi}$ with exactly m adjacencies. From the definition, $\mathbb{P}(\|\mathcal{I}_m\| = k \mid \tilde{\xi} = x) = P_{n,m,k}$ for any choice of $x \in S_n$, i.e. $\|\mathcal{I}_m\|$ is independent of $\tilde{\xi}$ and $\|\mathcal{I}_m\| \sim H(n-1, n-m, m)$. The rth moment of $\|\mathcal{I}_m\|$ is then given by

$$\mathbb{E}\|\mathcal{I}_m\|^r = \sum_{j=1}^{r} \frac{(n-m)_{[j]} m_{[j]}}{(n-1)_{[j]}} \left\{ {r \atop j} \right\}, \tag{1}$$

where $\left\{ {r \atop j} \right\}$ is the Stirling number of the second kind, and $n_{[j]}$ stands for the falling factorial, defined by $n_{[j]} = n!/(n-j)!$. It is clear from the definition that, for any k, the law of $\mathcal{I}_m$ distributes a uniform mass on $\mathscr{I}_{m,k}$. Therefore, defining $\mathcal{I}_{m,k}^{(n)}$ by $\mathbb{P}(\mathcal{I}_{m,k}^{(n)} = I) = \mathbb{P}(\mathcal{I}_m^{(n)} = I \mid \|\mathcal{I}_m^{(n)}\| = k)$ for $I \in \mathscr{I}_{m,k}^{(n)}$, we get $\mathbb{P}(\mathcal{I}_{m,k} = I) = 1/|\mathscr{I}_{m,k}|$. The distribution of $\mathcal{I}_m$ however is not uniform on $\mathscr{I}_m$. To see this we need the following lemma whose proof is provided in the Appendix. Denote by $\mathcal{R}_n(I) = \{x \in S_n : I \subset \mathcal{A}_x\}$.

Lemma 2. *Given a segment set $I \in \mathscr{I}_{m,k}^{(n)}$, the number of permutations in S_n containing I is given by $|\mathcal{R}_n(I)| = 2^k(n-m)!$.*

For any $I \in \mathscr{I}_m$, the distribution of $\mathcal{I}_m = \mathcal{I}_m^{(n)}$ can be obtained from Lemma 2

$$\mathbb{P}(\mathcal{I}_m = I) = \sum_{x \in \mathcal{R}_n(I)} \mathbb{P}(\mathcal{I}_m = I \mid \tilde{\xi} = x)\mathbb{P}(\tilde{\xi} = x)$$

$$= \frac{|\mathcal{R}_n(I)|}{n!\binom{n-1}{m}} = \frac{2^{\|I\|}(n-m)!}{n!\binom{n-1}{m}}. \tag{2}$$

By definition, for $I \in \mathscr{I}_{m,k}$, we get

$$\mathbb{P}(\mathcal{I}_{m,k} = I) = \frac{\mathbb{P}(\mathcal{I}_m = I)}{\mathbb{P}(\|\mathcal{I}_m\| = k)} = \frac{2^k(n-m)!}{n!\binom{n-m}{k}\binom{m-1}{m-k}}.$$

Since $\mathbb{P}(\mathcal{I}_{m,k} = I) = 1/|\mathscr{I}_{m,k}|$, for any $I \in \mathscr{I}_{m,k}$, we obtain

$$|\mathscr{I}_{m,k}| = \frac{n!\binom{n-m}{k}\binom{m-1}{m-k}}{2^k(n-m)!}.$$

Furthermore,

$$|\mathscr{I}_m| = \sum_{k=1}^{n-m} |\mathscr{I}_{m,k}| = \frac{n!}{(n-m)!} \sum_{k=1}^{n-m} \frac{\binom{n-m}{k}\binom{m-1}{m-k}}{2^k} = \frac{\binom{n-1}{m}n!}{(n-m)!} \mathbb{E}\left[2^{-\|\mathcal{I}_m\|}\right]. \quad (3)$$

This can also be written as

$$|\mathscr{I}_m| = \frac{n! \; {}_2F_1(1-m,1+m-n;2;0.5)}{2(n-m-1)!},$$

where ${}_2F_1$ is the hypergeometric function defined by

$${}_2F_1(a,b;c;z) = \sum_{r=0}^{\infty} \frac{a_{(r)}b_{(r)}}{c_{(r)}} \frac{z^r}{r!}; \; |z| < 1,$$

where $n_{(r)} = \Gamma(n+r)/\Gamma(n)$ is the rising factorial, and Γ is the gamma function.

3.2　Number of Adjacencies of Different Types

Consider a random permutation $\xi = \xi^{(n)}$ chosen uniformly from S_n. We assume that $\xi^{(n)}$ is independent of $\mathcal{I}_m^{(n)}$ and $\mathcal{I}_{m,k}^{(n)}$. Let $\alpha,\beta,\gamma,\delta$ be functions

$$\alpha,\beta,\gamma,\delta : \bigcup_{n\in\mathbb{N}} (S_n \times \mathscr{I}^{(n)}) \to \mathbb{Z}_+,$$

such that, for $x \in S_n$ and a segment set of S_n, namely I, let $\alpha(x,I)$, $\beta(x,I)$, $\gamma(x,I)$ and $\delta(x,I)$ be the number of 2-free-end adjacencies, 1-free-end adjacencies, trivial segments, and 0-free-end adjacencies of x w.r.t. I, respectively. In particular, let $\alpha_m^{(n)} := \alpha(\xi^{(n)},\mathcal{I}_m^{(n)})$, $\beta_m^{(n)} := \beta(\xi^{(n)},\mathcal{I}_m^{(n)})$, $\gamma_m^{(n)} := \gamma(\xi^{(n)},\mathcal{I}_m^{(n)})$ and $\delta_m^{(n)} := \delta(\xi^{(n)},\mathcal{I}_m^{(n)})$. Similarly, let $\alpha_{m,k}^{(n)} := \alpha(\xi^{(n)},\mathcal{I}_{m,k}^{(n)})$, $\beta_{m,k}^{(n)} := \beta(\xi^{(n)},\mathcal{I}_{m,k}^{(n)})$, $\gamma_{m,k}^{(n)} := \gamma(\xi^{(n)},\mathcal{I}_{m,k}^{(n)})$ and $\delta_{m,k}^{(n)} := \delta(\xi^{(n)},\mathcal{I}_{m,k}^{(n)})$. As before, we drop "$n$" from the superscripts when there is no risk of ambiguity. We have the following results whose proof will be given in the Appendix.

Theorem 1. *Let $m = m(n)$ and $k = k(n)$ be such that $0 < k \le m < n$, and let I be an arbitrary segment set in $\mathscr{I}_{m,k}$. Then*

$$\mathbb{E}[\alpha_{m,k}] = \mathbb{E}[\alpha(\xi,I)] = \frac{(n-m-k)(n-m-k-1)}{n},$$

$$\mathbb{E}[\beta_{m,k}] = \mathbb{E}[\beta(\xi,I)] = \frac{4k(n-m-k)}{n},$$

$$\mathbb{E}[\gamma_{m,k}] = \mathbb{E}[\gamma(\xi,I)] = \frac{2k(2k-1)}{n},$$

$$\mathbb{E}[\delta_{m,k}] = \mathbb{E}[\delta(\xi,I)] = \frac{(m-k)(2n-m+k-1)}{n}.$$

Furthermore,

$$\mathbb{E}[\alpha_m] = \frac{(n-m)(n-m-1)^2(n-m-2)}{n(n-1)(n-2)},$$

$$\mathbb{E}[\beta_m] = \frac{4m(n-m)(n-m-1)^2}{n(n-1)(n-2)},$$

$$\mathbb{E}[\gamma_m] = \frac{2m(n-m)(2m(n-m)+n)}{n(n-1)(n-2)},$$

$$\mathbb{E}[\delta_m] = \frac{m(m-1)(2n^2-6n-m^2+3m+2)}{n(n-1)(n-2)}.$$

Theorem 2. *Let $m = m(n)$ and $k = k(n)$ be such that $0 < k \le m < n$, and let I be an arbitrary segment set in $\mathscr{I}_{m,k}$. Then*

$$Var(\alpha_{m,k}) = Var(\alpha(\xi, I)) = \left(1 - \frac{m+k}{n}\right)^2 \left(\frac{m+k}{n}\right)^2 n + o(n),$$

$$Var(\beta_{m,k}) = Var(\beta(\xi, I)) = 4k\left(1 - \frac{m+k}{n}\right)\left(\frac{k}{n}\left(3 - \frac{4k}{n}\right) + \frac{m}{n}\left(1 - \frac{4k}{n}\right)\right) + o(n),$$

$$Var(\gamma_{m,k}) = Var(\gamma(\xi, I)) = 4\left(1 - \frac{2k}{n}\right)^2 \left(\frac{k}{n}\right)^2 n + o(n),$$

$$Var(\delta_{m,k}) = Var(\delta(\xi, I)) = \left(\frac{m-k}{n}\right)^2 \left(1 - \frac{m-k}{n}\right)^2 n + o(n).$$

$$Var(\alpha_m) = \left(1 - \frac{m}{n}\right)^4 \left(\frac{m}{n}\right)^2 \left(8 + \frac{m}{n}\left(\frac{5m}{n} - 12\right)\right) n + o(n),$$

$$Var(\beta_m) = 4\left(1 - \frac{m}{n}\right)^3 \left(\frac{m}{n}\right)^2 \left(8 - \frac{m}{n}\left(31 + \frac{4m}{n}\left(\frac{5m}{n} - 11\right)\right)\right) n + o(n),$$

$$Var(\gamma_m) = 4\left(1 - \frac{m}{n}\right)^2 \left(\frac{m}{n}\right)^2 \left[1 - \frac{4m}{n}\left(1 - \frac{m}{n}\right)\left(1 + \frac{5m}{n}\left(1 - \frac{m}{n}\right)\right)\right] n + o(n),$$

$$Var(\delta_m) = \left(\frac{m}{n}\right)^2 \left(1 - \left(\frac{m}{n}\right)^2\right)^2 \left(4 + \frac{m}{n}\left(\frac{5m}{n} - 8\right)\right) n + o(n).$$

We are ready to state a convergence theorem for all different types of adjacencies of $\xi^{(n)}$ w.r.t. $\mathcal{I}^{(n)}_{m(n),k(n)}$ or $\mathcal{I}^{(n)}_{m(n)}$, where m and k are such that $1 \le k(n) \le m(n) \le n - k(n)$, for any $n \in \mathbb{N}$. Let $(\hat{I}_n)_{n \in \mathbb{N}}$ be an arbitrary sequence of segment sets that $\hat{I}_n \in \mathscr{I}^{(n)}_{m(n),k(n)}$. Denote $\tilde{\alpha}_n := \alpha(\xi^{(n)}, \mathcal{I}^{(n)}_{m(n)})$ and $\bar{\alpha}_n := \alpha(\xi^{(n)}, \mathcal{I}^{(n)}_{m(n),k(n)})$. We similarly define $\tilde{\beta}_n, \tilde{\gamma}_n, \tilde{\delta}_n$, and $\bar{\beta}_n, \bar{\gamma}_n, \bar{\delta}_n$, for $n \in \mathbb{N}$.

A sequence of random variables (Z_n) converges to a random variable Z in L^2 if $\mathbb{E}[(Z_n - Z)^2] \to 0$, as $n \to \infty$. It is well known that convergence in L^2 implies convergence in probability of the Z_n to Z, denoted by $Z_n \xrightarrow{p} Z$, which means that for any $\varepsilon > 0$, as n goes to infinity, $\mathbb{P}(|Z_n - Z| > \varepsilon) \to 0$. See [1,5] for the definitions and relations. To prove the L^2-convergence of the $Z_n \in L^2$ to a given $z \in \mathbb{R}$, it suffices to show $\mathbb{E}Z_n \to z$ and $Var(Z_n) \to 0$, as $n \to \infty$. We can now state our main convergence result which is also illustrated in Fig. 2.

Theorem 3. *Suppose* $\frac{m(n)}{n} \to c$ *and* $\frac{k(n)}{n} \to c'$, *as* $n \to \infty$. *Then, as* $n \to \infty$

$$\frac{\tilde{\alpha}_n}{n} \xrightarrow{L^2,p} (1-c)^4, \qquad \frac{\bar{\alpha}_n}{n}, \frac{\alpha(\xi^{(n)}, \hat{I}_n)}{n} \xrightarrow{L^2,p} (1-c-c')^2,$$

$$\frac{\tilde{\beta}_n}{n} \xrightarrow{L^2,p} 4c(1-c)^3, \qquad \frac{\bar{\beta}_n}{n}, \frac{\beta(\xi^{(n)}, \hat{I}_n)}{n} \xrightarrow{L^2,p} 4c'(1-c-c'),$$

$$\frac{\tilde{\gamma}_n}{n} \xrightarrow{L^2,p} 4c^2(1-c)^2, \qquad \frac{\bar{\gamma}_n}{n}, \frac{\gamma(\xi^{(n)}, \hat{I}_n)}{n} \xrightarrow{L^2,p} 4c'^2,$$

$$\frac{\tilde{\delta}_n}{n} \xrightarrow{L^2,p} c^2(2-c)^2, \qquad \frac{\bar{\delta}_n}{n}, \frac{\delta(\xi^{(n)}, \hat{I}_n)}{n} \xrightarrow{L^2,p} (c-c')(2-c+c').$$

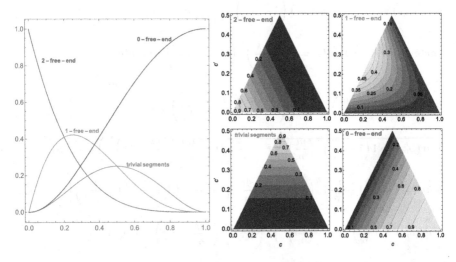

Fig. 2. Left: Plots illustrating the limits of $\tilde{\alpha}_n/n$ (in blue), $\tilde{\beta}_n/n$ (in orange), $\tilde{\gamma}_n/n$ (in green) and $\tilde{\delta}_n/n$ (in red) as functions of c, as given in Theorem 3. Right: Comparison of the limits of $\bar{\alpha}_n/n$ (top-left), $\bar{\beta}_n/n$ (top-right), $\bar{\gamma}_n/n$ (bottom-left) and $\bar{\delta}_n/n$ (bottom-right). Contours represent equal values of these quantities, with lighter colors indicating higher values; the horizontal axis represents the variable c, while the vertical axis represents c'. (Color figure online)

3.3   Necessary Conditions for the Sizes of the Segment Sets

For a segment set I in S_n, define $X_n(I)$ to be the set of all permutations x containing a segment set J such that $I \cap J = \emptyset$ and $I \cup J = A_\pi$ for some permutation π. In fact, J serves as $\bar{I}_\pi$. Let I be a segment set of $id = id^{(n)} \in S_n$. In order to construct a permutation $x \in X_n(I)$, we need to find a segment set of S_n, namely J, such that $I \cap J = \emptyset$ and $I \cup J = A_\pi$, for a permutation π. Then, x is constructed by completing the segment set J. Conversely, when a permutation $x \in X_n(I)$ is given, an easy observation shows that there exists at least one permutation π containing I such that $J = A_\pi \setminus I \subset A_x$ and all 2-free-end adjacencies of x are used in π (Lemma 3). For the moment, let us denote by J°, the segment set of x containing all 2-free-end adjacencies of x w.r.t. I, and note that we must have $J^\circ \subset J$. So in order to find the permutation π with the above property, we first take the segment set $I \cup J^\circ$. In fact, $A_\pi \setminus (I \cup J^\circ)$ should still be a segment set of x, and $n - 1 - |I| - |J^\circ|$ more adjacencies of x (1-free-end adjacencies and trivial segments) should be taken in order to complete $I \cup J^\circ$. To analyze this further, we define this formally as follows. Let F be a function

$$F : \bigcup_{n \in \mathbb{N}} (S_n \times \mathscr{I}^{(n)}) \to \mathscr{I}^{(n)},$$

where for any permutation $x \in S_n$ and any segment set $I \in \mathscr{I}^{(n)}$, $F(x, I)$ is the segment set of x containing all 2-free-end adjacencies of x w.r.t. I, that is

$$F(x, I) := \{\{l, l'\} \in A_x : \{l, l'\} \text{ is 2-free-end}\}.$$

Let

$$Q : \bigcup_{n \in \mathbb{N}} (S_n \times \mathscr{I}^{(n)}) \to \mathbb{Z}_+,$$

where for $(x, I) \in S_n \times \mathscr{I}^{(n)}$, $Q(x, I)$ is the number of adjacencies needed in order to complete $I \cup F(x, I)$ to a permutation π, that is, $Q(x, I) = n - 1 - |I| - |F(x, I)|$. The following theorem restricts the range of $Q(x, I)$, for $x \in X_n(I)$.

Theorem 4. *Let $I \in \mathscr{I}^{(n)}$, and $x \in X_n(I)$. Then $\|I\| - 1 \le Q(x, I) \le 2\|I\|$.*

Before proving the above theorem, we introduce a new concept. Let I be a segment set of S_n. The freedom factor of a point (number) $k \in [n]$, is 0 if $k \in Int(I)$. It is 1, if $k \in End(I)$. Finally, it is 2, if $k \in Ext(I)$. Similarly, the freedom factor of a segment $s = [v_1, ..., v_l] = [v_l, ..., v_1]$ is denoted by $u = <u_1, ..., u_l> = <u_l, ..., u_1>$, where for each $i \in [l]$, u_i is the freedom factor of v_i. A segment s, with the freedom vector u is called a u-segment. Also, for $\pi \in S_n$ and $i \in [n]$, the set of neighbours of i in π is defined by $\mathcal{N}_\pi(i) := \{j \in [n] : \{i, j\} \in A_\pi\}$. For an arbitrary segment set of S_n, namely I, in order that $x \in X_n(I)$, we need to find a segment set J contained in x such that $I \cup J = A_\pi$ and $I \cap J = \emptyset$. As we mentioned, J may not have all adjacencies of $F(x, I)$. For instance, let $I = [4, 5, 6, 7]$ and $x = 6\ 4\ 1\ 3\ 8\ 10\ 2\ 9\ 7\ 5$. Then $x \in X_{10}(I)$ and $J_1 = \{[3, 1, 4], [7, 9, 2, 10, 8]\}$ have the required property, while it does not contain the adjacency $\{3, 8\} \in F(x, I)$. However, even in this case, we see that there are segment sets $J_2 = [9, 2, 10, 8, 3, 1, 4]$ and $J_3 = [1, 3, 8, 10, 2, 9, 7]$ including all adjacencies of $F(x, I)$ both with the required properties. In fact, in the following lemma (proved in the Appendix) we can see that there are not many adjacencies of $F(x, I)$ that can be ignored in the construction of π from x and I.

Lemma 3. *Let I be a segment set in $\mathscr{I}^{(n)}$, and $x \in X_n(I)$.*

a) *Let $\pi \in S_n$ be such that $I \subset \mathcal{A}_\pi$, and $\mathcal{A}_\pi \setminus I \subset \mathcal{A}_x$. Then either $F(x, I) \subset \mathcal{A}_\pi \setminus I$, or there exists an adjacency of $F(x, I)$, namely $e \in F(x, I)$ such that $F(x, I) \setminus \{e\} \subset \mathcal{A}_\pi \setminus I$.*

b) *There always exists a permutation $\pi \in S_n$ such that $I \subset \mathcal{A}_\pi$, and $F(x, I) \subset \mathcal{A}_\pi \setminus I \subset \mathcal{A}_x$.*

Proof: (of Theorem 4). The left inequality holds, since when $F(x, I)$ is an empty segment set, we need at least $\|I\| - 1$ $<1, 1>$-segments (trivial segments) from x to complete π. To prove the right inequality, let π be a permutation such that $I \subset \mathcal{A}_\pi$ and $F(x, I) \subset \mathcal{A}_\pi \setminus I \subset \mathcal{A}_x$. From Lemma 3, we know that such π exists. As the freedom of every number in any segment of $F(x, I)$ is 2, for two segments of $F(x, I)$, say s_1, s_2, the segment of π that is located between them in π, say s, should necessarily contain at least one segment of I. In fact, the freedom of s cannot be $<2, 2, ..., 2>$ (since in that case $s_1 \cup s \cup s_2$ should be a segment of $F(x, I)$ that is not supposed so). Therefore there must be at least a number in the segment s with freedom 1, and this implies that a segment of I must be contained in s. This yields that two segments of $F(x, I)$ cannot be connected to each other in π without using at least a segment of I between them. On the other hand, let s_1, s_2 be two segments of I, and denote by s the segment of π located between them in π. If s does not contain a segment of $F(x, I)$ and does not contain a segment of I, then it must be either a $<1, 1>$-segment (i.e. a trivial segment) or a $<1, 2, 1>$-segment. Lastly, let s_1 be a segment of I and s_2 be a segment of $F(x, I)$ and let s be a segment of π that is located between s_1 and s_2 in π. If s does not contain a segment of I, it should be a $<1, 2>$-segment necessarily. Putting all these together, we conclude that between each pair of segments of I in π, say s_1, s_2, we may need either a $<1, 2, 1>$-segment of $\mathcal{A}_x \setminus F(x, I)$ or at most one segment of $F(x, I)$. In the latter case, for each end of this segment from $F(x, I)$, we need a $<1, 1>$-segment of $\mathcal{A}_x \setminus F(x, I)$ to connect it to s_1 and s_2. On the other hand, on the right-hand side (left-hand side) of the most right (most left) segment of I in π, we may place either a $<1, 2>$-segment ($<2, 1>$-segment) or a $<1, 2>$-segment ($<2, 1>$-segment) followed by a segment of $F(x, I)$ on its right (on its left). So in general, we need at most 2 adjacencies of $\mathcal{A}_x \setminus F(x, I)$ between each pair $s_1, s_2 \hat{\in} I$ which are neighbours with respect to π and in extremities we need at most one adjacency of $\mathcal{A}_x \setminus F(x, I)$. In other words, we need at most $2(\|I\| - 1) + 2 = 2\|I\|$ adjacencies of $\mathcal{A}_x \setminus F(x, I)$ in order to complete π. This finishes the proof. $\square$

Let $(\hat{I}_n)_{n \in \mathbb{N}}$ be an arbitrary sequence of segment sets with $|\hat{I}_n| = m(n)$ and $\|\hat{I}_n\| = k(n)$ for $n \in \mathbb{N}$. As we already saw, to have $x \in X_n(\hat{I}_n)$, it is necessary to have $k(n) \leq m(n) \leq n - k(n)$ and also by Theorem 4

$$\|\hat{I}_n\| - 1 \leq Q(x, \hat{I}_n) \leq 2\|\hat{I}_n\|.$$

Also, for $x \in X_n(\hat{I}_n)$, by definition we have $Q(x, \hat{I}_n) \leq \beta(x, \hat{I}_n) + \gamma(x, \hat{I}_n)$. Now suppose $m(n)/n \to c$ and $k(n)/n \to c'$, as $n \to \infty$, for $c, c' \in \mathbb{R}_+$. Then after normalizing, Theorem 3 implies that the right side of the above inequality converges to $4c' - 4cc'$, in L^2 and probability, as n goes to ∞. Similarly, the left side of the last inequality converges to $1 - c - (1 - c - c')^2$, in L^2 and probability, as $n \to \infty$. Now suppose c, c' is such that $1 - c - (1 - c - c')^2 > 4c' - 4cc'$. Let $0 < \varepsilon << 1 - c - (1 - c - c')^2 - 4c' + 4cc'$. Then

$$\mathbb{P}[\xi^{(n)} \in X_n(\hat{I}_n)] \leq \mathbb{P}[Q(\xi^{(n)}, \hat{I}_n) \leq \beta(\xi^{(n)}, \hat{I}_n) + \gamma(\xi^{(n)}, \hat{I}_n)]$$

$$\leq \mathbb{P}[|\frac{Q(\xi^{(n)}, \hat{I}_n)}{n} - (1 - c - (1 - c - c')^2)| > \varepsilon]$$

$$+ \mathbb{P}[|\frac{\beta(\xi^{(n)}, \hat{I}_n + \gamma(\xi^{(n)}, \hat{I}_n)}{n} - (4c' - 4cc')| > \varepsilon] \to 0,$$

as $n \to \infty$. So, to avoid this, we should assume $1 - c - (1 - c - c')^2 \leq 4c' - 4cc'$. Similarly, we derive the following necessary conditions (see also Fig. 3)

$$\begin{cases} 1 - c - (1 - c - c')^2 \leq 4c' - 4cc', \\ c' \leq 1 - c - (1 - c - c')^2 \leq 2c', \\ 0 < c' \leq c \leq 1 - c'. \end{cases} \tag{4}$$

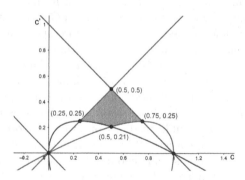

Fig. 3. The highlighted region includes all the points (c, c') satisfying (4).

4   Conclusion

In their simulation studies, Haghighi and Sankoff [2] showed that the breakpoint medians of $k \geq 3$ input genomes $x_1, \cdots, x_k$, picked independently and uniformly at random from the space of genomes S_n, tend to approach the corners (i.e. one of the input genomes). This indicates that most of the medians are uniformly distributed to small neighborhoods of $x_1, \cdots, x_k$, and medians have tendency to fall near the input genomes. Any of these corner medians reflects no information from the other $k - 1$ input genomes. Thus they are insignificant in reconstructing gene-order phylogenetic trees. Although there are a small proportion of the medians located far from the corners, it is hard to find them as the size of the genomes n gets large. The notion of "accessible median genomes" introduced by Jamshidpey et al. [4] helps to find these non-trivial medians located far from the corners $x_1, \cdots, x_k$. The construction of these accessible medians significantly relies on finding non-trivial geodesic points of x_i and x_j for $i, j \in \{1, \cdots, k\}$.

In other words, to be able to find such non-trivial medians, we need to find a geodesic point M for any given pair of genomes G and G' s.t. $d(M, G) \approx \alpha n \approx n - 1 - d(M, G')$, with $\alpha \in (0, 1)$. This is equivalent to finding $M \in S_n$ with approximately $(1 - \alpha)n$ adjacencies selected from G and αn adjacencies selected from G'. Selecting a segment set I from a genome G places certain restrictions on the usability of the gene adjacencies of another genome G', in constructing an intermediary genome M for which $I \subset \mathcal{A}_M$ and $J := \mathcal{A}_M \setminus I \subset \mathcal{A}_{G'}$. A genome M with this property is almost a geodesic point of G and G', if the size of $\mathcal{A}_{G,G'}$ is small. Note that, to be a definite geodesic point, M must also contain all gene adjacencies of $\mathcal{A}_{G,G'}$ [4]. To successfully identify a segment set J from G' with the specified property, it is imperative to delicately select the gene adjacencies from G that form I. In fact, for a substantial range of choices for I, the challenge arises as we encounter difficulty in finding a corresponding segment set J such that $I \cup J$ constitutes a valid genome. According to their specific usage in J, we categorized the gene adjacencies of G' w.r.t. I into four distinct classes. We proved a strong L^2-convergence theorem for the frequencies of each class. This induces certain conditions for I, in terms of its number of segments and adjacencies, necessary to find a segment set J from G' whose union with I forms an intermediate genome between G and G'. More specifically, to find such J from a genome G' picked uniformly at random from S_n, we showed that $\|I\|/n$ and $|I|/n$ must asymptotically satisfy (4), as n gets large. The conditions (4) are not sufficient but significantly restrict the shape and size of an appropriate choice of I, thus aiding in understanding how to construct approximate geodesic points of G and G' that are far from both. This brings us one step closer to discovering non-trivial accessible medians of three or more genomes, i.e. the near medians far from the corners. While our results are specifically presented for unsigned linear unichromosomal genomes, it is crucial to emphasize that our methodology readily extends to all genome types. Hence, analogous results hold true for signed, unsigned, unichromosomal and multichromosomal genomes with linear and/or circular chromosomes.

Appendix A: Proofs

Here we provide the proofs of Lemma 1, Lemma 2, Theorem 1, Theorem 2, Theorem 3 and Lemma 3.

Proof of Lemma 1. Consider a segment set $I = \{s_1, ..., s_k\}$, with k non-empty segments and m adjacencies that is contained in $x \in S_n$. Then $|\|\bar{I}_x\| - k| \leq 1$, and therefore we represent the segments of $\bar{I}_x$ by $s'_1, ..., s'_{k+1}$, where s'_j is non-empty for $2 \leq j \leq k$, and s'_1 and s'_{k+1} may be empty. Note that $\sum_{i=1}^{k} |s_i| = m$ and $\sum_{j=1}^{k+1} |s'_j| = n - 1 - m$ with $|s_i| \geq 1$ for $1 \leq i \leq k$ and $|s'_j| \geq 1$ for $2 \leq j \leq k$. Hence, the number of solutions for these two equations is equal to:

$$\binom{m - k + (k - 1)}{k - 1} \binom{n - 1 - m - (k - 1) + (k + 1 - 1)}{(k + 1 - 1)} = \binom{m - 1}{k - 1} \binom{n - m}{k}$$

In other words, that is the number of ways we can choose k segments with m adjacencies of x. $\qquad \square$

Proof of Lemma 2. As the segment set I has m adjacencies and k segments, each permutation containing I has $n - m - k$ external points with respect to I. Therefore, noting that segments have two directions, we have $2^k (k + (n - m - k))!$ permutations containing I. $\qquad \square$

Proof of Theorem 1. As $\alpha_{m,k}$ is independent of $\mathcal{I}_{m,k}$ and $\mathcal{L}(\mathcal{I}_{m,k}) = \mathcal{L}(\mathcal{I}_m \mid \|\mathcal{I}_m\| = k)$, we have

$$\mathbb{E}[\alpha_m \mid \|\mathcal{I}_m\| = k] = \mathbb{E}[\alpha_{m,k}] = \mathbb{E}[\alpha_{m,k} \mid \mathcal{I}_{m,k} = I] = \mathbb{E}[\alpha(\xi, I)].$$

So for the first part, we only need to compute $\mathbb{E}[\alpha_m \mid \|\mathcal{I}_m\| = k]$. To this end, note that there are $n - m - k$ external points (gens), $m - k$ internal points, and $2k$ end points in any segment set with m adjacencies and k segments. Sampling a random adjacency from $\mathcal{I}_m$, conditional on $\|\mathcal{I}_m\| = k$, the chance to have a 2-free-end, 1-free-end, trivial segment adjacency, is respectively

$$\frac{(n - m - k)(n - m - k - 1)}{n(n - 1)}, \quad \frac{4k(n - m - k)}{n(n - 1)}, \quad \frac{2k(2k - 1)}{n(n - 1)},$$

while the chance to have a 0-free-end adjacency is given by

$$\frac{2(m - k)(n - m + k) + (m - k)(m - k - 1)}{n(n - 1)} = \frac{(m - k)(2n - m + k - 1)}{n(n - 1)}.$$

Now, for $i = 1, ..., n - 1$, let $\hat{\alpha}_{m,i}$ be a random variable such that $\hat{\alpha}_{m,i} = 1$ if the ith adjacency of ξ, i.e. $\{\xi_i, \xi_{i+1}\}$, is 2-free-end w.r.t. $\mathcal{I}_m$ and $\hat{\alpha}_{m,i} = 0$ otherwise. Then, for every $i = 1, ..., n - 1$, we have

$$\mathbb{P}(\hat{\alpha}_{m,i} = 1 \mid \|\mathcal{I}_m\| = k) = \frac{(n - m - k)(n - m - k - 1)}{n(n - 1)},$$

implying that $\mathbb{E}[\alpha_m \mid \|\mathcal{I}_m\| = k]$ is equal to

$$\sum_{i=1}^{n-1} \mathbb{P}(\hat{\alpha}_{m,i} = 1 \mid \|\mathcal{I}_m\| = k) = \frac{(n - m - k)(n - m - k - 1)}{n}.$$

The other conditional expected values in the statement of the theorem are computed similarly. For the second part of the theorem, averaging over the possible number of segment sets, we have

$$\mathbb{E}[\alpha_m] = \frac{1}{n}\mathbb{E}\left[(n - m - \|\mathcal{I}_m\|)(n - m - \|\mathcal{I}_m\| - 1)\right]$$

$$= \frac{(n - m)(n - m - 1)}{n} + \frac{2m - 2n + 1}{n}\mathbb{E}\|\mathcal{I}_m\| + \frac{1}{n}\mathbb{E}\|\mathcal{I}_m\|^2.$$

Since $\|\mathcal{I}_m\| \sim H(n - 1, m - n, m)$, its moments are given in (1). Therefore, after some simplification, we obtain

$$\mathbb{E}[\alpha_m] = \frac{(n - m)(1 + m - n)^2(n - m - 2)}{(n - 2)(n - 1)n}.$$

Similarly,

$$\mathbb{E}[\beta_m] = \frac{1}{n}\mathbb{E}\left[4\|\mathcal{I}_m\|(n - m - \|\mathcal{I}_m\|)\right] = \frac{4m(n - m)(1 + m - n)^2}{(n - 2)(n - 1)n},$$

$$\mathbb{E}[\gamma_m] = \frac{1}{n}\mathbb{E}\left[2\|\mathcal{I}_m\|(2\|\mathcal{I}_m\| - 1)\right] = \frac{2m(n - m)(2m(n - m) + n)}{(n - 2)(n - 1)n},$$

and

$$\mathbb{E}[\delta_m] = \frac{1}{n}\mathbb{E}\left[(m - \|\mathcal{I}_m\|)(2n - m + \|\mathcal{I}_m\| - 1)\right]$$

$$= \frac{m(m - 1)(2n^2 - 6n - m^2 + 3m + 2)}{n(2 - 3n + n^2)}.$$

$\square$

Proof of Theorem 2. There are two options for choosing two adjacencies of ξ. They are either consecutive, $\{\xi_i, \xi_{i+1}\}, \{\xi_{i+1}, \xi_{i+2}\}$, or nonconsecutive, $\{\xi_i, \xi_{i+1}\}$, $\{\xi_j, \xi_{j+1}\}$ for $i + 1 < j$. If we select two consecutive adjacencies of ξ at random, the chances that both are 2-free end, both are 1-free end, and both are trivial segment adjacencies are respectively given by

$$\frac{(n - m - k)_{[3]}}{n_{[3]}}, \quad \frac{2k(n - m - k)_{[2]} + (n - m - k)(2k)_{[2]}}{n_{[3]}}, \quad \frac{(2k)_{[3]}}{n_{[3]}},$$

while the chance that both are 0-free end is

$$\frac{(m - k)_{[3]} + 3(n - m + k)(m - k)_{[2]} + (m - k)(n - m + k)_{[2]}}{n_{[3]}}.$$

Similarly, if we pick two nonconsecutive adjacencies of ξ at random, the chances that both are 2-free end, both are 1-free end, and both are trivial segment adjacencies are respectively given by

$$\frac{(n - m - k)_{[4]}}{n_{[4]}}, \quad \frac{4(n - m - k)_{[2]}(2k)_{[2]}}{n_{[4]}}, \quad \frac{(2k)_{[4]}}{n_{[4]}},$$

and finally the chance that both are 0-free end is readily obtained

$$\frac{(m-k)_{[4]} + 4(n-m+k)(m-k)_{[3]} + 4(n-m+k)_{[2]}(m-k)_{[2]}}{n_{[4]}}.$$

Now, for the first part of the theorem, as before we only need to compute the left of

$$Var(\alpha_m \mid \|\mathcal{I}_m\| = k) = Var(\alpha_{m,k}) = Var(\alpha(\xi, I)).$$

For $i = 1, \ldots, n-1$, recall the definition of $\hat{\alpha}_{m,i}$ from the proof of Theorem 1, and let $\hat{\alpha}_{m,k,i}$ be random variable such that $\hat{\alpha}_{m,k,i} = 1$ if the ith adjacency of ξ, i.e. $\{\xi_i, \xi_{i+1}\}$, is 2-free-end w.r.t. $\mathcal{I}_{m,k}$ and $\hat{\alpha}_{m,k,i} = 0$ otherwise. Then, for every $i = 1, \ldots, n-1$

$$\mathbb{E}[\alpha_{m,k}^2] = \sum_i \mathbb{E}[\hat{\alpha}_{m,k,i}^2] + 2\sum_{i>j} \mathbb{E}[\hat{\alpha}_{m,k,i}\hat{\alpha}_{m,k,j}]$$

$$= \sum_i \mathbb{P}(\hat{\alpha}_{m,k,i}^2 = 1) + 2\sum_{i>j} \mathbb{P}(\hat{\alpha}_{m,k,i}\hat{\alpha}_{m,k,j} = 1)$$

$$= \sum_i \mathbb{P}(\hat{\alpha}_{m,k,i} = 1) + 2\sum_{i>j} \mathbb{P}(\hat{\alpha}_{m,k,i}\hat{\alpha}_{m,k,j} = 1)$$

$$= \mathbb{E}[\alpha_{m,k}] + 2\sum_{i>j} \mathbb{P}(\hat{\alpha}_{m,k,i}\hat{\alpha}_{m,k,j} = 1).$$

Note that

$$\sum_{i>j+1} \mathbb{P}(\hat{\alpha}_{m,k,i}\hat{\alpha}_{m,k,j} = 1) = \sum_{i>j+1} \frac{(n-m-k)_{[4]}}{n_{[4]}} = \frac{(n-m-k)_{[4]}}{2n(n-1)},$$

and

$$\sum_{i=j+1} \mathbb{P}(\hat{\alpha}_{m,k,i}\hat{\alpha}_{m,k,j} = 1) = \sum_{i=j+1} \frac{(n-m-k)_{[3]}}{n_{[3]}} = \frac{(n-m-k)_{[3]}}{n(n-1)}.$$

Hence,

$$Var(\alpha_{m,k}) = \mathbb{E}[\alpha_{m,k}](1 - \mathbb{E}[\alpha_{m,k}]) + \frac{(n-m-k)_{[3]}(n-m-k-1)}{n(n-1)}.$$

Exactly the same calculations give $Var(\alpha(\xi, I))$. Similarly we can compute $Var(\beta_{m,k}) = Var(\beta(\xi, I))$, $Var(\gamma_{m,k}) = Var(\gamma(\xi, I))$ and $Var(\delta_{m,k}) = Var(\delta(\xi, I))$. Now to compute $Var(\alpha_m)$, write $\mathbb{E}[\alpha_m^2]$ as

$$\sum_i \mathbb{E}[\hat{\alpha}_{m,i}^2] + 2\sum_{i>j} \mathbb{E}[\hat{\alpha}_{m,i}\hat{\alpha}_{m,j}] = \sum_i \mathbb{P}(\hat{\alpha}_{m,i}^2 = 1) + 2\sum_{i>j} \mathbb{P}(\hat{\alpha}_{m,i}\hat{\alpha}_{m,j} = 1)$$

$$= \sum_i \mathbb{P}(\hat{\alpha}_{m,i} = 1) + 2\sum_{i>j} \mathbb{P}(\hat{\alpha}_{m,i}\hat{\alpha}_{m,j} = 1) = \mathbb{E}[\alpha] + 2\sum_{i>j} \mathbb{P}(\hat{\alpha}_{m,i}\hat{\alpha}_{m,j} = 1).$$

Letting $A_{m,k} = A_{m,k}^{(n)} := \{\|\mathcal{I}_m^{(n)}\| = k\}$, note that

$$\sum_{i>j+1} \mathbb{P}(\hat{\alpha}_{m,i} \cdot \hat{\alpha}_{m,j} = 1) = \sum_{i>j+1} \sum_{k=1}^{m} \frac{(n-m-k)_{[4]}}{n_{[4]}} \mathbb{P}(A_{m,k})$$

$$= \sum_{k=1}^{m} \frac{(n-m-k)_{[4]}}{2n(n-1)} \mathbb{P}(A_{m,k}),$$

and

$$\sum_{i=j+1} \mathbb{P}(\hat{\alpha}_{m,i}\hat{\alpha}_{m,j} = 1) = \sum_{i=j+1} \sum_{k=1}^{m} \frac{(n-m-k)_{[3]}}{n_{[3]}} \mathbb{P}(A_{m,k})$$

$$= \sum_{k=1}^{m} \frac{(n-m-k)_{[3]}}{n(n-1)} \mathbb{P}(A_{m,k}).$$

Therefore from (1)

$$Var(\alpha_m) = \mathbb{E}[\alpha_m](1-\mathbb{E}[\alpha_m]) + \frac{1}{n(n-1)}\mathbb{E}[(n-m-\|\mathcal{I}_m\|)_{[3]}(n-m-\|\mathcal{I}_m\|-1)]$$

$$= \mathbb{E}[\alpha_m](1-\mathbb{E}[\alpha_m]) + \frac{(n-m)_{[4]}(n-m-1)_{[2]}\left((n-m)^2 - 5n + 4 + 7m\right)}{n_{[5]}(n-1)}$$

$$= \left(1-\frac{m}{n}\right)^4 \left(\frac{m}{n}\right)^2 \left(8 + \frac{m}{n}\left(5\frac{m}{n} - 12\right)\right)n + o(n).$$

In the same way, we can show that

$$Var(\beta_m) = \mathbb{E}[\beta_m](1-\mathbb{E}[\beta_m])$$

$$+ \frac{1}{n(n-1)}\mathbb{E}[16\|\mathcal{I}_m\|^2(n-m-\|\mathcal{I}_m\|)^2 + 4\|\mathcal{I}_m\|^3 - 4\|\mathcal{I}_m\|(n-m)^2]$$

$$= \mathbb{E}[\beta_m](1-\mathbb{E}[\beta_m]) + \left(\frac{4m(m-n)(m-n+1)^2}{n_{[5]}(n-1)}\right) \times$$

$$\{(1-4m)n^3 + (4m(3m+5) - 3)n^2$$

$$- (m+1)(3m(4m+11)+1)n + 4(m+1)^2(m(m+4)+1)\}$$

$$= 4\left(1-\frac{m}{n}\right)^3 \left(\frac{m}{n}\right)^2 \left(8 - \frac{m}{n}\left(31 + \frac{4m}{n}\left(\frac{5m}{n} - 11\right)\right)\right)n + o(n),$$

$$Var(\gamma_m) = \mathbb{E}[\gamma_m](1-\mathbb{E}[\gamma_m]) + \frac{1}{n(n-1)}\mathbb{E}[(2\|\mathcal{I}_m\|)_{[3]}(2\|\mathcal{I}_m\| - 1)]$$

$$= \mathbb{E}[\gamma_m](1-\mathbb{E}[\gamma_m])$$

$$+ \frac{4m_{[2]}(n-m)_{[2]}}{n_{[5]}(n-1)} \times \{m^4 - 8m^3 n + 4m^2\left(n^2 + n + 3\right)$$

$$- 4mn(n+3) + n(n+9) - 4\}$$

$$= 4\left(1-\frac{m}{n}\right)^2 \left(\frac{m}{n}\right)^2 \left[1 - \frac{4m}{n}\left(1-\frac{m}{n}\right)\left(1 + \frac{5m}{n}\left(1-\frac{m}{n}\right)\right)\right]n + o(n),$$

and finally,

$$Var(\delta_m) = \mathbb{E}[\delta_m](1 - \mathbb{E}[\delta_m])$$

$$+ \frac{1}{2n(n-1)}\mathbb{E}[(m - \|\mathcal{I}_m\|)_{[2]}(2n - m + \|\mathcal{I}_m\| - 2)_{[2]}] + \frac{(n-2)}{n}\mathbb{E}[m - \|\mathcal{I}_m\|]$$

$$= \mathbb{E}[\delta_m](1 - \mathbb{E}[\delta_m]) + \frac{m_{[2]}}{n_{[5]}(n-1)} \times$$

$$\{(m-5)m\left(m\left(m^3 - 10m^2 + m + 40\right) + 4\right)$$

$$+ 4(m-4)(m+1)n^4 + 2(9 - 23(m-3)m)n^3$$

$$+ 2(m(m(51 - 2(m-8)m) - 235) + 50)n^2$$

$$+ 2m(m(13(m-8)m + 121) + 170)n + 2n^5 - 152n + 48\}$$

$$= \left(\frac{m}{n}\right)^2\left(1 - \left(\frac{m}{n}\right)^2\right)^2\left(4 + \frac{m}{n}\left(\frac{5m}{n} - 8\right)\right)n + o(n).$$

$\square$

Proof of Theorem 3. First observe that, by Theorem 1, as $n \to \infty$,

$$\mathbb{E}[\frac{\tilde{\alpha}_n}{n}] \to (1-c)^4, \quad \mathbb{E}[\frac{\bar{\alpha}_n}{n}], \; \mathbb{E}[\frac{\alpha(\xi^{(n)}, \hat{I}_n)}{n}] \to (1 - c - c')^2,$$

$$\mathbb{E}[\frac{\tilde{\beta}_n}{n}] \to 4c(1-c)^3, \quad \mathbb{E}[\frac{\bar{\beta}_n}{n}], \; \mathbb{E}[\frac{\beta(\xi^{(n)}, \hat{I}_n)}{n}] \to 4c'(1 - c - c'),$$

$$\mathbb{E}[\frac{\tilde{\gamma}_n}{n}] \to 4c^2(1-c)^2, \quad \mathbb{E}[\frac{\bar{\gamma}_n}{n}], \; \mathbb{E}[\frac{\gamma(\xi^{(n)}, \hat{I}_n)}{n}] \to 4c'^2,$$

$$\mathbb{E}[\frac{\tilde{\delta}_n}{n}] \to c^2(2-c)^2, \quad \mathbb{E}[\frac{\bar{\delta}_n}{n}], \; \mathbb{E}[\frac{\delta(\xi^{(n)}, \hat{I}_n)}{n}] \to (c - c')(2 - c + c').$$

Also, following Theorem 2, the variances of all these sequences converge to 0. Hence, the convergence in L^2 and in probability holds. $\square$

Proof of Lemma 3. Suppose $\{a, b\} \in F(x, I) \setminus \mathcal{A}_\pi$. As $a, b \in Ext(I)$ and therefore the neighbours of a in π should be from set $\mathcal{N}_x(a) \setminus \{b\}$ and the neighbours of b in π should be from set $\mathcal{N}_x(b) \setminus \{a\}$, we have $|\mathcal{N}_\pi(a)|, |\mathcal{N}_\pi(b)| \le 1$. But $|\mathcal{N}_\pi(a)|$ and $|\mathcal{N}_\pi(b)|$ cannot be 0, since in that case a or b cannot be connected to the rest of the numbers to construct π, and therefore $|\mathcal{N}_\pi(a)| = |\mathcal{N}_\pi(b)| = 1$ which means that a and b are extremities of permutation π, i.e. $\{\pi_1, \pi_n\} = \{a, b\}$. In other words, there may exist at most one adjacency $\{a, b\} \in F(x, I) \setminus \mathcal{A}_\pi$. This proves part (a). For part (b), suppose $\pi' \in [id, x]$ and there exists adjacency $\{a, b\}$ such that $\{a, b\} \in F(x, I) \setminus \mathcal{A}_{\pi'}$. As we showed above $\{\pi_1', \pi_n'\} = \{a, b\}$. Also, as a and b are connected in π' through a segment of π' containing at least one segment of I and this means that there exists at least one $<1, 2>$-adjacency ($<1, 2>$-segment) in the segment of π' connecting a to b, namely e, and hence e is not in $F(x, I)$. Therefore, we can construct a new permutation π by cutting e in π' and joining a to b. This proves part (b). $\square$

References

1. Billingsley, P.: Probability and Measure, 3r edn. John Wiley & Sons, New York (1995)
2. Haghighi, M., Sankoff, D.: Medians seek the corners, and other conjectures. BMC Bioinform. **13**(19), S5 (2012)
3. Jamshidpey, A.: Population dynamics in random environment, random walks on symmetric group, and phylogeny reconstruction. Ph.D. thesis, Université d'Ottawa/University of Ottawa (2016)
4. Jamshidpey, A., Jamshidpey, A., Sankoff, D.: Sets of medians in the non-geodesic pseudometric space of unsigned genomes with breakpoints. BMC Genomics **15**(6), S3 (2014)
5. Kallenberg, O.: Foundations of Modern Probability. Springer, Cham (2006)
6. Larlee, C.A., Zheng, C., Sankoff, D.: Near-medians that avoid the corners; a combinatorial probability approach. BMC Genomics **15**(6), S1 (2014)
7. Sankoff, D., Blanchette, M.: The median problem for breakpoints in comparative genomics. In: Jiang, T., Lee, D.T. (eds.) COCOON 1997. LNCS, vol. 1276, pp. 251–263. Springer, Heidelberg (1997). https://doi.org/10.1007/BFb0045092
8. Sankoff, D., Blanchette, M.: Multiple genome rearrangement and breakpoint phylogeny. J. Comput. Biol. **5**(3), 555–570 (1998)

Genome Evolution

Transcription Factors Across the *Escherichia coli* Pangenome: A 3D Perspective

Gabriel Moreno-Hagelsieb[(✉)] ⓘD

Wilfrid Laurier University, Waterloo, ON N2L 3C5, Canada
`gmorenohagelsieb@wlu.ca`

Abstract. Identification of complete sets of transcription factors (TFs) is a foundational step in the inference of genetic regulatory networks. The availability of high-quality predictions of protein three-dimensional structures (3D), has made it possible to use structural similarities for the inference of homology beyond what is possible from sequence analyses alone. This work explores the potential to use such 3D structures for the identification of TFs in the *Escherichia coli* pangenome. Comparisons between predicted structures and their experimental counterparts confirmed the high-quality of predicted structures, with most 3D structural alignments showing TM-scores well above established structural similarity thresholds, though the quality seemed slightly lower for TFs than for other proteins. As expected, structural similarity decreased with sequence similarity, though most TM-scores still remained well above the structural similarity threshold. This was true regardless of the aligned structures being experimental or predicted. Results at the lowest sequence identity levels revealed potential for 3D structural comparisons to extend homology inferences below the "twilight zone" of sequence-based methods. The body of predicted 3D structures covered 99.7% of available proteins from the *E. coli* pangenome, missing only two of those matching TF domain sequence profiles. After cleaning out potential false positives, structural comparisons increased the number of inferred TFs in the pangenome by 50% over those obtained with sequence profiles alone, which translated into around 20% more TFs per genome.

Keywords: Tertiary structure · Homology · Structural alignment · Transcription factors · Alphafold · DNA-binding domains

1 Introduction

Genome sequences continue to accumulate at enormous rates [22, 25], constantly adding to the pile of proteins to characterize. Identifying proteins within newly sequenced genomes, as well as inferring their potential functions, depends mainly

Supported by the Natural Sciences and Engineering Research Council of Canada (NSERC).

C. Scornavacca and M. Hernández-Rosales (Eds.): RECOMB-CG 2024, LNBI 14616, pp. 213–225, 2024.
https://doi.org/10.1007/978-3-031-58072-7_11

on the inference of homology, commonly achieved either by sequence [5,8,24], or by comparison against databases of protein family profiles [17,18,20], which offer improved sensitivity and specificity over pairwise sequence alignments [14]. Both methods for inferring homology are limited by the insufficient information available from dealing with character sequences.

Ultimately, protein three dimensional (3D) structures might offer the most information for the inference of homology and function [13,16,23]. However, the task has been hindered by the slow accumulation of 3D structures relative to the accumulation of sequences. The need for structures for improved clues to homology and function inspired the Structural Genomics project [6,7,23], aiming at solving representative structures for each inferred structural class, which would then be used for predicting, by homology, the structures of all the protein family members. While the project accelerated the pace of experimental solution of 3D structures, it still remained too slow compared to the pace of sequence accumulation.

Predictions of 3D structures, founded on artificial intelligence, has revolutionised the field [2,27]. The recent methods for 3D predictions have produced structures that seemingly compete with experimental 3D structure solutions [2,27]. Not only that, the Alphafold project offered to predict the structures of all proteins available at the UniProt knowledgebase [25], thus releasing more than 214,000,000 3D structures [28]. Therefore, we count today with a very rich database for experimentation, specifically, for the theme of this work, about the inference of homology and function beyond what is possible using sequence information alone.

Because of our interest on the identification of transcription factors (TFs), proteins that play an important role in modulating gene expression [9,21], here we explore the inference of homology by 3D structural comparisons across the *Escherichia coli* pangenome. To this end, we use experimentally confirmed TFs from RegulonDB [26], a collection of TF protein family profiles from the Pfam database [18,21], as well as databases of 3D structures, both experimental [4] and predicted [28].

2   Methods

2.1   Sequence Data and Alignment

Genome data were downloaded from NCBI's RefSeq genome database [10] by the end of August 2023. To retrieve the *E. coli* pangenome, genome distances were estimated between all genomes classified as Enterobacterales using mash [19]. Genomes were clustered and cut into groups at a Mash-distance threshold of 0.048, a threshold previously found to correspond to species-level grouping among these bacteria [11]. The pangenome consisted of the group containing all the genomes labeled as *E. coli*.

The sequences of experimentally available three-dimensional (3D) structures, were retrieved from the PDB [4] by the beginning of December 2023. The sequences for all Alphafold 3D structures were downloaded from the specialized database [27,28], by February 2023 (the data has been stable since October 2022 as version 4).

Protein sequence comparisons were run using diamond v. 2.1.8 [5], with an E-value threshold of 1e-6, and soft-masking [12].

2.2 Transcription Factors and Protein Family Profiles

To identify protein sequences of experimentally know transcription factors (TFs), we downloaded the list of TFs from *E. coli* K12 MG1655 from RegulonDB [26]. Only TFs with strong evidence were considered. Cross-checking against the genome of this strain resulted in 86 TFs sequences.

To identify further TFs, the protein sequences from the *E. coli* pangenome, from the PDB, and from the SwissProt subset of Alphafold, were compared against all available Pfam profiles [18,20] (v. 36), using MMseqs2 v. 15-6f452 [24]. Pfam TFs were proteins matching Pfam domains characteristic of TF in Bacteria and Archaea [21]. MMseqs2 was run with an E value threshold of 1e-3 (NCBI' CDD uses an E-value threshold of 1e-2 for profile matching). To try and keep matches more likely to retain the domain's function, the results were filtered for a minimum coverage of 60% of the Pfam models, with no more than 15% overlap between matched domains, keeping the match with the highest score.

2.3 Protein 3D Structures and Alignment

Protein structures were downloaded from the PDB [4], and from Alphafold [28] as necessary. Structural alignments were performed using US-align (v. 20220626) [30], and foldseek (v. 8-ef4e960) [15], both with default parameters, plus options for producing tab-separated tables.

Information about clusters of Alphafold structures were downloaded from the AFDB [3].

3 Results and Discussion

3.1 The *E. Coli* Pangenome Contained 2,878 Genomes and 718,581 Unique Proteins

Clustering all Enterobacteriaceae genomes, and cutting the resulting hierarchy at the mash-distance threshold appropriate for species grouping [11] (see Methods), produced one group containing all but one of the genomes labeled as *E. coli*, plus all the genomes under the genus *Shigella*, as expected (Table 1) [11]. One *E. coli* genome clustered with an *Escherichia ruysiae* genome, apart from any other genome under the genus *Escherichia* or *Shigella*. Thus, we ignored this genome as a potentially mislabelled one. The resulting pangenome contained 2,878 genomes (Table 1).

The total number of protein sequences annotated in the *E. coli* pangenome, was 13,265,057. After filtering for identical sequences, the dataset reduced to 718,581 unique proteins.

Table 1. The *Escherichia coli* pangenome.

Species	Count
Escherichia coli	2703
Shigella flexneri	93
Shigella sonnei	37
Shigella dysenteriae	24
Shigella boydii	14
Escherichia sp	7

3.2 Pfam Profiles Found 31,295 Unique Transcription Factors (TFs) in the *E. Coli* Pangenome

The transcription factor dataset from RegulonDB contained 86 transcription factors (TFs) with strong evidence. Cross-checking the TF names, synonyms, and annotations, against the corresponding annotations in the *E. coli* K12 MG1655 genome, allowed us to find the protein identifiers for different collections, such as SwissProt and NCBI.

To identify potential TFs in the pangenome, we compared the unique proteins dataset against all Pfam profiles using MMseqs2 (see Methods). These results were filtered to keep matches covering 60% or more of the Pfam profile in the alignment, with no more than 15% overlap between profile matches. After filtering, 568,578, or 71.1%, proteins matched at least one Pfam profile.

Seventy-three of the 86 RegulonDB TFs matched at least one Pfam TF profile. Of these, 71 matched Pfam profiles typical of TFs in bacteria [21], while the other two matched one typical of Archaea [21]. Therefore, to identify TFs by Pfam analysis, we included every protein matching Pfam TF profiles listed as either from Bacteria or from Archaea. The list of bacterial TF domains contained 123 Pfam identifiers, while that of archaeal TFs contained 43 [21]. The union of these two lists produced 155 Pfam identifiers. A total of 31,282 of the 568,578 unique *E. coli* proteins found above, matched at least one of these Pfam TF profiles. These Pfam-inferred TFs (Pfam TFs) matched 101 of the 155 TF Pfam profiles, 19 of them unique to the list of archaeal TF profiles. Combined with the RegulonDB TFs, we obtained a total of 31,295 TFs.

3.3 Alignments Against Experimental 3D Structures Confirmed the High-Quality of Structural Predictions

To identify predicted 3D structures corresponding to experimental ones, we compared the protein sequences from PDB against those from Alphafold corresponding to the *E. coli* K12 MG1655 model organism (4,363 proteins). The PDB dataset was filtered to eliminate redundant sequences found in the same PDB structural file. Equivalent pairs were defined as at least 95% identical, with 95% coverage of both protein sequences in the alignment. The total number of proteins with both a PDB and an Alphafold structure was 1,319 (30.2%). Among

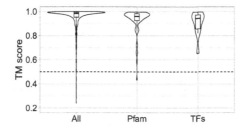

Fig. 1. Alignment between *E. coli* K12 MG1655 predicted and experimental 3D protein structures. The overall quality of predicted structures was high. Though the quality for both, experimentally confirmed transcription factors (TFs), as well as proteins containing TF Pfam domains (Pfam), seems lower, their structural alignments still resulted in TM scores above the 0.5 threshold, normally associated with shared structures [29].

them, 65 were Pfam TFs, and 20 were RegulonDB TFs, with 16 of the latter intersecting with the Pfam TFs.

To check the quality of Alphafold predicted 3D structures, we aligned them with their equivalent, experimental, 3D structures using US-align (see methods). Most of these alignments produced TM-scores well above the structural similarity threshold of 0.5 (Fig. 1), a threshold previously shown to correspond to structures sharing the same fold [29]. While also well above the structural similarity threshold, the TM-scores were slightly lower for both, RegulonDB TFs and Pfam TFs.

3.4 Structural Similarity Between TFs Decreased with Sequence Identity

To identify 3D structures corresponding to transcription factors (TFs), we compared all PDB sequences, as well as sequences in the Alphafold SwissProt subset, against all Pfam profiles using MMseqs2 (see methods). These results were filtered to ensure coverage of the Pfam profile equal or above 60% and no more than 15% overlap between Pfam matching segments. TFs were proteins matching Pfam domains characteristic of bacterial and archaeal TFs [21]. The procedure yielded 3,901 PDB and 13,175 Alphafold/SwissProt Pfam TFs.

PDB and Alphafold Pfam TF sequences were compared against each other using diamond (see methods). Sequence alignment results were filtered to keep those showing a minimum of 95% coverage of the shortest protein in the alignment. An *ad hoc* perl script was used to select these proteins, and organise them into identity bins at 10% identity steps. The program also sampled randomly for a maximum of 250 pairs per bin. The same perl script downloaded the corresponding structures and ran pairwise structural alignments using US-align, producing a table with results that included the TM-scores obtained from the structural comparisons. The program selected the TM-scores calculated on the basis of the shortest protein.

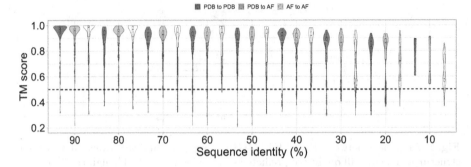

Fig. 2. Sequence divergence *versus* structural alignments between Pfam TFs. Regardless of the compared 3D structures being experimentally determined (PDB), or predicted by Alphafold (AF), all structural alignments decreased in quality with sequence divergence, with the tendency being more obvious at the lowest sequence identity levels for comparisons between predicted structures (AF to AF). The apparent double distributions at the lowest levels of identity might indicate increased proportions of false positive sequence alignments. Still, most structural alignments had TM-scores above the 0.5 threshold (dashed line), normally associated with shared structures [29].

Table 2. Transcription factor inference

Dataset	Aligned	90/90	Pfam	AFDB	foldseek
Pangenome	716,730	693,862	31,074	33,576	46,541
Alphafold	348,386	340,753	14,756	16,368	22,595

As expected, structural similarity had a tendency to decrease with sequence divergence. Still, most results remained well above the 0.5 TM-score threshold (Fig. 2). However, results at the lowest level of sequence identity showed a tendency towards double distributions, perhaps revealing higher proportions of false positive sequence alignments. Problems at lower levels of divergence were more obvious for comparisons between predicted structures than any comparisons involving experimental ones.

3.5   Alphafold Matched 99.7% Proteins of the *E. Coli* Pangenome

After sequence comparisons, 718,581 unique proteins of the *E. coli* pangenome, 373,702 (52%) found matches against proteins with experimentally determined 3D structures (PDB). Of these, only 150,343, 40.2% (or 20.1% of the total) were "90/90" alignments, consisting of those with identities above 90%, as well as sequence coverage of the *E. coli* protein above 90% (Fig. 3, Table 2). Overall PDB matches across the *E. coli* pangenome ranged from 55.1% to 68.2%, with a median of 62.2%.

When compared against Alphafold protein sequences, 716,730, or 99.7%, of the unique pangenome proteins found a match. Of these, 693,862, or 96.8%

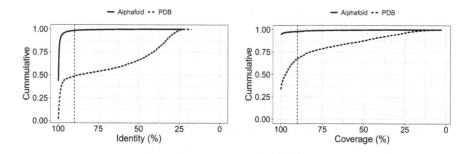

Fig. 3. Statistics for best sequence alignment matches. Predicted (Alphafold) structures had much better sequence-based matches against the unique protein dataset annotated in the *E. coli* pangenome, than experimental (PDB) ones. More than 95% of Alphafold matches had sequence coverage and identity values above 90% (vertical dashed line).

(96.6% of the total), constituted 90/90 matches (Fig. 3, Table 2). Overall, Alphafold structures matched between 91.8% and 99.4% of the proteins annotated in the genomes analysed, with a median of 99.1%. It thus seems unlikely that Alphafold missed proteins of wide interest for the *E. coli* scientific community. Thus, Alphafold should provide enough material to try and infer further TFs by structural similarity.

3.6 Structural Analyses Identified Around 1.5 Times as Many TFs as Pfam

AFDB is a database containing clusters of Alphafold structures [3], or structural families, which the authors obtained based on alignments performed with the foldseek fast 3D structural alignment program [15]. Thus, as a first approach towards inferring TFs by structural similarity, we explored the use of the AFDB structural families.

As a starting point for this analysis, we selected Alphafold proteins that could represent the pangenome's Pfam TFs. To select these proteins, we found alignments with at least 90% identity, though, instead of 90% overall coverage of the *E. coli* protein, we selected the cases where the alignments covered the segment of the *E. coli* protein that matched the Pfam TF profile. This procedure yielded 14,756 Alphafold proteins that aligned with 31,074 pangenome Pfam TFs (Table 2). Of these 14,756 Alphafold proteins, 14,636 matched a Pfam TF profile directly.

We cross-checked the 14,756 Alphafold TF identifiers obtained above against the cluster information found at the AFDB. Of these 14,756 proteins, 14,708 were found in 696 AFDB clusters or structural families. The complete set of structure identifiers within the 696 structural families, represented a total of 5,069,754 protein structures, including the 14,708 Alphafold TFs. With this list, we checked

Table 3. Transcription factor (TF) GO terms

GO Identifier	Description
TF specific	
GO:0000976	F:transcription cis-regulatory region binding
GO:0000987	F:cis-regulatory region sequence-specific DNA binding
GO:0001046	F:core promoter sequence-specific DNA binding
GO:0001216	F:DNA-binding transcription activator activity
GO:0001217	F:DNA-binding transcription repressor activity
GO:0003700	F:DNA-binding transcription factor activity
GO:0005667	C:transcription regulator complex
GO:0006351	P:DNA-templated transcription
GO:0006355	P:regulation of DNA-templated transcription
GO:0045892	P:negative regulation of DNA-templated transcription
GO:0045893	P:positive regulation of DNA-templated transcription
GO:2000142	P:regulation of DNA-templated transcription initiation
GO:2000143	P:negative regulation of DNA-templated transcription initiation
GO:2000144	P:positive regulation of DNA-templated transcription initiation
More generic:	
GO:0003677	F:DNA binding
GO:0003681	F:bent DNA binding
GO:0003690	F:double-stranded DNA binding
GO:0032993	C:protein-DNA complex
GO:0043565	F:sequence-specific DNA binding

back the unique *E. coli* proteins *versus* Alphafold sequence alignments. This way we identified 17,371 Alphafold structures that both belong with the 696 TF structural families mentioned above, and found closely related sequences in the *E. coli* pangenome. This represents a potential addition of 2,615, or 15%, TFs to the Pfam-based inferences. These AFDB TFs matched three of the thirteen RegulonDB TFs that did not match a Pfam TF profile.

To further explore the quality of these results, we checked UniProt annotations. UniProt runs a battery of analyses to try and assign functions to most of the protein sequences in the "knowledgebase" [25]. Since the Alphafold 3D structure collection consists of predicted structures for UniProt proteins [28], finding the corresponding annotations is straightforward.

In the case of Alphafold proteins matching Pfam TFs, we found UniProt annotations for 8,058 of the 14,756 proteins. Among these, 5,771, or 71.6%, contained Gene Ontology (GO) [1] keys found in the RegulonDB TFs, whose descriptions match TF activities (Table 3, top). Adding more generic GO keys to the search (Table 3, bottom), found 6,883 matches, covering 85.4% of the pro-

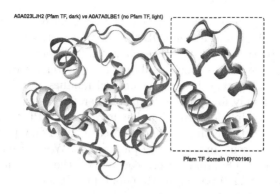

A0A023LJH2 (Pfam TF, dark) *vs* A0A7A0LBE1 (no Pfam TF, light)

Pfam TF domain (PF00196)

Fig. 4. Structural inference of transcription factors. Comparing structures whose sequences matched Pfam domains, against structures lacking such matches, while ensuring that the structural alignments covered the appropriate segments, allowed for the inference of transcription factors by structural similarity.

teins with UniProt annotations. Thus, a majority of Pfam TF-matching proteins contained consistent GO keys.

In contrast, of the 2,515 Alphafold proteins added by AFDB analyses, potentially new TFs that did not match a Pfam TF profile, 1,296 had UniProt annotations. Among these, 342, or 26.4%, had GO keys proper of TF activities (Table 3), while generic Go keys complemented that number to 721. Thus, at most 55.6% of these results had annotations consistent with TFs, a noticeable lower proportion than the Pfam TF-matching proteins described above. While this result could be interpreted as inferences that could not be done by UniProt's battery of analyses, and although AFDB structural families are produced with a coverage threshold of 90% [3], it is still possible that some members did not contain the Pfam TF domain(s), and thus represent false positives.

To select potential Alphafold TFs from the AFDB analyses that actually had a TF domain, we alignment them, structurally, against the ones matching Pfam TF profiles using foldseek [15]. We then checked if the segments of the proteins matching Pfam TF profiles were covered, in the structural alignment, by the other proteins (Fig. 4). This way we found that only 1,612 (64.1%) of the 2,515 Alphafold proteins that did not match a Pfam TF profile, aligned, structurally, with regions corresponding to such profiles in the Pfam TF-matching proteins. Of these, 849 had UniProt annotations, 335 matching GO keys appropriate for TFs (39.5%), 702 (82.7%) if we add the more generic terms. These results considered only structural alignments with TM-scores above 0.5 that covered completely the regions matching a Pfam TF profile. The proportion of proteins with appropriate annotations is much closer to that obtained for Alphafold Pfam TFs above, thus confirming that being in the same AFDB structural family is not enough to infer that all proteins in the family share all their domains. Up to this point, using the AFDB allowed us to increment the number of inferred TFs from 14,756 to 16,368, which translated into 33,576 TFs among the unique proteins of the *E. coli* pangenome (Table 2). A modest increment of 7.5% over Pfam-TF profiles. The three TFs from RegulonDB that did not match a Pfam TF domain directly, rescued by the initial AFDB results, were still found in this subset.

The results above suggest the possibility that proteins with DNA-binding domains might also be found in other structural families and be missed by the AFDB structural family analyses performed. Therefore, we also ran structural alignments between Alphafold proteins that matched TF Pfam profiles against the whole dataset of those that did not match a TF Pfam profile, but were close homologs to proteins in the *E. coli* pangenome. We then checked whether the regions matching TF profiles were covered by the structural alignments. This way we found 17,439 potential Alphafold TFs. That is, 17,439 Alphafold proteins that did not match, directly, a TF Pfam profile, yet, when aligned structurally, covered regions of other proteins that did match a TF Pfam profile. These included the 1,612 proteins analysed in the paragraph above. If we trusted all these proteins to correspond to TFs, adding these proteins to the 14,756 Alphafold Pfam TFs we obtained 32,195 total potential TFs, more than twice the TFs inferred by Pfam analyses alone (2.18 times, or 218%). Of these 17,439 additional proteins, 7,090 had UniProt annotations, 1,236 of them matching GO keys associated with TFs (17.4%), or 3,184 (44.9%) allowing for more generic GO keys. Given that these proteins cover, by structural alignments, Pfam-TF domains, we need further analyses to try and confirm if these might be truly newly discovered TFs, or at least DNA-binding proteins.

To further explore the quality of these results, we checked whether the sections aligning with these Pfam TF domains were covered by other Pfam domains, thus suggesting false positives. A great majority of these 17,439 proteins, 17,419, had matches with Pfam. Among these, only 7,839 did not have other Pfam domains competing for the position where the Pfam TF domains were projected from the structural alignments. Of these 2,718 (34.7%) had UniProt annotations, 987 (36.3%) with TF-consistent GO keys, 2,009 (73.9%) if considering more generic GO keys. These numbers are closer to those obtained for Pfam TF. If we considered the rest as false positives, since domains other than Pfam TF domains overlapped the potential TF domains, these results reduced the increments above from 32,195 to 22,595, so, instead of 218% the Alphafold TF dataset would have increased by 53.1% Table 3. When matched against the *E. coli* unique proteins, these translated into a total of 46,541 TFs, 49.7% more than those obtained using Pfam alone (Table 3). Of the thirteen RegulonDB TFs that did not match a Pfam domain directly, nine, including the three found by AFDB analyses, were found by these analyses. Only four TFs from RegulonDB could not be found by any of the analyses performed.

Mapping these inferred TFs onto each of the genomes in our dataset resulted in an average increment of 21.4% (Fig. 5), as well as into more evenly distributed proportions of TFs. These results suggests that most of the inferences complement the datasets of particular genomes, rather than TFs belonging to the core genome. Thus, most of the newly found TFs might be horizontally transferred. Studying these TFs might shed further light into the role of TFs in the success of horizontal gene transfer.

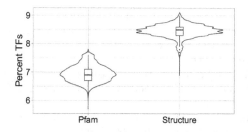

Fig. 5. Effect of structural inferences in the proportion of TFs across the *E. coli* pangenome. Though we found close to 50% more TFs than would be produced from Pfam analyses alone, these translated into an average of 21% extra TFs per genome, suggesting that most of the additional TFs are particular to a few genomes.

That the projected Pfam TF domains did not have competing domains overlapping them, does not mean that they are indeed DNA-binding domains. Also, given that the Pfam domain overlap analyses reduced the total number of proteins with appropriate GO keys (specific: 1,236 to 987; generic: 3,184 to 2,009), it is possible that further examination of the ones eliminated by competing domains might be necessary to confirm that this overlap is enough evidence of false positives, rather than evidence of unsuspected DNA-binding family "clans," families sharing similar domains [18]. In the meantime, the results show high potential for the identification of TFs beyond what is possible using sequence profiles alone.

4   Conclusion

In this work we have confirmed the high-quality of 3D structures predicted by Alphafold. Though the quality of 3D structural alignments were less reliable for predicted structures than for experimental ones, when the proteins have low sequence identity, they still offered potential for the inference of homology beyond what is possible using sequence family profiles. Our results allowed us to increase our database of inferred TFs in the *E. coli* pangenome by around 50%, with several of the extra TFs being specific to a few genomes. Further research might be necessary to confirm the quality of these inferences.

References

1. Aleksander, S.A., et al.: The gene ontology knowledgebase in 2023. Genetics **224**(1), iyad031 (2023). https://doi.org/10.1093/genetics/iyad031
2. Baek, M., et al.: Accurate prediction of protein structures and interactions using a three-track neural network. Science **373**(6557), 871–876 (2021). https://doi.org/10.1126/science.abj8754
3. Barrio-Hernandez, I., et al.: Clustering-predicted structures at the scale of the known protein universe. Nature 1–9 (2023). https://doi.org/10.1038/s41586-023-06510-w

4. Bittrich, S., et al.: RCSB protein data bank: efficient searching and simultaneous access to one million computed structure moddels alongside the pdb structures enabled by architectural advances. J. Mol. Biol. 167994 (2023). https://doi.org/10.1016/j.jmb.2023.167994

5. Buchfink, B., Reuter, K., Drost, H.G.: Sensitive protein alignments at tree-of-life scale using DIAMOND. Nat. Methods **18**(4), 366–368 (2021). https://doi.org/10.1038/s41592-021-01101-x

6. Burley, S.K.: An overview of structural genomics. Nat. Struct. Biol. **7**(Suppl 11), 932–934 (2000). https://doi.org/10.1038/80697

7. Burley, S.K., et al.: Structural genomics: beyond the human genome project. Nat. Genet. **23**(2), 151–157 (1999). https://doi.org/10.1038/13783

8. Camacho, C., et al.: BLAST+: architecture and applications. BMC Bioinform. **10**(1), 421 (2009). https://doi.org/10.1186/1471-2105-10-421

9. Freyre-González, J.A., Treviño-Quintanilla, L.G., Valtierra-Gutiérrez, I.A., Gutiérrez-Ríos, R.M., Alonso-Pavón, J.A.: Prokaryotic regulatory systems biology: common principles governing the functional architectures of Bacillus subtilis and Escherichia coli unveiled by the natural decomposition approach. J. Biotechnol. **161**(3), 278–286 (2012). https://doi.org/10.1016/j.jbiotec.2012.03.028

10. Haft, D.H., Badretdin, A., et al.: RefSeq and the prokaryotic genome annotation pipeline in the age of metagenomes. Nucleic Acids Res. **52**(D1), D762–D769 (2023). https://doi.org/10.1093/nar/gkad988

11. Hernández-Salmerón, J.E., Irani, T., Moreno-Hagelsieb, G.: Fast genome-based delimitation of Enterobacterales species. PLoS ONE **18**(9), e0291492 (2023). https://doi.org/10.1371/journal.pone.0291492

12. Hernández-Salmerón, J.E., Moreno-Hagelsieb, G.: Progress in quickly finding orthologs as reciprocal best hits: comparing blast, last, diamond and MMseqs2. BMC Genom. **21**(1), 741 (2020). https://doi.org/10.1186/s12864-020-07132-6

13. Illergård, K., Ardell, D.H., Elofsson, A.: Structure is three to ten times more conserved than sequence—a study of structural response in protein cores. Proteins: Struct. Funct. Bioinform. **77**(3), 499–508 (2009). https://doi.org/10.1002/prot.22458

14. Johnson, L.S., Eddy, S.R., Portugaly, E.: Hidden Markov model speed heuristic and iterative HMM search procedure. BMC Bioinform. **11**(1), 431 (2010). https://doi.org/10.1186/1471-2105-11-431

15. Kempen, M.V., et al.: Fast and accurate protein structure search with Foldseek. Nat. Biotechnol. 1–4 (2023). https://doi.org/10.1038/s41587-023-01773-0

16. Leman, J.K., et al.: Sequence-structure-function relationships in the microbial protein universe. Nat. Commun. **14**(1), 2351 (2023). https://doi.org/10.1038/s41467-023-37896-w

17. Lu, S., et al.: CDD/SPARCLE: the conserved domain database in 2020. Nucleic Acids Res. **48**(D1), D265–D268 (2019). https://doi.org/10.1093/nar/gkz991

18. Mistry, J., et al.: Pfam: the protein families database in 2021. Nucleic Acids Res. **49**(D1), gkaa913 (2020). https://doi.org/10.1093/nar/gkaa913

19. Ondov, B.D., et al.: Mash: fast genome and metagenome distance estimation using MinHash. Genome Biol. **17**(1), 132 (2016). https://doi.org/10.1186/s13059-016-0997-x

20. Paysan-Lafosse, T., et al.: InterPro in 2022. Nucleic Acids Res. **51**(D1), D418–D427 (2022). https://doi.org/10.1093/nar/gkac993

21. Sanchez, I., Hernandez-Guerrero, R., Mendez-Monroy, P.E., Martinez-Nuñez, M.A., Ibarra, J.A., Pérez-Rueda, E.: Evaluation of the abundance of DNA-binding

transcription factors in prokaryotes. Genes **11**(1), 52 (2020). https://doi.org/10.3390/genes11010052

22. Sayers, E.W., et al.: Database resources of the national center for biotechnology information in 2023. Nucleic Acids Res. **51**(D1), gkac1032 (2022). https://doi.org/10.1093/nar/gkac1032

23. Skolnick, J., Fetrow, J.S., Kolinski, A.: Structural genomics and its importance for gene function analysis. Nat. Biotechnol. **18**(3), 283–287 (2000). https://doi.org/10.1038/73723

24. Steinegger, M., Söding, J.: MMseqs2 enables sensitive protein sequence searching for the analysis of massive data sets. Nat. Biotechnol. **35**(11), 1026–1028 (2017). https://doi.org/10.1038/nbt.3988

25. The UniProt Consortium, et al.: UniProt: the universal protein knowledge base in 2023. Nucleic Acids Res. **51**(D1), D523–D531 (2022). https://doi.org/10.1093/nar/gkac1052

26. Tierrafría, V.H., et al.: RegulonDB 11.0: comprehensive high-throughput datasets on transcriptional regulation in Escherichia coli K-12. Microb. Genom. **8**(5), mgen000833 (2022). https://doi.org/10.1099/mgen.0.000833

27. Varadi, M., et al.: AlphaFold protein structure database: massively expanding the structural coverage of protein-sequence space with high-accuracy models. Nucleic Acids Res. **50**(D1), D439–D444 (2021). https://doi.org/10.1093/nar/gkab1061

28. Varadi, M., et al.: AlphaFold protein structure database in 2024: providing structure coverage for over 214 million protein sequences. Nucleic Acids Res. **52**(D1), D368–D375 (2023). https://doi.org/10.1093/nar/gkad1011

29. Xu, J., Zhang, Y.: How significant is a protein structure similarity with TM-score = 0.5? Bioinformatics **26**(7), 889–895 (2010). https://doi.org/10.1093/bioinformatics/btq066

30. Zhang, C., Shine, M., Pyle, A.M., Zhang, Y.: US-align: universal structure alignments of proteins, nucleic acids, and macromolecular complexes. Nat. Methods **19**(9), 1109–1115 (2022). https://doi.org/10.1038/s41592-022-01585-1

Revisiting the Effects of MDR1 Variants Using Computational Approaches

Tal Gutman[1] and Tamir Tuller[1,2(✉)] (iD)

[1] Department of Biomedical Engineering, The Engineering Faculty, Tel-Aviv University,
69978 Tel Aviv, Israel
tamirtul@tauex.tau.ac.il
[2] The Sagol School of Neuroscience, Tel-Aviv University, 6997801 Tel Aviv, Israel

Abstract. P-glycoprotein, encoded by the MDR1 gene, is an ATP-dependent pump that exports various substances out of cells. Its overexpression is related to multi drug resistance in many cancers. Numerous studies explored the effects of MDR1 variants on p-glycoprotein expression and function, and on patient survivability. T1236C, T2677G and T3435C are prevalent MDR1 variants that are the most widely studied, typically in-vitro and in-vivo, with remarkably inconsistent results. In this paper we perform computational, data-driven analyses to assess the effects of these variants using a different approach. We use knowledge of gene expression regulation to elucidate the variants' mechanism of action. Results indicate that T1236C is correlated with worse patient prognosis. Additionally, examination of MDR1 folding strength suggests that T3435C potentially modifies co-translational folding. Furthermore, all three variants reside in potential translation bottlenecks and likely cause increased translation rates. These results support several hypotheses suggested by previous studies. To the best of our knowledge, this study is the first to apply a computational approach to examine the effects of MDR1 variants.

Keywords: MDR1 · synonymous variants · gene expression · cancer evolution · mRNA folding selection

1 Introduction

P-glycoprotein (p-gp) was first discovered in 1976, where it was shown to alter drug permeability in Chinese hamster ovary cells [1]. Ensuing extensive studies unveiled ample information about the structure and function of this transmembrane protein; P-gp is a member of the ATP-binding cassette transporter superfamily [2]. It is a 170 KDa molecule that is comprised of two halves with high homology, each containing six transmembrane domains (TMD) and a single nucleotide-binding domain (NBD) [3]. P-gp is interleaved in the cell's membrane and functions as an ATP-dependent efflux pump, exporting diverse substances out of the cell. Lipids, steroids, xenobiotics, various drugs and chemotherapy agents are some of the molecules transferred by this protein [4]. P-gp is normally highly expressed in the adrenal glands, kidneys, liver, the gastro-intestinal

C. Scornavacca and M. Hernández-Rosales (Eds.): RECOMB-CG 2024, LNBI 14616, pp. 226–247, 2024.
https://doi.org/10.1007/978-3-031-58072-7_12

tract and the blood-brain barrier [5]. It protects the body from deleterious substances by excreting them to the gut lumen, urine and bile, and by reducing permeability of sensitive tissues [6]. P-gp is also highly expressed in many cancer cells. Its ability to export substrates that vary greatly in chemical structure and size makes it fundamental for the multi-drug resistance (MDR) mechanism [7] of cancer cells.

P-gp is encoded by the MDR1 gene, also frequently called ABCB1. It is located on chromosome 7 and encodes for a protein of 1,280 amino acids [8]. MDR1 is highly polymorphic, with over 50 single nucleotide polymorphisms (SNP) currently discovered in its coding region [9]. Three variants are the most extensively studied - T1236C (rs1128503), T2677G (rs2032582) and T3435C (rs1045642). T1236C is a synonymous variant that occurs in the first NBD (at position 1236 of the coding sequence), changing the codon GGT to GGC (both encode for Glycine). T2677G is a non-synonymous variant that occurs between the tenth and eleventh TMDs. It changes the codon TCT to GCT, resulting in Serine being replaced by Alanine. Finally, T3435C is a synonymous variant that occurs in the second NBD, causing the codon ATT to be changed to ATC but keeping the encoded amino acid Isoleucine [10]. All three variants occur in intracellular parts of the protein [8]. Though the frequencies vary for different ethnicities, they are common in the general population; In fact, the alternative allele is present in 57%, 55% and 49% of genotypes for T1236C, T2677G and T3435C respectively [11]. The three also make a common haplotype, occurring in 33% of individuals [12].

Variants, particularly ones defined as "silent" such as synonymous alterations, can modify the gene expression process. They can affect transcription through changing transcription-factor binding sites (TFBS) [13]; they can impact translation and co-translational folding (CTF) through changing codon usage [14, 15]; they can also accelerate mRNA degradation by changing affinity to miRNA binding sites [16] or lead to mis-spliced proteins by modifying the nucleotides in the vicinity of donor or acceptor sites [17]. Altogether, they can impact all phases of gene expression. Though these three variants are widely studied, conclusions regarding their effects have been exceptionally inconsistent [18]. Some studies suggest they change mRNA or protein expression [19–23] while others claim they do not [24–29] and associate their presence with modifications in protein structure [24, 29]. Some find them to significantly impact patients' response to chemotherapy and survivability while others do not [30–46]. Remarkably, controversy is also found among those who claim that the variants do affect survivability [47]; whether the variants are detrimental or advantageous for patient survival remains a matter of disagreement.

The effects of these MDR1 variants on p-gp expression, function and on chemotherapy resistance have been studied mostly in clinical trials and in-vitro cell line experiments. Each method has inherent disadvantages; results of clinical trials are affected by factors such as cohort size, patient ethnicity, tumor type and conducted chemotherapy regimen. In-vitro experiments cannot fully mimic the symbiosis of a cell with its natural environment and substantially differ from the corresponding cell type in-vivo [48]. Even though the effects of the variants may vary under different conditions, it is possible that some of the contradictions emerge not due to biological complexity but rather due to impediments in scientific methods. The objective of this paper is to examine, for the first time, the effects of T1236C, T2677G and T3435C on all phases of the gene expression process using various computational approaches. We aim to gain a better understanding of the mechanism of action of this three MDR1 variants, endeavoring to propose plausible explanations and validate some of the previously proposed effects.

2   Methods

2.1   Data Sources

For performing the various analyses, we utilized data of several known databases. Single Nucleotide Variants (SNV) data, mRNA expression data and clinical data of TCGA [49, 50] projects was downloaded from the GDC [51] (https://gdc.cancer.gov/) on November 2021. The complete human CDS was downloaded from Ensembl [52, 53] (https://ftp.ens embl.org/pub/release-109/fasta/homo_sapiens/cds/) on 2020. Human protein expression measurements were downloaded from PaxDb [54] on 2020 (https://pax-db.org/dataset/9606/1502934799/).

2.2   MDR1 Expression

MDR1 Expression Change of TCGA Patients. SNVs and expression data of TCGA patients were used to examine the effects of T1236C, T2677G, T3435C and the haplotypes on MDR1 expression. For each mutation or haplotype, the cohort was split to carriers and non-carriers groups. The non-carriers group was randomly sampled 100,000 times such that each sampled group contained the same amount of patients as the carriers group. The average MDR1 expression was calculated for each sampled group and used to create a distribution of the MDR1 expression for non-carriers. The average MDR1 expression of the carriers group was compared to the distribution and an empirical p-value was calculated.

MDR1 Expression Change According to Enformer. Enformer [55] is a state-of-the-art model that predicts gene expression and transcription regulation based on an extremely long input sequence (hundreds of thousands of nucleotides). Enformer was used to predict MDR1 expression levels both for the wild type and the mutated sequence, for all three variants. As the output of Enformer is made of predictions for multiple genomic tracks that differ in tissues and measure of gene expression, we used several different approaches to obtain a single score from the output. For example, some of the scores only considered tracks related to tissues where MDR1 is highly expressed, others only considered CAGE tracks, some accounted all tracks. The same post-processing was performed for random variants with similar characteristics (explained in the "Empirical p-values" section) and the scores of the original and random variants were compared. If the score of the variant was larger than the scores of 95% of the random variants it was deemed to significantly change MDR1 expression.

2.3   Splicing Events

To examine the effect of T1236C, T2677G, T3435C on the occurrence of splicing events we used SpliceAI [56], a model that, given a genomic sequence, predicts the probabilities of each site to be a donor or acceptor site. We perform the prediction for both the wildtype and mutated sequence, centered around each of the three variants. For each position in the prediction range, we calculated the difference in the probability of it being a donor/acceptor site that was caused by the variant. We searched for positions for which the probability changed by more than 50%. Positions exhibiting an increment of more than 50% are considered new prospective donor/acceptor sites, whereas positions for which the probability decreases by more than 50% are regarded as potentially abolished donor/acceptor sites.

2.4 Translation Rates

To examine the effect of T1236C, T2677G and T3435C on translation rates we utilized several measures positively correlated with it - MFE, CAI, FPTC and tAI. We calculated these measures for both the wildtype and mutated MDR1 CDS sequences and examined the difference in these measures at the position of the variant.

MFE. A per-position MFE score was computed using ViennaRNA [57]; first, a sliding window (length = 39 nucleotides, stride = 1 nucleotide) was used to obtain a per-window MFE score. Then, the MFE score of a specific position in the CDS was set as the average of all MFE scores of the windows that the position is in.

CAI. Human CAI weights were computed as suggested in the original paper [58]. The set of highly expressed genes (15% most highly expressed) was curated using human protein expression levels from PAXdb [54].

FPTC. Human FPTC weights were downloaded from the Kazusa website [59].

tAI. tAI tissue-specific weights were taken from Hernandez-Alias et al. [60] and the s weights were optimized as depicted in Sabi et al. [61].

2.5 Co-translational Folding

To examine the effect of T1236C, T2677G and T3435C on the CTF of p-gp, a computational model that assesses which positions in the CDS are important for correct protein folding was deployed. The model is yet to be published and therefore we will provide much detail about the model's methodology. The analysis is based on the basic assumption that CTF is governed by local translation rates, and that this rate is evolutionary conserved for a position that is crucial for correct folding [62]. Thus, it searches for positions with both evolutionarily conserved low and evolutionarily conserved high MFE (a measure correlated with translation rate) across orthologous versions of a gene. All MDR1 orthologous CDSs (n = 383) were downloaded from Ensembl (https://rest.ens embl.org/documentation/info/homology_ensemblgene) and were aligned using Clustal Omega [63]. The MFE score per nucleotide position was calculated for all sequences in the multiple sequence alignment (MSA). Then, we calculated the average MFE at each CDS position across the different organisms. To find the positions were the MFE score is conserved as significantly lower or higher than expected by chance, we created two kinds of permuted versions of the MSA. In the first method (named "vertical permutation"), we shuffle synonymous codons within the same column of the MSA. In the second method (named "horizontal" permutation), we horizontally swap between synonymous codons of pairs of columns. Both methods affect the MFE scores while keeping basic characteristics of the MSA such as the amino-acid, codon and nucleotide content. One hundred permutations of the MSA are created for each kind. We calculate the per-position MFE score averaged across orthologs for each permuted MSA in the same manner as was done for the original one. At this point, for each CDS position we have a single true MFE score and one hundred scores from each of the permuted versions. We utilized the permutations to calculate a z-score for each position. Finally, we intersect the results to get positions that had significantly low or high MFE when compared to both kinds of permuted versions of the MSA.

2.6 Survivability of TCGA Patients

To examine the effect of T1236C, T2677G, T3435C and their haplotypes on the survival of cancer patients we used TCGA SNV and clinical data. We used the vital status of the patients and a matching time-stamp in order to create Kaplan-Meier survival curves [64] for the carriers and non-carriers groups. The time-stamp was derived from the maximum value of the following attributes- "Days from the initial diagnosis to current follow-up", "Days from the initial diagnosis to the current confirmation of vital status", "Days from the initial diagnosis to patient death". The logrank test [65] was used to assess whether the survival curves of the carriers and non-carriers group significantly differ.

To assert that the results are not caused by differences in tumor mutational burden (TMB), we repeated the analysis when controlling for this confounder; the balance of the TMB between the mutated and control groups was tested using a two sample KS test [66].

2.7 Stratification of TCGA Patients

For several analyses TCGA patients were stratified according to their cancer types as suggested by Gao et al. [67] (see Table 1). The patients were assigned to one of three groups–metabolic cancers, which are associated with altered metabolic pathways, proliferative cancers, which are associated with dysregulated cell proliferation, and inflammatory cancers, which are associated with immune system dysregulation. Because cancers in these categories dysregulate different pathways, it is likely that drug resistance patterns also differ between these categories. For example, metabolic cancers could dysregulate drug-metabolizing enzymes or drug efflux pumps, while proliferative cancers have rapid division rates and could lead to emergence of drug resistant clones through acquisition of new mutations.

2.8 Empirical p-Values

To infer the significance of the changes caused by T1236C, T2677G and T3435C in several of the analyses, they were compared to changes caused by random variants with similar characteristics. T2677G is a non-synonymous variant and therefore its effect was compared to other, randomly sampled, T -> G non-synonymous variants in the MDR1 CDS. T1236C and T3435C are synonymous variants and therefore were both compared to randomly sampled synonymous T -> C variants in the MDR1 CDS. Each original variant was compared to one-hundred randomized sequences.

2.9 TCGA Variants Correlated with T1236C

T1236C was found as correlated with worse survival. To investigate the possibility of a causal relationship, we examined mutations that are highly correlated with T1236C and their effect on patient survival, searching for other potential effects. Highly correlated variants were defined as variants that are detected in the genomes of more than 75% of T1236C positive patients.

3 Results

Diverse computational models were deployed aiming to assess whether T1236C, T2677G and T3435C could modify different phases of the gene expression process. Additionally, genetic, clinical and expression data from The Cancer Genome Atlas (TCGA) were analyzed for this purpose. It is important to mind the difference between a germline variant, a genetic variation that is inherited and may be prevalent in the population, and a somatic variant which is a genetic variation that is acquired during one's lifetime. While most of the previous research regarding MDR1 analyses germline variants, our study examines both germline and somatic ones; when analyzing TCGA data we strictly evaluate the effect of somatic variants in cancerous tissues, whereas the rest of the computational analyses that do not utilized data from TCGA can be used to interpret the effects of both somatic and germline variants, as they simply analyze the effects caused by the nucleotide change and are "blind" to the type of variant.

3.1 T1236C, T2677G and T3435C Do not Seem to Significantly Affect MDR1 Expression Levels

MDR1 abundance was compared between TCGA patients that acquired any of the three somatic variants and patients who did not (see Methods). Results, shown in Fig. 1, indicate a marginal association between T1236C and MDR1 expression ($p = 0.06$) and no association between T2677G and T3435C with MDR1 expression levels. Differences in mRNA abundance between carriers of the haplotypes and non-carriers were also found non-significant ($0.11 < p < 0.24$). When stratifying patients according to cancer types (Table 1), no significant changes in expression were found as well.

Additionally, we used Enformer [55] to computationally assess whether the variants are likely to cause a change in MDR1 expression. Enformer is a transformer-based neural network that is trained on thousands of epigenetic and transcriptional datasets to predict gene expression. It identifies complicated genomic patterns related to gene expression regulation such as TSSs, TFBSs, histone modifications sites and miRNA binding sites. With an input sequence of 393,216 nucleotides, it is currently the gene expression predictor with the widest receptive field. For each variant we ran Enformer for the reference sequence and for the mutated sequence (see Methods). The predicted change in MDR1 expression was not significant for any of the variants, including T1236C.

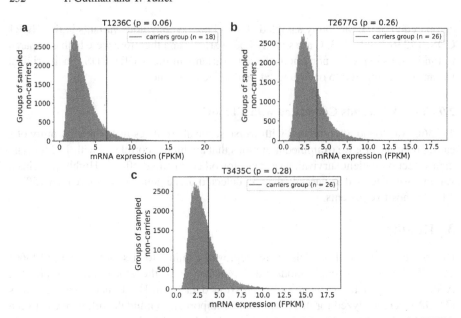

Fig. 1. MDR1 expression levels of carriers vs. non-carriers of the three variants in the TCGA database. a: T1236C; b: T2677G; c: T3435C.Comparison of the mean MDR1 expression of the carriers group (vertical line) to the mean MDR1 expression levels of 100,000 groups of randomly chosen non-carriers (distribution). Size of the carriers' group and non-carriers' groups are the same.

3.2 T1236C, T2677G and T3435C Do not Seem to Affect mRNA Splicing

SpliceAI [56] was used to assess whether the variants change MDR1 splicing. Given a sequence, the output of spliceAI is a matrix of probabilities, indicating the probability of each position in the sequence to be a donor or acceptor site (see Methods). For each variant we ran SpliceAI for the reference sequence and for the mutated sequence. Using the difference between probabilities we searched for canceled or newly generated donor and acceptors sites, as these events could lead to alternative splicing. None of the variants were found to cause significant changes.

3.3 T1236C, T2677G and T3435C Potentially Increase p-gp Levels Through Raising Global Translation Rates

The rate of translation elongation is a key component that shapes protein abundance. A change in the translation rate of a single codon can impact protein abundance if it is located in a translation bottleneck. The translation rate varies throughout the mRNA sequence and is dependent on multiple cellular conditions and factors such as accessibility of the mRNA to the ribosome, codon usage bias and availability of tRNAs. In this section we examine several measures correlated with the translation rate and evaluate whether the three variants are expected to change it, locally or globally.

T1236C, T2677G and T3435C All Decrease mRNA Folding Strength. T1236C and T3435C Potentially Increase Global Translation Rates Due to this Decrease. Minimum Free Energy (MFE) is a measure derived from predictions of the secondary structure of an mRNA sequence; a lower MFE value indicates a structure that is tightly packed and less accessible for translation. Therefore, lower MFE values are correlated with lower translation rates [68]. We computed MFE scores for the region surrounding each of the three variants (see Methods) and predicted the instigated change in MFE. The results (Fig. 2) demonstrate that all three variants cause an increase in MFE. This increase is significantly larger (p = 0.01, p = 0.02 and p = 0.04 for T1236C, T2677G and T3435C respectively) than the increase caused by random variants with similar characteristics in the MDR1 gene (see Methods). Moreover, positions 1236 and 3435 (with the wild type alleles "T") have relatively low MFE scores compared to the rest of the coding sequence (CDS) positions (19th and 12th percentile respectively) and therefore constitute possible translation bottlenecks (Fig. 7), suggesting that T1236C and T3435C could also increase global translation rates.

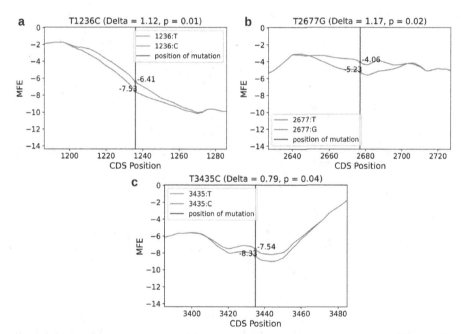

Fig. 2. Effect of the three variants on MFE in their vicinity. a: T1236C; b: T2677G; c: T3435C. x axis: the nucleotide position in the coding sequence (CDS); y axis: MFE score. The vertical line indicates the position of the variant, and the curves depict the MFE scores of the nucleotides proximal to the position of the variant, with (pink) or without it (green). (Color figure online)

T1236C, T2677G and T3435C Increase the Optimality of Codon Usage. T1236C Potentially Increases Global Translation Rates Due to This Increase. We compute Codon Adaptation Index (CAI) and Frequency Per 1000 codons (FPTC) to measure the change in codon usage bias (CUB) caused by the three variants, where CAI is computed for the synonymous variants and FPTC for the non-synonymous variant. Higher CUB suggests better adaptation of the sequence to the cellular translation machinery and resources and is correlated with faster translation [69]. Results (Fig. 3) indicate that all three variants cause a less prevalent codon to be replaced by a more prevalent codon, suggesting an increase in local translation rate. Moreover, T1236C substitutes the least prevalent codon of Glycine to the most common one. When comparing to random variants with similar characteristics in the MDR1 gene, the increase in CAI caused by T1236C is marginally significant (p = 0.08). Also, the CAI score of codon 412 (in which position 1236 resides) is in the 16th percentile of CAI scores compared to all the codons in the CDS, further strengthening the possibility that it is a translational bottleneck and that T1236C could increase global translation rate.

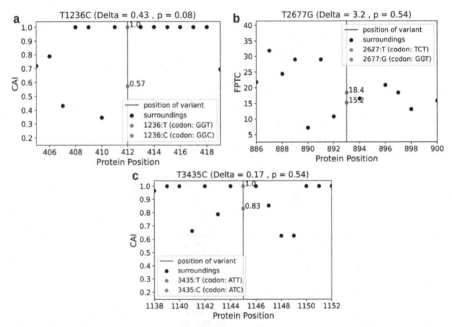

Fig. 3. Effect of the three variants on CUB. a: T1236C; b: T2677G; c: T3435C. x axis: the amino-acid position in the protein sequence; y axis: the CUB score (CAI/FPTC). The vertical line indicates the position of the variant. Black dots indicate the CUB scores in the vicinity of the variant and the dots on the vertical line indicate the CUB score of the mutated position, before (green) and after (pink). (Color figure online)

T1236C and T2677G Improve Adaptation to the tRNA Pool in Tissues Where MDR1 is Expressed. T2677G Potentially Increases Global Translation Rates Due to the Improvement in Adaptation The tRNA Adaptation Index (tAI) is a measure of translational efficiency which considers the intracellular concentration of tRNA molecules and the efficiencies of each codon–anticodon pairing [70]. Higher tAI scores are given to codons with better tRNA availability and are correlated with a higher translation rate [71]. The tRNA availability changes substantially between different tissues and organs. We examine the effect of the variants on the tAI profile in tissues where the MDR1 gene is typically expressed (see Methods) – liver, kidney, colon and brain tissues. tAI scores for the adrenal glands were not available. Figure 4 demonstrates the effect of the variants on the tAI profile in the liver tissue, but it is similar for all other examined tissues (more data will be report in the full version of this paper). The results of this analysis show that both T1236C and T2677G cause an increase in tAI. Moreover, T2677G replaces a codon with an extremely low tRNA availability to a codon with much higher tRNA availability, in all examined tissues. When comparing the tAI change caused by T2677G to changes caused by random variants with similar characteristics in the MDR1 gene, the former is found either significantly or marginally significantly larger in all relevant tissues (p < = 0.1). Moreover, the tAI score of the codon affected by T2677G is in the 1st–5th (depending on the tissue) percentile of the tAI scores of all the codons in

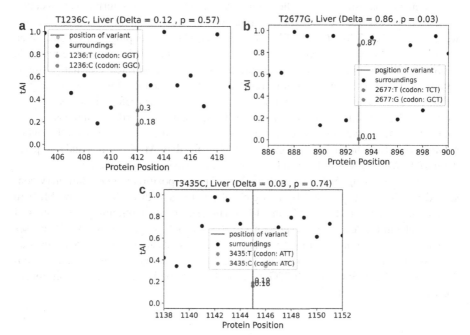

Fig. 4. Effect of the three variants on tAI. a: T1236C; b: T2677G; c: T3435C.x axis: the amino-acid position in the protein sequence; y axis: the tAI score. The vertical line indicates the position of the variant. Black dots indicate the tAI scores in the vicinity of the variant and the dots on the vertical line indicate the tAI score of the mutated position, before (green) and after (pink). (Color figure online)

the CDS, strengthening the indication that the position could be a translation bottleneck and that T2667G could increase global translation speed.

To conclude, the variants are in positions that are potential translation bottlenecks as they all obtain scores that are close to the minimal score of the entire CDS in at least one of the examined measures. All three variants lead to an increase of MFE and CUB scores, the increase in MFE being significantly larger than expected by chance. Additionally, T1236C and T2677G cause an increase in tAI scores in tissues where MDR1 is expressed, with T2677G leading to an extreme and significant tAI change. Notably, a variant can cause an increase of one factor while decreasing the others and vise-versa. In fact, when examining all synonymous variants across TCGA we find a negative correlation between the effect of variants on MFE and CAI ($p = -0.13$, $p < 10^{-323}$) and between the effect on MFE and tAI ($p = -0.18$, $p < 10^{-323}$). The increase in MFE, CAI and tAI caused by all three variants provides accumulating evidence and suggest that T1236C, T2677G and T3435C likely increase local translation rates in their vicinity, and possibly increase global translation rates and protein levels.

3.4 T3435C Potentially Modifies Co-translational Protein Folding Through Raising Local Translation Rate in a Conserved Slowly Translated Region

CTF is the mechanism in which the nascent protein begins to fold during its translation [72]. This process was shown to be governed by local translation rates [62]. To examine whether T1236C, T2677G and T3435C influence CTF we deployed a model which detects positions that have evolutionary conserved extreme MFE scores. The rationale being that MFE scores can be used as proxies for translation rates and that positions with evolutionary conserved extreme translation rates (especially slow rates) are largely conserved as such due to their importance for optimal CTF. Therefore, it is possible that variants in these positions interfere with the CTF mechanism. In order to find positions in the CDS of MDR1 with conserved extreme MFE scores, the model utilizes orthologous MDR1 genes from hundreds of organisms, calculates their MFE profiles and compares them to the MFE profiles of permuted versions of these genes (see Methods).

Model output (Fig. 5) indicates that T3435C is in a position with evolutionary conserved low MFE, surrounded by a stretch of positions with conserved low MFE (6 nucleotides upstream of T3435C and 17 downstream of it). Combining the results of this model and the results of the previous analysis which suggests that T3435C causes a local increase in translation rate, we deduce that it is possible that T3435C causes an increase in translation rate in a region of conserved low translation rates, and thus modifies CTF. Both T1236C and T2677G were not found to be in a position of evolutionary conserved extreme MFE.

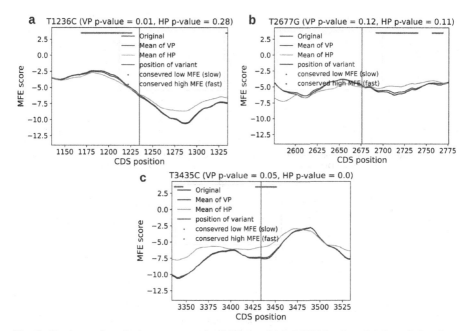

Fig. 5. Regions of evolutionary conserved low or high MFE in the vicinity of the three variants. a: T1236C; b: T2677G; c: T3435C. x axis: the nucleotide position on the CDS of the MDR1 gene. y axis: the MFE score, which is used by this model as the proxy for translation rate. The purple curve indicates the mean MFE score of the true MDR1 CDS across 383 orthologs. The orange and brown curves indicate the mean MFE score of a permuted MDR1 CDS, using two different kinds of permutations – vertical permutations (VP) and horizontal permutations (HP). Green dots are nucleotide positions for which the MFE of the true MDR1 was significantly higher than the MFE of both permutated versions, across organisms. Red dots are nucleotide positions for which the MFE of the true MDR1 was significantly lower than the MFE of both randomized versions, across organisms.

3.5 T1236C Potentially Decreases Patient Survivability

Clinical data of TCGA patients were used to assess the variants' effect on survivability. Kaplan Meier curves [64] were compared between patients that acquired one or more of the three variants in their cancerous tissue and those who did not (see Methods). Results (Fig. 6 and Fig. 8 for the haplotypes) suggest that T1236C decreases survival probability (logrank $T = 5.22$, $p = 0.02$. Bonferroni corrected $p = 0.06$) whereas T2677G and T3435C do not. Other TCGA variants that are highly correlated with the presence of T1236C (see Methods) are not correlated with decreased survivability (Fig. 9), enhancing the likelihood that T1236C is a causal variant.

When stratifying patients according to cancer types (Table 1), we find a significant negative impact ($p = 0.002$) on the survival of patients with inflammatory cancers (Fig. 10).

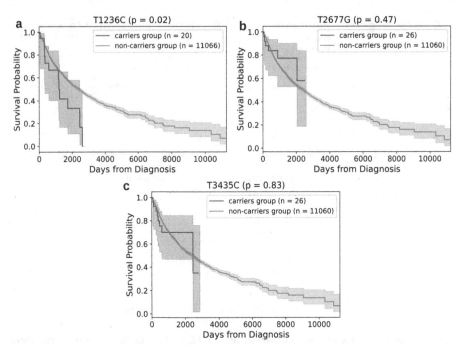

Fig. 6. Effect of the three variants on patient survivability. a: T1236C; b: T2677G; c: T3435C. Comparison of the Kaplan-Meier survival curves of the carriers' group and the non-carriers group of the three variants.

4 Discussion

Since the discovery of MDR1 and its many variants, numerous in-vitro and in-vivo studies have attempted to understand their mechanisms of action and contribution to chemotherapy resistance and cancer prognosis. Due to contradictory findings and the lack of success of MDR1 inhibitors in clinical trials [73], research of MDR1 and its variants has gradually diminished. However, the current surge in genomic data and evolving technology enables the examination of this case from a new, data-driven, perspective.

Our analyses aid to paint a clearer image, providing evidence of accumulating effects. T1236C causes an increase in CAI, tAI and MFE, all positively correlated with translation rates. Because it is in a region of low MFE and CAI, this increase could lead to a rise in global translation rates and p-gp over-expression. Moreover, T1236C was found to be correlative with worse patient prognosis, which can be explained by overexpression of the protein. T2677G also causes an increase in measures correlated in translation rates and most significantly effects tAI. It causes a codon that is extremely rare in tissues where MDR1 is expressed to be replaced with a common codon. T3435C increases MFE and CAI as well, in a region of very low MFE. Additionally, it possibly modifies p-gp structure through changing local translation rate. Altogether, when exploring the mechanism of action of T1236C, T2677G and T3435C, it is much more likely that they cause an increase in p-gp expression rather than a decrease, providing an advantage for the cancer cell. Due to these findings, we expected patients with the CGC haplotype to have significantly worse survival, but it was not detected (Fig. 8). Perhaps the combination of the three variants leads to a more complexed effect than the accumulated impact of each one separately. Alternatively, perhaps a significant effect on survival was not observed due to the small size of the cohort and confounders such as patients' cancer type and treatment regime.

Though the findings of previous research on the matter are inconsistent, our analyses support the conclusions of several major studies. Kimchey-Zarfaty at el. Demonstrated that T3435C, combined with either T1236C or T2677G, changes MDR1 substrate specificity [24]. They hypothesized that T3435C modifies CTF (and therefore protein conformation) because it changes the local translation rate in a slowly translated region, important for correct CTF. They supported their hypothesis by identifying a cluster of non-prevalent codons in the vicinity of T3435C (as can be seen in Fig. 3c). Our model, which utilizes MFE and assesses which positions are evolutionarily conserved for slow or fast translation, indeed detects 3435 in a region that is likely conserved for slow translation, supporting their findings. Moreover, we demonstrated that T3435C increases both MFE and CAI scores, suggesting that it could increase the local translation rate in this slowly translated region and thus modify CTF. Kimchey-Zarfaty et al. also examined mRNA expression, protein expression and aberrant splicing events in wild-type and mutated p-gp and reports that no changes were caused by the variants. To the best of our knowledge, no prior study has suggested that T1236C, T2677G or T3435C cause alternative splicing; our results, obtained using SpliceAI [56], support the consensus. As for mRNA and protein expression, most in-vitro studies do not report a significant change, while some in-vivo studies do. This could be due to a complicated mechanism of action that cannot be detected out of the natural cellular environment. Nevertheless, reports of the in-vivo studies are also controversial, perhaps due to the tremendous variability in MDR1 expression between patients, small cohorts and difference in patient ethnicities, genetic backgrounds and tumor types. Combining the results of our analyses, we assess that if the variants do modify protein levels, they should cause an increase rather than a decrease. In another comprehensive study, Johnatty SE et al. [74] examine 4,616 ovarian cancer patients from the Ovarian Cancer Association Consortium (OVAC) and TCGA, that have received chemotherapy treatments. They found a marginal association of T1236C with worse overall survival and did not find association between T2677G or T3435C and survival parameters. Additionally, Chen et al. [75] performed a meta-analysis involving 3,320 patients across 15 studies and concluded that a TT genotype in position 1236 is associated with better overall survival. Both these findings are supported by our pan-cancer survival analysis on TCGA data, shown in Fig. 6.

When reviewing the analyses described in this study, we must remember the limitations of in-silico models as well; A major limitation of this study is the small cohorts of mutated patients; our carriers data is comprised of 20, 26 and 20 patients with T1236C, T2677G and T3435C respectively. In the future, once more genomic data is available, we suggest measuring both mRNA and protein levels of MDR1 of cancer patients that carry these mutations. Also, a significantly larger cohort could allow us to split patients according to their specific cancer types, ethnicities and received chemotherapy regimens while keeping the required statistical power. As MDR1 is related to the response to chemotherapy, it is likely that the effect of the variants on patient survivability is dependent on the patient's received treatment; not all chemotherapeutic agents are substrates of p-gp, and the effect of the variant among different substrates could also vary, making this analysis highly important. Another limitation is the difficulty to create models that capture all relevant processes and interactions in highly complex biological mechanisms such as gene expression regulation. Moreover, some of the models were trained on data obtained from wet lab experiments and thus are also subjected to noise. Nonetheless, computational models have many advantages; they are often less expensive, faster and can incorporate larger quantities of data than in-vitro experiments. We believe the incorporation of computational, data-driven models is both essential and highly beneficial when analyzing the complex effects of variants on gene expression regulation, generally and specifically for the case of MDR1.

Acknowledgements. We thank Alma Davidson, Yoram Zarai, Sanjit Batra and Yun S. Song for their contributions in different analyses of this study.

Disclosure of Interests. The authors have no competing interests to declare that are relevant to the content of this article.

Appendix

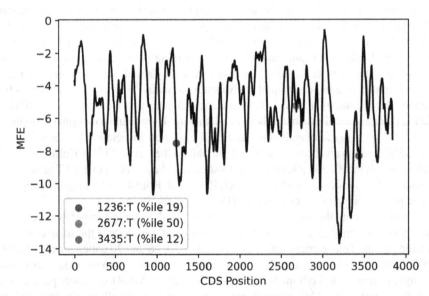

Fig. 7. MFE profile of the unmutated CDS sequence of MDR1. The positions of the variants are denoted in blue (1236), orange (2677) and green (3435). The percentile of the MFE score of the positions of the variants (when unmutated) is denoted in the legend. (Color figure online)

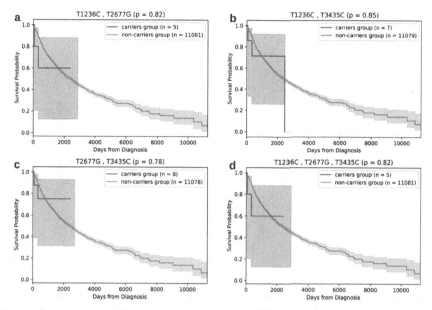

Fig. 8. Kaplan-Meier survival curves of the carriers and non-carriers of the combinations of the three variants, from the TCGA database. a: T1236C & T2677G; b: T1236C & T3435C; c: T2677G & T3435C;. d: T1236C, T2677G & T3435C.

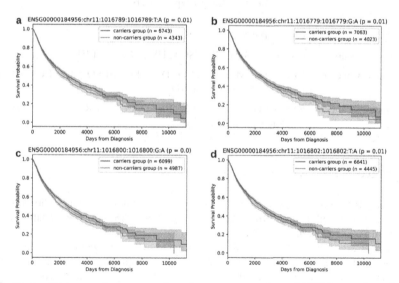

Fig. 9. Kaplan-Meier survival curves of carriers vs. non-carriers of TCGA mutations that are highly correlated with T1236C (present in the genomes of over 75% of T1236C positive patients)

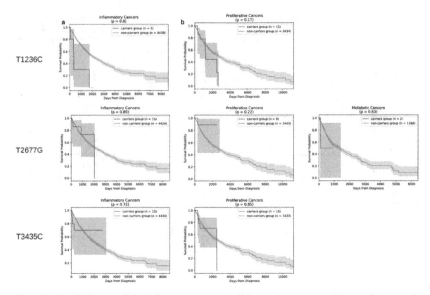

Fig. 10. Kaplan-Meier survival curves of carriers vs. non-carriers of the three variants, stratified cancer types. Left column – inflammatory cancers. Middle column- proliferative cancers. Right column- metabolic cancers. Categories are described in Table 1.

Table 1. Cancer type clusters

Cluster	Cancer Types
Metabolic Cancers	LAML, UCS, HNSC, ESCA, UVM, CHOL, LIHC
Proliferative Cancers	BLCA, SKCM, SARC, COAD, UCEC, MESO, ACC, LUAD, KIRC, KIRP, DLBC, TGCT, PRAD
Inflammatory Cancers	PAAD, LGG, CESC, GBM, READ, LUSC, BRCA, STAD, THCA, OV, KICH, PCPG, THYM

References

1. Juliano, R.L., Ling, V.: A surface glycoprotein modulating drug permeability in Chinese hamster ovary cell mutants. BBA - Biomembranes **455**(1), 152–162 (1976). https://doi.org/10.1016/0005-2736(76)90160-7
2. Allikmets, R., Gerrard, B., Hutchinson, A., Dean, M.: Characterization of the human ABC superfamily: isolation and mapping of 21 new genes using the expressed sequence tags database. Hum. Mol. Genet. **5**(10), 1649–1655 (1996). https://doi.org/10.1093/hmg/5.10.1649
3. Li, Y., Yuan, H., Yang, K., Xu, W., Tang, W., Li, X.: The structure and functions of P-Glycoprotein. Curr. Med. Chem. **17**(8), 786–800 (2010). https://doi.org/10.2174/092986710790514507

4. Sakaeda, T., Nakamura, T., Okumura, K.: MDR1 genotype-related pharmacokinetics and pharmacodynamics. Biol. Pharm. Bull. **25**(11), 1391–1400 (2002). https://doi.org/10.1248/bpb.25.1391

5. Thiebaut, F., Tsuruo, T., Hamada, H., Gottesman, M.M., Pastan, I.R.A., Willingham, M.C.: Cellular localization of the multidrug-resistance gene product P-glycoprotein in normal human tissues. Proc. Nat. Acad. Sci. **84**(21), 7735–7738 (1987). https://doi.org/10.1073/pnas.84.21.7735

6. Tanigawara, Y.: Role of P-glycoprotein in drug disposition. Ther. Drug Monit. **22**(1), 137–140 (2000). https://doi.org/10.1097/00007691-200002000-00029

7. Kartner, N., Evernden-Porelle, D., Bradley, G., Ling, V.: Detection of P-glycoprotein in multidrug-resistant cell lines by monoclonal antibodies. Nature **316**(6031), 820–823 (1985). https://doi.org/10.1038/316820a0

8. Fung, K.L., Gottesman, M.M.: A synonymous polymorphism in a common MDR1 (ABCB1) haplotype shapes protein function. Biochimica et Biophysica Acta (BBA)-Proteins Proteomics, **1794**(5), 860–871 (2009). https://doi.org/10.1016/j.bbapap.2009.02.014

9. Wang, L.-H., Song, Y.-B., Zheng, W.-L., Jiang, L., Ma, W.-L.: The association between polymorphisms in the MDR1 gene and risk of cancer: a systematic review and pooled analysis of 52 case-control studies. Cancer Cell Int. **13**, 46 (2013). https://doi.org/10.1186/1475-2867-13-46

10. Panczyk, M., Balcerczak, E., Piaskowski, S., Jamroziak, K., Pasz-Walczak, G., Mirowski, M.: ABCB1 gene polymorphisms and haplotype analysis in colorectal cancer. Int. J. Colorectal Dis. **24**(8), 895–905 (2009). https://doi.org/10.1007/s00384-009-0724-0

11. Sherry, S.T., et al.: DbSNP: the NCBI database of genetic variation. Nucleic Acids Res. **29**(1), 308–311 (2001). https://doi.org/10.1093/nar/29.1.308

12. Spooner, W., et al.: Haplosaurus computes protein haplotypes for use in precision drug design. Nat. Commun. **9**(1), 4128 (2018). https://doi.org/10.1038/s41467-018-06542-1

13. Wang, S.Y., et al.: A synonymous mutation in IGF-1 impacts the transcription and translation process of gene expression. Mol. Ther.-Nucleic Acids **26**, 1446–1465 (2021). https://doi.org/10.1016/j.omtn.2021.08.007

14. Tarrant, D., Von Der Haar, T.: Synonymous codons, ribosome speed, and eukaryotic gene expression regulation. Cell. Mol. Life Sci. **71**(21), 4195–4206 (2014). https://doi.org/10.1007/s00018-014-1684-2

15. Walsh, I.M., Bowman, M.A., Soto Santarriaga, I.F., Rodriguez, A., Clark, P.L.: Synonymous codon substitutions perturb cotranslational protein folding in vivo and impair cell fitness. Proc. Nat. Acad. Sci. **117**(7), 3528–3534 (2020). https://doi.org/10.1073/pnas.1907126117

16. Gu, W., Wang, X., Zhai, C., Xie, X., Zhou, T.: Selection on synonymous sites for increased accessibility around miRNA binding sites in plants. Mol. Biol. Evol. **29**(10), 3037–3044 (2012). https://doi.org/10.1093/molbev/mss109

17. Mueller, W.F., Larsen, L.S., Garibaldi, A., Hatfield, G.W., Hertel, K.J.: The silent sway of splicing by synonymous substitutions. J. Biol. Chem. **290**(46), 27700–27711 (2015). https://doi.org/10.1074/jbc.M115.684035

18. Robey, R.W., Pluchino, K.M., Hall, M.D., Fojo, A.T., Bates, S.E., Gottesman, M.M.: Revisiting the role of ABC transporters in multidrug-resistant cancer. Nat. Rev. Cancer **18**(7), 452–464 (2018). https://doi.org/10.1038/s41568-018-0005-8

19. He, H., et al.: Association of ABCB1 polymorphisms with prognostic outcomes of anthracycline and cytarabine in Chinese patients with acute myeloid leukemia. Eur. J. Clin. Pharmacol. **71**(3), 293–302 (2015). https://doi.org/10.1007/s00228-014-1795-6

20. Hemauer, S.J., Nanovskaya, T.N., Abdel-Rahman, S.Z., Patrikeeva, S.L., Hankins, G.D., Ahmed, M.S.: Modulation of human placental P-glycoprotein expression and activity by MDR1 gene polymorphisms. Biochem. Pharmacol. **79**(6), 921–925 (2010). https://doi.org/10.1016/j.bcp.2009.10.026

21. Hoffmeyer, S., et al.: Functional polymorphisms of the human multidrug-resistance gene: multiple sequence variations and correlation of one allele with P-glycoprotein expression and activity in vivo. Proc. Nat. Acad. Sci. **97**(7), 3473–3478 (2000). https://doi.org/10.1073/pnas.97.7.3473

22. Song, P., et al.: G2677T and C3435T genotype and haplotype are associated with hepatic ABCB1 (MDR1) expression. J. Clin. Pharmacol. **46**, 373–379 (2006). https://doi.org/10.1177/0091270005284387

23. Pang, L., et al.: ATP-binding cassette genes genotype and expression: a potential association with pancreatic cancer development and chemoresistance? Gastroenterol. Res. Pract. **2014**, 414931 (2014). https://doi.org/10.1155/2014/414931

24. Kimchi-Sarfaty, C., et al.: A "silent" polymorphism in the MDR1 gene changes substrate specificity. Science **315**(5811), 525–528 (2007). https://doi.org/10.1126/science.1135308

25. Kroetz, D.L., et al.: Sequence diversity and haplotype structure in the human ABCB1 (MDR1, multidrug resistance transporter) gene. Pharmacogenet. Genomics **13**(8), 481–494 (2003). https://doi.org/10.1097/00008571-200308000-00006

26. Gow, J.M., Hodges, L.M., Chinn, L.W., Kroetz, D.L.: Substrate-dependent effects of human ABCB1 coding polymorphisms. J. Pharmacol Exp. Ther. **325**(2), 435–442 (2008). https://doi.org/10.1124/jpet.107.135194

27. Hung, C.C., Chen, C.C., Lin, C.J., Liou, H.H.: Functional evaluation of polymorphisms in the human ABCB1 gene and the impact on clinical responses of antiepileptic drugs. Pharmacogenet. Genomics **18**(5), 390–402 (2008). https://doi.org/10.1097/FPC.0b013e3282f85e36

28. Salama, N.N., Yang, Z., Bui, T., Ho, R.J.: MDR1 haplotypes significantly minimize intracellular uptake and transcellular P-gp substrate transport in recombinant LLC-PK1 cells. J. Pharm. Sci. **95**(10), 2293–2308 (2006). https://doi.org/10.1002/jps.20717

29. Fung, K.L., et al.: MDR1 synonymous polymorphisms alter transporter specificity and protein stability in a stable epithelial monolayer. Cancer Res. **74**(2), 598–608 (2014). https://doi.org/10.1158/0008-5472.CAN-13-2064

30. Ni, L.-N., et al.: Multidrug resistance gene (MDR1) polymorphisms correlate with imatinib response in chronic myeloid leukemia. Med. Oncol. **28**, 265–269 (2011). https://doi.org/10.1007/s12032-010-9456-9

31. Lu, Y., et al.: Host genetic variants of ABCB1 and IL15 influence treatment outcome in paediatric acute lymphoblastic leukaemia. Br. J. Cancer **110**(6), 1673–1680 (2014). https://doi.org/10.1038/bjc.2014.7

32. Zheng, Q., et al.: ABCB1 polymorphisms predict imatinib response in chronic myeloid leukemia patients: a systematic review and meta-analysis. Pharmacogenomics J. **15**(2), 127–134 (2015). https://doi.org/10.1038/tpj.2014.54

33. Chu, Y.-H., et al.: Association of ABCB1 and FLT3 polymorphisms with toxicities and survival in Asian patients receiving sunitinib for renal cell carcinoma. PLoS ONE **10**(8), e0134102 (2015). https://doi.org/10.1371/journal.pone.0134102

34. Munisamy, M., et al.: Pharmacogenetics of ATP binding cassette transporter MDR1 (1236C>T) gene polymorphism with glioma patients receiving Temozolomide-based chemoradiation therapy in Indian population. Pharm. J. **21**(2), 262–272 (2021). https://doi.org/10.1038/s41397-021-00206-y

35. Li, J.Z., Tian, Z.Q., Jiang, S.N., Feng, T.: Effect of variation of ABCB1 and GSTP1 on osteosarcoma survival after chemotherapy. Genet. Mol. Res. **13**(2), 3186–3192 (2014). https://doi.org/10.4238/2014.April.25.3

36. Olarte Carrillo, I., García Laguna, A.I., De la Cruz Rosas, A., Ramos Peñafiel, C.O., Collazo Jaloma, J., Martínez Tovar, A.: High expression levels and the C3435T SNP of the ABCB1 gene are associated with lower survival in adult patients with acute myeloblastic leukemia in

Mexico City. BMC Med. Genomics **14**(1), 1–9 (2021). https://doi.org/10.1186/s12920-021-01101-y

37. Balcerczak, E., Panczyk, M., Piaskowski, S., Pasz-Walczak, G., Sałagacka, A., Mirowski, M.: ABCB1/MDR1 gene polymorphisms as a prognostic factor in colorectal cancer. Int. J. Colorectal Dis. **25**(10), 1167–1176 (2010). https://doi.org/10.1007/s00384-010-0961-2

38. Caronia, D., et al.: Effect of ABCB1 and ABCC3 polymorphisms on osteosarcoma survival after chemotherapy: a pharmacogenetic study. PLoS ONE **6**(10), e26091 (2011). https://doi.org/10.1371/journal.pone.0026091

39. Wu, H., et al.: Roles of ABCB1 gene polymorphisms and haplotype in susceptibility to breast carcinoma risk and clinical outcomes. J. Cancer Res. Clin. Oncol. **138**(9), 1449–1462 (2012). https://doi.org/10.1007/s00432-012-1209-z

40. Knez, L., et al.: Predictive value of ABCB1 polymorphisms G2677T/A, C3435T, and their haplotype in small cell lung cancer patients treated with chemotherapy. J. Cancer Res. Clin. Oncol. **138**(9), 1551–1560 (2012). https://doi.org/10.1007/s00432-012-1231-1

41. Vivona, D., et al.: ABCB1 haplotypes are associated with P-gp activity and affect a major molecular response in chronic myeloid leukemia patients treated with a standard dose of imatinib. Oncol. Lett. **7**(4), 1313–1319 (2014). https://doi.org/10.3892/ol.2014.1857

42. Li, W., et al.: ABCB1 3435TT and ABCG2 421CC genotypes were significantly associated with longer progression-free survival in Chinese breast cancer patients. Oncotarget, **8**(67), 111041 (2017).https://doi.org/10.18632/oncotarget.22201

43. Gregers, J., et al.: Polymorphisms in the ABCB1 gene and effect on outcome and toxicity in childhood acute lymphoblastic leukemia. Pharm. J. **15**(4), 372–379 (2015). https://doi.org/10.1038/tpj.2014.81

44. Xiaohui, S., Aiguo, L., Xiaolin, G., Ying, L., Hongxing, Z., Yilei, Z.: Effect of ABCB1 polymorphism on the clinical outcome of osteosarcoma patients after receiving chemotherapy. Pak. J. Med. Sci. **30**(4), 886–890 (2014). https://doi.org/10.12669/pjms.304.4955

45. Liu, S., Yi, Z., Ling, M., Shi, J., Qiu, Y., Yang, S.: Predictive potential of ABCB1, ABCC3, and GSTP1 gene polymorphisms on osteosarcoma survival after chemotherapy. Tumor Biol. **35**(10), 9897–9904 (2014). https://doi.org/10.1007/s13277-014-1917-x

46. Zmorzynski, S., et al.: The relationship of ABCB1/MDR1 and CYP1A1 variants with the risk of disease development and shortening of overall survival in patients with multiple myeloma. J. Clin. Med. **10**(22), 5276 (2021). https://doi.org/10.3390/jcm10225276

47. Chen, Q., et al.: Prognostic value of two polymorphisms, rs1045642 and rs1128503, in ABCB1 following taxane-based chemotherapy: a meta-analysis. Asian Pac. J. Cancer Prev. **22**(1), 3–10 (2021). https://doi.org/10.31557/APJCP.2021.22.1.3

48. Graudejus, O., Wong, R., Varghese, N., Wagner, S., Morrison, B.: Bridging the gap between in vivo and in vitro research: reproducing in vitro the mechanical and electrical environment of cells in vivo. Front. Cell Neurosci. **12** (2018). https://doi.org/10.3389/conf.fncel.2018.38.00069

49. Tomczak, K., Czerwinska, P., Wiznerowicz, M.: Review the cancer genome atlas (TCGA): an immeasurable source of knowledge. Contemp. Oncol. (Pozn) **19**, A68–A77 (2015). https://doi.org/10.5114/wo.2014.47136

50. Chang, K., et al.: The cancer genome atlas pan-cancer analysis project. Nat. Genet. **45**, 1113–1120 (2013). https://doi.org/10.1038/ng.2764

51. Grossman, R., et al.: Toward a shared vision for cancer genomic data. New. Engl. J. Med. **375**, 1109–1112 (2016). https://doi.org/10.1056/NEJMp1607591

52. Cunningham, F., et al.: Ensembl 2022. Nucleic Acids Res. **50**(D1), D988–D995 (2022). https://doi.org/10.1093/nar/gkab1049

53. Hunt, S.E., et al.: Ensembl variation resources. Database, **2018**, bay119 (2018). https://doi.org/10.1093/database/bay119

54. Wang, M., et al.: PaxDb, a database of protein abundance averages across all three domains of life. Mol. Cell. Proteomics **11**, 492–500 (2012). https://doi.org/10.1074/mcp.O111.014704

55. Avsec, Ž, et al.: Effective gene expression prediction from sequence by integrating long-range interactions. Nat. Methods **18**, 1196–1203 (2021). https://doi.org/10.1038/s41592-021-01252-x

56. Jaganathan, K., et al.: Predicting splicing from primary sequence with deep learning. Cell **176**, 535–548 (2019). https://doi.org/10.1016/j.cell.2018.12.015

57. Hofacker, I., et al.: Automatic detection of conserved RNA structure elements in complete RNA virus genomes. Nucleic Acids Res. **26**, 3825–3836 (1998). https://doi.org/10.1093/nar/26.16.3825

58. Sharp, P., Li, W.-H.: The codon adaptation Index—a measure of directional synonymous codon usage bias, and its potential applications. Nucleic Acids Res. **15**, 1281–1295 (1987). https://doi.org/10.1093/nar/15.3.1281

59. Nakamura, Y., Gojobori, T., Ikemura, T.: Codon usage tabulated from international DNA sequence databases: status for the year 2000. Nucleic Acids Res. **28**(1), 292 (2000). https://doi.org/10.1093/nar/28.1.292

60. Hernandez-Alias, X., Benisty, H., Schaefer, M.H., Serrano, L.: Translational adaptation of human viruses to the tissues they infect. Cell Rep. **34**(11), 108872 (2021). https://doi.org/10.1016/j.celrep.2021.108872

61. Sabi, R., Tuller, T.: Modelling the efficiency of codon–tRNA interactions based on codon usage bias. DNA Res. **21**(5), 511–526 (2014). https://doi.org/10.1093/dnares/dsu017

62. Yu, C.-H., et al.: Codon usage influences the local rate of translation elongation to regulate co-translational protein folding. Mol. Cell **59**(5), 744–754 (2015). https://doi.org/10.1016/j.molcel.2015.07.018

63. Sievers, F., et al.: Fast, scalable generation of high-quality protein multiple sequence alignments using clustal omega. Mol. Syst. Biol. **7**, 539 (2011). https://doi.org/10.1038/msb.2011.75

64. Kaplan, E.L., Meier, P.: Nonparametric estimation from incomplete observations. In: Kotz, S., Johnson, N.L. (eds.) Breakthroughs in Statistics. Springer Series in Statistics, pp. 319–337. Springer, New York (1992). https://doi.org/10.1007/978-1-4612-4380-9_25

65. Mantel, N.: Evaluation of survival data and two new rank order statistics arising in its consideration. Cancer Chemother. Rep. 50(3), 163–170 (1966). http://europepmc.org/abstract/MED/5910392

66. Karson, M.: Handbook of Methods of Applied Statistics. Volume I: Techniques of Computation Descriptive Methods, and Statistical Inference. Volume II: Planning of Surveys and Experiments. I. M. Chakravarti, R. G. Laha, and J. Roy, New York, John Wiley; 1967, $9.00. J Am. Stat. Assoc. 63(323), 1047–1049 (1968). https://doi.org/10.1080/01621459.1968.11009335

67. Gao, H., et al.: Clustering cancers by shared transcriptional risk reveals novel targets for cancer therapy. Mol. Cancer **21**(1), 116 (2022). https://doi.org/10.1186/s12943-022-01592-y

68. Kudla, G., Murray, A.W., Tollervey, D., Plotkin, J.B.: Coding-sequence determinants of gene expression in Escherichia coli. Science **324**(5924), 255–258 (2009). https://doi.org/10.1126/science.1170160

69. Futcher, B., Latter, G.I., Monardo, P., McLaughlin, C.S., Garrels, J.I.: A sampling of the yeast proteome. Mol. Cell Biol. **19**(11), 7357–7368 (1999). https://doi.org/10.1128/MCB.19.11.7357

70. Dos Reis, M., Wernisch, L., Savva, R.: Unexpected correlations between gene expression and codon usage bias from microarray data for the whole Escherichia coli K-12 genome. Nucleic Acids Res. **31**(23), 6976–6985 (2003). https://doi.org/10.1093/nar/gkg897

71. Waldman, Y.Y., Tuller, T., Shlomi, T., Sharan, R., Ruppin, E.: Translation efficiency in humans: tissue specificity, global optimization and differences between developmental stages. Nucleic Acids Res. **38**(9), 2964–2974 (2010). https://doi.org/10.1093/nar/gkq009

72. Hardesty, B., Tsalkova, T., Kramer, G.: Co-translational folding. Curr. Opin. Struct. Biol. **9**(1), 111–114 (1999). https://doi.org/10.1016/S0959-440X(99)80014-1

73. Binkhathlan, Z., Lavasanifar, A.: P-glycoprotein inhibition as a therapeutic approach for overcoming multidrug resistance in cancer: current status and future perspectives. Curr. Cancer Drug Targets **13**(3), 326–346 (2013). https://doi.org/10.2174/15680096113139990076

74. Johnatty, S.E., et al.: ABCB1 (MDR1) polymorphisms and ovarian cancer progression and survival: a comprehensive analysis from the ovarian cancer association consortium and the cancer genome atlas. Gynecol. Oncol. **131**(1), 8–14 (2013). https://doi.org/10.1016/j.ygyno.2013.07.107

75. Chen, Q., et al.: Prognostic value of two polymorphisms, rs1045642 and rs1128503, in ABCB1 following taxane-based chemotherapy: a meta-analysis. Asian Pac. J. Cancer Prev. **22**(1), 3 (2021). https://doi.org/10.31557/APJCP.2021.22.1.3

Evidence of Increased Adaptation of Omicron SARS-CoV-2 Codons to Humans

Alma Davidson[1], Marina Parr[2], Franziska Totzeck[2], Alexander Churkin[3], Danny Barash[4], Dmitrij Frishman[2], and Tamir Tuller[1,5(✉)]

[1] Department of Biomedical Engineering, Tel Aviv University, 6997801 Tel Aviv, Israel
tamirtul@tauex.tau.ac.il
[2] Department of Bioinformatics, School of Life Sciences, Technical University of Munich, Freising, Germany
[3] Department of Software Engineering, Sami Shamoon College of Engineering, 84100 Beer-Sheva, Israel
[4] Department of Computer Science, Ben-Gurion University, 8410501 Beer-Sheva, Israel
[5] Sagol School of Neuroscience, Tel Aviv University, 6997801 Tel Aviv, Israel

Abstract. Viruses are highly dependent on their hosts to carry out cellular mechanisms and cause productive infection. Thus, they undergo extensive adaptations to the host intracellular machinery, which occur over the evolution of the virus, and during the emergence of new viral strains with different properties. One aspect of viral adaptation is related to the efficiency of recruiting the host's gene expression machinery and specifically the translation machinery. This process can be partially detected using measures of codon usage bias (CUB).

While previous studies in the field suggested that there is an adaptation of codons in the viral genome to the host, none of them studied these adaptations among the different strains of the same virus over time. Thus, in this study, we focused on the SARS-CoV-2 and demonstrated for the first time that the omicron strain has an increased codon usage adaptation to humans in the early gene ORF1ab compared to previous strains. In addition, our findings indicate that the observed differences in CUB scores were primarily attributed to non-synonymous mutations. This conclusion holds for additional human-infecting viruses.

Keywords: SARS-CoV-2 · Codon usage bias · Genome evolution · Genomic variation · Gene expression

1 Introduction

Viruses are dependent on their host cell for different purposes such as particle assembly, genome replication, and gene expression to achieve an efficient infection and spreading. They have co-evolved for over millions of years with their hosts in order to escape from its immune system on the one hand and establish the optimal conditions for their specific needs on the other [1–5]. The evolution from the virus side occurs rapidly, due to short generation times and large population sizes that lead to high mutation rates. Thus, the virus could rapidly generate *de-novo* mutations in a short period of time,

© The Author(s), under exclusive license to Springer Nature Switzerland AG 2024
C. Scornavacca and M. Hernández-Rosales (Eds.): RECOMB-CG 2024, LNBI 14616, pp. 248–270, 2024.
https://doi.org/10.1007/978-3-031-58072-7_13

creating many viral variants that could form in large amounts and enable a selection towards the more adaptive ones. In the process of natural selection, the more adaptive mutations are the ones that are preferred by the host organism and can enhance viral replication and increase viral protein synthesis efficiency [6–8]. Therefore, by having frequent interaction between viruses and their hosts, it could lead to the emergence of common patterns and codon usage in both organisms that would be preserved, optimized, and tracked over the viral evolution [9–11].

Codon usage bias (CUB) is a common factor to determine viral adaptation to the host and could be exhibited in many viruses [12–18]. For example, one study conducted a comprehensive analysis of human RNA viruses, comparing their CUB to the expected codon frequency. The results revealed a correlation between codon bias and viral structure in aerosol-transmitted viruses, suggesting the presence of translational selection within these viruses. Another study explored the impact of dinucleotide content on CUB and its implications for translational bias, further supporting the hypothesis [19, 20].

Viral evolution exhibits variation across different virus types, leading to significant differences over time. This evolutionary process is influenced by various factors, including population density, way of transmission, and the availability of susceptible hosts [21, 22]. These factors contribute to rapid viral evolution and the emergence of new strains that may possess enhanced transmissibility or altered pathogenicity. Additionally, selective pressures imposed by antiviral treatments or vaccines play a crucial role [23]. Viruses exposed to antiviral drugs can develop mutations that confer resistance, enabling their survival and propagation. Similarly, widespread vaccination can drive the selection of viral variants capable of evading immune responses triggered by the vaccine [24, 25]. Furthermore, it is important to consider the timescale of viral evolution, as it can vary significantly among different viruses. While certain viruses maintain relatively stable genomes over extended periods, RNA viruses like SARS-CoV-2 undergo rapid genetic changes due to their high mutation rates and short replication cycles. Continual monitoring and research on viral evolution offer valuable insights into the emergence of novel strains, facilitating the mitigation of future outbreaks and pandemics [24, 26, 27].

The latest SARS-CoV-2 pandemic has led to a breakthrough in terms of data availability. As the SARS-CoV-2 pandemic affected people all over the world, it involved collaborative efforts among professionals from different disciplines to generate and share data rapidly and in large amounts. Those efforts have advanced our understanding of the SARS-CoV-2 structure, and the role for each of its components [28]. Having high infection rate and constant interaction with its host, have enabled tracking the evolution of SARS-CoV-2 over a relatively short period of time and gaining insights regarding viral selection [29, 30]. Therefore, when comparing it to its former versions such as SARS-CoV and MERS-CoV, SARS-CoV-2 has been highly adapted to its environment and went through a strict selection pressure with rapidly expanding virus lineages. Some of those lineages have resulted in the emergence of new variants that were designated as variants of concern (VOC), which have some shared mutations, as well as new ones: alpha, beta, gamma, delta, omicron [31–33].

Recent evidence suggests that the omicron VOC may be more transmissible and potentially able to evade some immune responses, which has raised a great concern compared to other variants. The omicron VOC has many accumulated mutations in the spike protein that bind to human cells during the infection process and has been the target in SARS-CoV-2 vaccines. With more than 30 mutations in the spike protein and more than 50 mutations throughout its entire genome, those changes have increased the association between the variant and its host, which potentially increased the transmission rate [34–36]. The main example is the binding of the spike protein to the angiotensin-converting enzyme-2 (ACE2), which is localized in various human organs and naturally involved in a few processes within the human body that regulate a variety of factors such as hormones and cytokines. On the other hand, it also plays a role in the viral infection process, when it mediates the entry of the virus by binding to the receptor-binding domain (RBD) in the spike protein [37–40]. Thus, by acquiring mutations that would increase the affinity of the spike protein for ACE2, as well as enhance other stages during the viral infection, the omicron variant has adapted to the human host and increased its ability to spread and cause disease.

However, mutations observed in the virus are not limited to the spike protein. Other genes, such as the late genes N, M and E, and the early gene ORF1ab, have also undergone mutations that may contribute to the increased infection and transmissibility of the omicron variant. For example, in ORF1ab gene, which encodes into multiple nonstructural proteins (NSP) that play various roles in viral replication and immune escape, multiple mutations observed in the omicron variant can affect their activity and potentially enhance those processes [41, 42] (see Fig. 1).

Nevertheless, while most studies have investigated the effect of non-synonymous mutations that change the amino acid sequence on the viral protein's structure, there is no consideration towards the adaptation of virus in the codon-level to the host's environment. Furthermore, those mutations could affect the codon usage in the viral genome, which in turn can affect the protein expression, folding and function [7, 43]. More specifically, in SARS-CoV-2, missense mutations were shown to be the most common type of mutation in the virus, and it was suggested it might be associated with increased virus replication due to changes in the secondary structure of the viral RNA. In the omicron variant, the distribution of those mutations was shown to be significantly different, suggesting some contribution to the viral fitness and spreading [29, 44, 45].

In this study we have performed a comprehensive computational analysis of the patterns and codon usage of SARS-CoV-2 omicron variant comparing to other variants. The analysis considered a large-scale dataset of coronavirus variants and their genes, as well as available patterns and codon usage information of their human host. Based on comparison between early gene and late genes, we suggest that the early gene within the SARS-CoV-2 tends to undergo a significant codon selection with optimized codons exhibited in the omicron variant. We analyze these changes in different mutation types and argue that those changes are mainly due to non-synonymous mutations which are more frequent. In addition to our analysis of various viral proteins in SARS-CoV-2, we have also conducted similar investigations on other human viruses. We examine genetic changes and evolutionary patterns, aiming to understand the impact of mutations on their adaptability to environmental and host factors. Finally, we discuss some possible explanations for our findings, and present some possible applications to antiviral therapy.

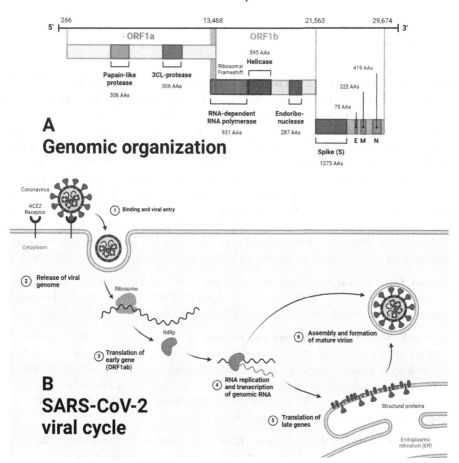

Fig. 1. The figure shows the structure of SARS-CoV-2 and its genome. (A) Shows the genes that compose the coronavirus. The first two-thirds include the open reading frame (ORF) 1ab, an early gene which contains two overlapping ORFs 1a and 1b and encodes to a polyprotein composed of nonstructural proteins (NSP). These NSPs include conserved areas such as RNA-dependent RNA polymerase (RdRp) that plays a crucial role in the viral replication process by copying the viral RNA genome. Another example is Papain-like protease (PLpro) that helps in cleaving process of viral polyproteins and evading the host's immune system [73, 74].On the last third, there are the late genes that encode to the structural components of the virus – Spike (S), Envelope (E), Membrane (M) and Nucleocapsid (N) that are encoded to proteins essential for the virus's structure, and accessory proteins between them. (B) Shows an overview of the SARS-CoV-2 viral cycle, focused on the expression of early genes and late genes. After the virus binds and enters the host cell (1), the viral genome is released (2), and the early gene is translated by the host's ribosome (3). The timing of gene expression is tightly regulated as the early gene is expressed first to establish the viral replication machinery. Once the RdRp is translated, specific factors enhance the expression of late genes, and the late genes are replicated, transcribed (4) and translated (5). Later, the viral components, including the viral genome, are being assembled to form a new virion (6). Adapted from "Coronavirus Replication Cycle" and "Genomic organization of SARS-CoV-2", by BioRender.com (2020). Retrieved from https://app.biorender.com/biorender-templates

2 Materials and Methods

2.1 Analyzed Data

SARS-CoV-2 Viral Data Extraction and Processing. Genomes of SARS-CoV-2 variants were downloaded from the National Center for Biotechnology Information (NCBI) database [46]. Genomes which are not complete and without a pangolin classification were filtered, leaving a total of 1.5 million genomes as of October 2022. Those genomes consist of the whole viral sequence and the genes' location, and the genes ORF1ab, S, N, M and E were spliced. Five datasets – one for each gene – were created and consisted of the following information: variant accession, protein accession, pangolin, and the spliced sequence. Later, the sequences were classified to 5 variants: alpha, beta, gamma, delta and omicron, based on the following pangolin: alpha (B.1.1.7 and Q lineages), beta (B.1.351 and descendent lineages), gamma (P.1 and descendent lineages), delta (B.1.617.2 and descendant lineages) and omicron (B.1.1.529 and descendant lineages). The variant-pangolin matching was based on the classification made in CoVariants [47].

Human Data Extraction. Human transcripts data was downloaded from Human Genome Resources at NCBI (RefSeq Transcripts, GRCh38 edition) [48]. Codon Usage was downloaded from Codon and Codon Pair Usage Tables (CoCoPUTs) [49].

Human Viruses' Data Extraction and Processing. Genomes of 114 different viruses were obtained from NCBI. All the genomes were complete, collected from human host and with up to 30 ambiguous characters. The data retrieved for each virus contained its genes, coding sequences, Baltimore classification, and collection time of each sample. The functional annotation for each of the viral proteins was retrieved from VOGdb (https://vogdb.org, release number vog218).

tRNA Copy Numbers and s-Values. tRNA copy numbers aim to quantify the number of tRNA molecules which take part in the translation process. Those values were downloaded from GtRNAdb database for Homo sapiens (GRCh38/hg38) edition, which contains tRNA gene predictions made by tRNAscan-SE [50, 51]. s-values are selective constraints on the efficiency of the codon–anticodon coupling, which can vary and be optimized among different organisms. Human specific s-values were obtained from the optimized values in [52].

2.2 Indices Calculation

We calculated the weights of Codon Adaptation Index (CAI) as defined in [53]. The adaptability of the viral genes to the tRNA pool was estimated using the tRNA Adaptation Index (tAI) and was computed as in [54]. The Supply-to-Demand adaptation index (SDA) was computed as in [55] and is based on normalizing the translational efficiency [56]. Those CUB weights are calculated per codon, rather than a whole gene, enabling the tracking of a single mutation and its adaptation scores.

2.3 Mutations Quantification

Data Processing. A dataset for each of the variants (alpha, beta, gamma, delta, omicron) was created, based on the datasets for each of the genes (ORF1ab, S, N, M, E) that were generated. The genes themselves were composed out of nucleotides. Any sequence with ambiguous character was removed.

Comparing between Genes from Different Variants. In order to compare genes taken from different datasets of variants, we wanted to find for a sequence in the dataset of variant X the closest sequence that exhibited in the dataset of variant Y. Therefore, we built a similarity matrix as follows: For each comparison between variant X and variant Y, a $M \times N$ sized matrix was built, where M and N are the number of sequences within the datasets of variants X and Y respectively. In each position (m, n) in the matrix, we calculated the similarity between the sequence in row m of variant X and sequence in column n of variant Y. The similarity in each position (m, n) was considered as the optimal global alignment score between the sequence in row m and the sequence in column n. The two sequences were translated from nucleotides to amino acids, and we performed global alignment using Needleman-Wunsch algorithm [57]. After the matrix was complete, we looked for the closest sequence for each sequence within the dataset of variant X, that was defined as the one with maximum alignment score when applying Needleman-Wunsch algorithm.

Mutation Classification. We took each couple of sequences that were defined as the closest based on the previous section, translated them to their amino acid sequences and aligned them using Needleman-Wunsch algorithm. Then, we returned each sequence to its nucleotide composition and compared the codons in each column of the alignment. The mutations were classified based on the following definitions:

- Silent – codon change without amino acid change.
- Missense – codon change with amino acid change.
- Deletion – the codon is missing in the sequence from row m.
- Insertion – the codon is missing in the sequence from column n.

2.4 Over Time Randomization Models and Statistical Analysis

Variant Comparison Analysis. To test the hypothesis regarding selection towards optimal codons in the omicron variant, we used the following model: we generated dataset of sequences when half of them were taken from an omicron classified gene, and the other half belonged to the same group of another variant: alpha, beta, gamma, delta or all the variants together. The sequences for the dataset were chosen randomly from the datasets that were built in Sect. 2.3. We performed multiple sequence alignment (MSA) on the sequences using Clustal Omega [58]. Then, we assigned each of the codons within each of the columns in the MSA a CUB weight (see Sect. 2.2), which enabled us to have a numerical representation of each codon's adaptability. Using those numerical values, we separated each column into two arrays of scores, one for each group compared. Columns with indels were excluded. We calculated the mean weights value for each variant array and calculated the difference between the mean values of the two groups. The difference between two variants within each of the columns in the MSA was defined as follows:

$$Diff(IDX_i) = mean(IDX[X]_i) - mean(IDX[Y]_i) \qquad (1)$$

When:

i–the number of the column in the MSA

$IDX[X]_i$, $IDX[Y]_i$–an array of indices scores (CAI/SDA/tAI) of variant X and Y in column i

X and Y–the two variants that were considered in the analysis

Later, we calculated the mean value of all the columns, excluding the start and stop codons, and defined:

$$mean(Diff(IDX)) = \frac{1}{N-2} \sum_{i=2}^{N-1} Diff(IDX_i) \qquad (2)$$

When:

N - the number of columns in the MSA.

$Diff(IDX_i)$–the calculated difference between two variants within each of the columns, as defined in Eq. (1).

We hypothesized that there would be a significant increase in the CUB scores for genes in their later version compared to the earlier one. Therefore, we applied a randomization model per gene for each of the categories described later in Sect. 2.4. We performed 1,000 permutations where we randomly flipped the sign of $Diff(IDX_i)$ with 50% chance. As each column holds two components, magnitude and directionality, the permutation enables to examine the change due to a sign flip.

Later, we compared between the 1,000 permutations to the original case (no permutations), and calculated an empirical p-value as follows [59]:

$$empirical\ p - value = \frac{r+1}{n+1} \qquad (3)$$

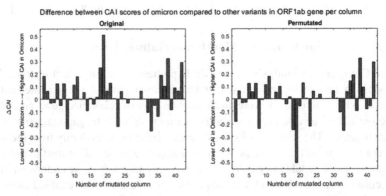

Fig. 2. A bar plot representing the original case versus a permutated case in the randomization model. The original case (left) shows the differences between omicron and other variants' CAI scores, while the permutated case (right) shows the differences with the same magnitude and a 50% chance of sign flip.

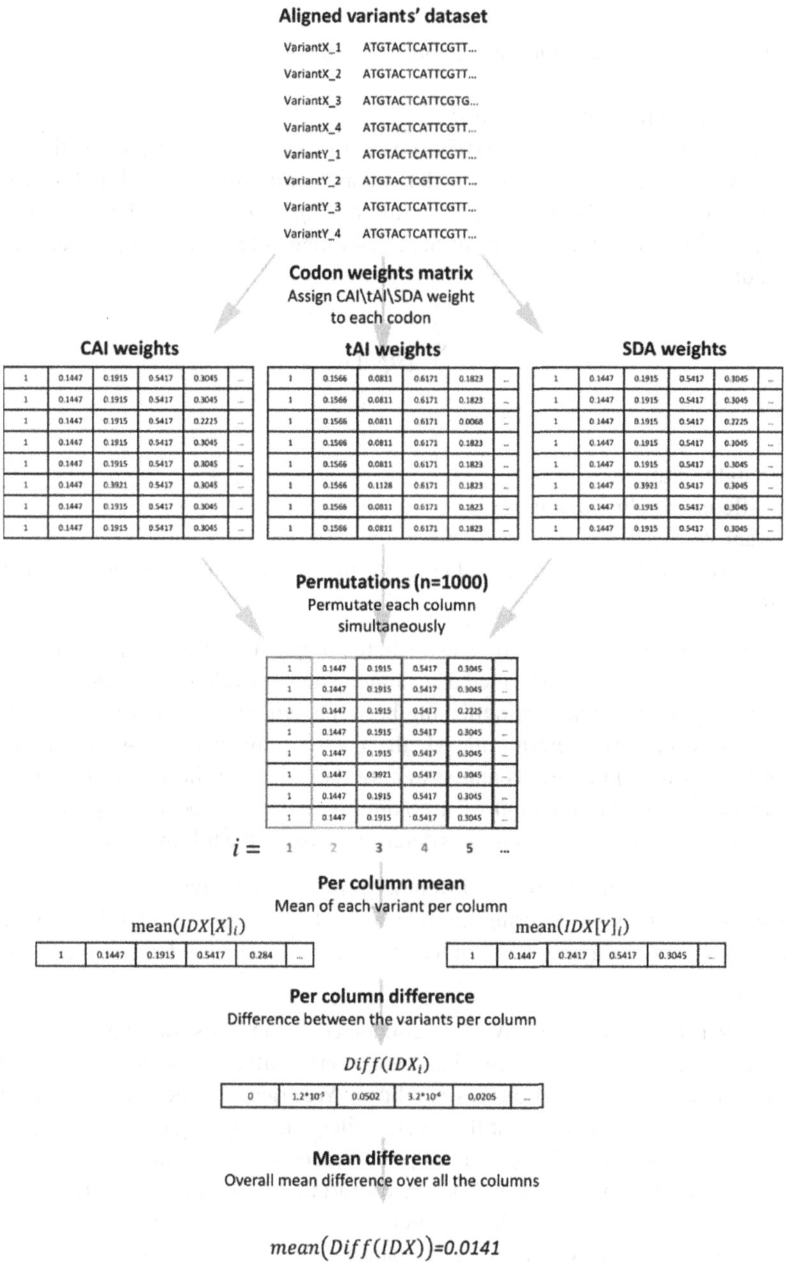

Aligned variants' dataset

VariantX_1	ATGTACTCATTCGTT...
VariantX_2	ATGTACTCATTCGTT...
VariantX_3	ATGTACTCATTCGTG...
VariantX_4	ATGTACTCATTCGTT...
VariantY_1	ATGTACTCATTCGTT...
VariantY_2	ATGTACTCGTTCGTT...
VariantY_3	ATGTACTCATTCGTT...
VariantY_4	ATGTACTCATTCGTT...

Codon weights matrix
Assign CAI\tAI\SDA weight
to each codon

CAI weights

1	0.1447	0.1915	0.5417	0.3045	...
1	0.1447	0.1915	0.5417	0.3045	...
1	0.1447	0.1915	0.5417	0.2225	...
1	0.1447	0.1915	0.5417	0.3045	...
1	0.1447	0.1915	0.5417	0.3045	...
1	0.1447	0.3921	0.5417	0.3045	...
1	0.1447	0.1915	0.5417	0.3045	...
1	0.1447	0.1915	0.5417	0.3045	...

tAI weights

1	0.1566	0.0811	0.6171	0.1823	...
1	0.1566	0.0811	0.6171	0.1823	...
1	0.1566	0.0811	0.6171	0.0068	...
1	0.1566	0.0811	0.6171	0.1823	...
1	0.1566	0.0811	0.6171	0.1823	...
1	0.1566	0.1128	0.6171	0.1823	...
1	0.1566	0.0811	0.6171	0.1823	...
1	0.1566	0.0811	0.6171	0.1823	...

SDA weights

1	0.1447	0.1915	0.5417	0.3045	...
1	0.1447	0.1915	0.5417	0.3045	...
1	0.1447	0.1915	0.5417	0.2225	...
1	0.1447	0.1915	0.5417	0.3045	...
1	0.1447	0.1915	0.5417	0.3045	...
1	0.1447	0.3921	0.5417	0.3045	...
1	0.1447	0.1915	0.5417	0.3045	...
1	0.1447	0.1915	0.5417	0.3045	...

Permutations (n=1000)
Permutate each column
simultaneously

1	0.1447	0.1915	0.5417	0.3045	...
1	0.1447	0.1915	0.5417	0.3045	...
1	0.1447	0.1915	0.5417	0.2225	...
1	0.1447	0.1915	0.5417	0.3045	...
1	0.1447	0.1915	0.5417	0.3045	...
1	0.1447	0.3921	0.5417	0.3045	...
1	0.1447	0.1915	0.5417	0.3045	...

$$i = \quad 1 \quad 2 \quad 3 \quad 4 \quad 5 \quad ...$$

Per column mean
Mean of each variant per column

$mean(IDX[X]_i)$

| 1 | 0.1447 | 0.1915 | 0.5417 | 0.284 | ... |

$mean(IDX[Y]_i)$

| 1 | 0.1447 | 0.2417 | 0.5417 | 0.3045 | ... |

Per column difference
Difference between the variants per column

$Diff(IDX_i)$

| 0 | $1.2*10^5$ | 0.0502 | $3.2*10^4$ | 0.0205 | ... |

Mean difference
Overall mean difference over all the columns

$$mean\big(Diff(IDX)\big)=0.0141$$

Fig. 3. The pipeline that was used to calculate the mean difference between each gene variant. First, a codon weights matrix is created using CUB weights for three indices, CAI, tAI and SDA, where each row represents each aligned sequence from the MSA. Next, columns within each matrix are permutated together, and the mean weight value for each variant within the i-th column is calculated separately to create two arrays – one for each variant. These arrays are subtracted from each other to produce a single array, and the mean value over this array is calculated to determine the final mean difference between the gene variants.

where:

r–the number of permutations that are equal to or more extreme than the observed value (original)

n–the total number of permutations

We used a significance level of 0.05 in our analysis. This empirical test, estimating the mean change within the columns, was defined as **mean-change p-value.** Furthermore, in order to examine all the differences within the *Diff* (*IDX $_i$*) altogether, we performed per gene for each of the categories in Sect. 2.4 a one-sided Wilcoxon signed-rank test on each difference array as follows:

$$W = \sum_{i=1}^{N} [sign(x_i) \cdot R_i] \tag{4}$$

where:

N–length of array

x_i–difference value in column *i*

R_i–rank of *i*

Then, we calculated the p-value based on this test, which is defined as **directional p-value.**

Mutation-Specific Analysis. To investigate the impact of different types of mutations between two variants, we performed a separate analysis, evaluating columns within the MSA with a mutation, and distinguishing between synonymous and non-synonymous mutations. It was done by determining whether each column in the MSA has synonymous (or non-synonymous) mutations and assigning the values of those specific columns to each category. The classification of a column as having a specific type of mutation (synonymous or non-synonymous) was determined using the following diagram (Fig. 4):

Then, we created three separate difference arrays for synonymous, non-synonymous and overall mutations, by filtering the relevant columns from *Diff* (*IDX $_i$*). The mean-change p-value and directional p-value were calculated based on the equations in Sect. 2.4.

Viruses Over Time Analysis. We have conducted the analysis for other human viruses mentioned in Sect. 2.1 by separating their evolution to initial versus progressive stages in their evolution. The classification was as follows: We organized the samples of each gene in each virus according to the year they were collected, arranging them in chronological order. Next, we established a separation point based on the middle year among all the collection years. Samples collected before or during the middle year were categorized as initial stages, while those collected after were classified as progressive stages. This classification enables us to conduct a comparative study between earlier and later viral populations, offering insights into the evolutionary process even in the absence of explicit information on the variant's emergence.

Afterwards, we performed the same analysis that was described earlier in Sect. 2.4, when we defined the initial and progressive stages as the two groups aligned and randomized.

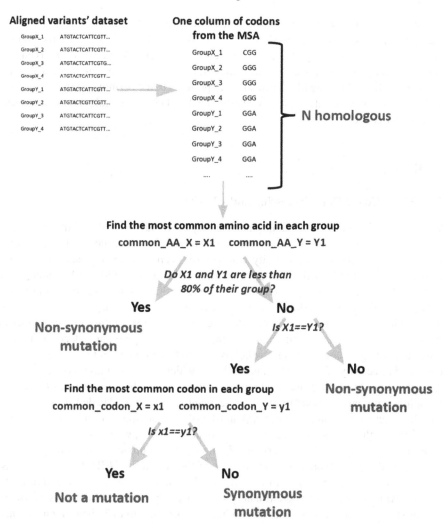

Fig. 4. A decision tree summarizing the classification process of columns into the types of mutations. For each column within the MSA, we first separate the column into two arrays based on its variant's classification. Next, we translate the codons into their corresponding amino acids and calculate the ratio of the most common amino acid within each of the arrays. If the most common amino acid constitutes less than 80% of the total, it indicates a non-synonymous mutation. Otherwise, if the most common amino acids are the same, we compare the two common amino acids directly. If they are unequal, it signifies a non-synonymous mutation. In case they are equal, we revert to the codon representation and compare the most common codons – if the codons are identical, it implies no mutation whereas non-equality denotes a synonymous mutation.

FDR Randomization Model. To estimate the FDR of viral proteins showing significant trends in CUB over time, we employed a randomized model. This involved generating difference vectors, as defined in Eq. (1), for each of the viral proteins. These difference vectors were then subjected to permutation, as illustrated in Fig. 2, to compare the initial and progressive stages as outlined in Sect. 2.4. Afterwards, we determined the number of viral proteins showing significant CUB trends in the original (non-permuted) dataset and compared it to the number of significant viral proteins in the permutated dataset. By calculating the percentage of significant viral proteins in the permutated dataset relative to the non-permuted dataset, we could estimate the FDR of the viral proteins exhibiting CUB patterns over time.

3   Results

3.1   SARS-CoV-2 Data Processing and CUB Trend

We conducted an analysis to quantify and classify the mutations in different genes of the omicron variant compared to four other variants. First, by using the pipeline presented in Sect. 2.3, we generated similarity matrices for each gene to compare the variants. In the S gene, which consists of approximately 1,273 amino acids (AAs), there were between 25 missense mutations (compared to delta) to 30 missense mutations (compared to beta). We observed a predominant presence of missense mutations, which were found to be up to ten times more frequent than other types. The variability in the number of missense mutations among different sequences suggests a potential selection pressure towards altering protein structure, which is known to be related to the process of viral entry and infectivity. In the early gene ORF1ab, which consists of approximately 7,096 AAs, the rates of missense and silent mutations were both high with between 14 missense mutations (compared to beta) to 21 missense mutations (compared to delta), and between 11 silent mutations (compared to beta) and 16 silent mutations (compared to gamma). Interestingly, the S gene showed almost no silent mutations, whereas the ORF1ab gene exhibited relatively similar proportions of both silent and missense mutations.

The analysis was conducted for the rest of the structural SARS-CoV-2 genes (N, M.E) as well to examine the number and types of mutations in each of the genes. The N gene, which consists of approximately 419 AAs, exhibited between 2 to 4 missense mutations when comparing omicron to other variants, and up to 3 silent mutations and insertion mutations. In the M gene, with 222 AAs, the mean missense mutation number was up to 3 mutations per gene, while silent mutations rate was relatively low and with no insertion or deletion mutations. In E gene, with 75 AAs, the mean number of missense mutations was 2 mutations per gene sequence, with no other mutations detected. Those relatively low numbers of mutations are mainly related to the short length of each of the genes.

Later, we calculated three different CUB scores – CAI, tAI and SDA, for every sequence within the variants' datasets built in Sect. 2.3. Then, we compared the five VOC by analyzing the distribution of these scores among different genes in the different variants, estimating their mean and standard deviation (SD). Omicron has a higher mean score than the rest of the variants in each of the indices for gene S (see Fig. 5a for CAI, similar results were observed for tAI and SDA and will appear in the full version of this paper) and for gene ORF1ab (see Fig. 5b for CAI, similar results were observed for tAI and SDA and will appear in the full version of this paper), which might suggests that those crucial genes within the omicron variant tend to be more adapted to their human host by choosing more optimized codons when compared to the rest of the variants. In the genes N, M and E, differences between the variants were minor for tAI and SDA indices, with almost no difference at all between omicron and other variants. However, using CAI scores, omicron variant was not detected to be the one with the highest scores – instead delta variant has shown the highest scores in N gene and E gene. These results could suggest that synonymous and non-synonymous mutations in ORF1ab may contribute to a more efficient translation process, while potentially causing structural changes in the resulting NSPs, that could affect the replication efficiency of the late genes S, N, M and E.

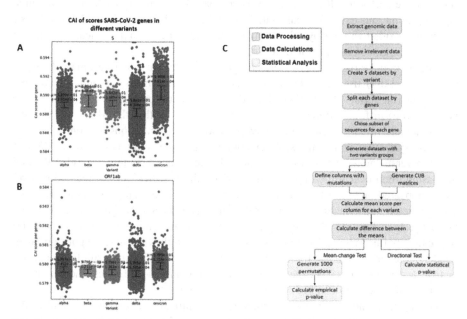

Fig. 5. a) and b) CAI scores of S gene and ORF1ab gene in different SARS-CoV-2 variants. c) The flowchart of the analysis conducted. Blue denotes preprocessing stages performed on the raw data. Orange describes the calculation performed on the processed data in order to evaluate CUB adaptability. Yellow is the statistical analysis performed on the data.

We developed a pipeline to assess the adaptability of the omicron variant compared to previously identified variants as depicted in Fig. 5c. The process involved extracting and filtering genomic data based on predefined characteristics outlined in the Methods section. Then, we created five datasets based on variant annotation and divided them by genes. From these datasets, we randomly selected 5,000 sequences, considering their frequency in the overall SARS-CoV-2 population. After performing MSA, the sequences were transformed into a matrix where each element represented a codon. In addition, the columns in the MSA were classified according to the type of mutation they contained, and each codon within these columns was converted into its representation using codon usage bias (CUB) weights. The resulting matrix was then divided into two groups based on the omicron versus non-omicron annotation. For each group, the mean codon weight score was calculated independently for each column. Afterwards, the differences in each column were calculated, and these differences underwent statistical analysis.

3.2 The Cumulative CUB Weights of All Codons Within the Omicron Genes Suggest an Overall Selection Towards More Optimal Human Codons in SARS-CoV-2 Over Time

We performed a MSA for all the genes (ORF1ab, S, M, N, E) based on 10,000 genes (5,000 from each variant compared) when comparing omicron to other variants. In addition, we performed the same analysis comparing omicron to each of the variants: alpha, gamma and delta. We calculated the difference between the CUB weights of omicron and other variants by calculating the difference between their means (Sect. 2.4) in mutated columns, as illustrated in Fig. 3.

The omicron variant's genes were compared to other variants to evaluate the trend that occurred over time in SARS-CoV-2 and the transition from non-omicron variants to the omicron variant. We evaluated the differences between each group (omicron vs. non-omicron) in four different categories: All columns within the MSA, all columns exhibiting any mutation, only columns with synonymous mutations and only columns with non-synonymous mutations. When comparing all the columns within the MSA of omicron versus non-omicron variants, and only the mutated columns, we identified an increasing trend over time. Figure 6 summarizes all the differences in the codon-level which appeared on each of the columns classified as having mutation in gene ORF1ab. We conducted statistical tests to evaluate directional and mean-change p-values (detailed in Sect. 2.4) and got a significant signal in most tests for the indices (p-value < 0.05) for gene ORF1ab. When analyzing those p-values on all the columns, we got significant results in the directional test for all indices, but only one significant result for the mean-change test in tAI index. In the analysis of the remaining categories, there were no significant differences observed for synonymous mutations.

However, for nonsynonymous mutations, a significant increase was found in the omicron variant, primarily in the CAI score (see Fig. 7). These findings suggest that the notable differences in the ORF1ab gene are mainly associated with nonsynonymous codon alterations to enhance codon efficiency during the translation process. However, assessing the tAI and SDA scores suggests that the adaptability lies in the modulation of tRNA utilization towards better adaptability to the host's tRNA pool, rather than being linked to a specific type of mutation. Thus, these mutations could potentially have implications not only to the protein structure due to the change in the resulting amino acid chain, but also to the process of translation efficiency and adaptation.

In the analysis of the S gene, no significant differences were observed when considering all mutations within the S gene codon matrix. However, when examining all the columns in the matrix, the directional tests yielded mainly significant results, but with higher non-significant mean-change p-values (Table 1). Furthermore, when separating the mutations into synonymous and non-synonymous categories, no significant differences were found in non-synonymous mutations, while the number of synonymous ones was too small to perform any statistical analysis. This observation could be explained

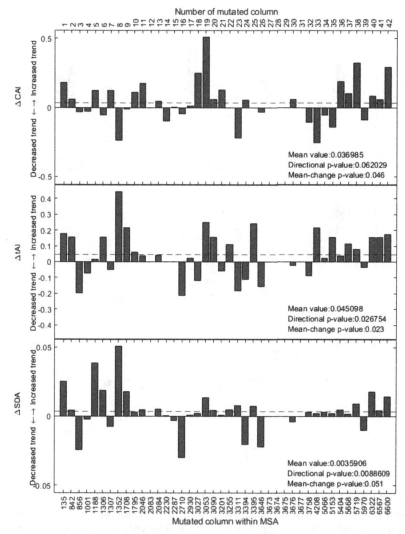

Fig. 6. Changes in CUB scores within each column in the MSA of ORF1ab which has a mutation. The upper subplot shows the CAI difference, the middle subplot shows the tAI difference, and the bottom subplot shows the SDA difference. Mean-change and Directional p-values were calculated based on the Eqs. (3) and (4) respectively. The mean of the bar values is scattered with the dashed horizontal line in each subplot.

by multiple columns that, despite not meeting our mutation criteria, had changes in their CUB over time, consequently influencing the analysis across all columns.

As for the genes N, M and E, due to their short length, the number of detected mutations were not sufficient in order to perform any statistical test. When examining all the codons within the genes, the p-values were divergent, as we were able to get a significant directional p-value, but non-significant mean-change p-value in genes N and

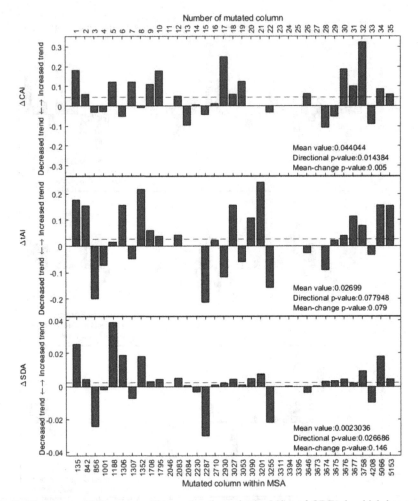

Fig. 7. Changes in CUB scores within each column in the MSA of ORF1ab which has a non-synonymous mutation. The upper subplot shows the CAI difference, the middle subplot shows the tAI difference, and the bottom subplot shows the SDA difference. Mean-change and Directional p-values were calculated based on the Eqs. (3) and (4) respectively. The mean of the bar values is scattered with the dashed horizontal line in each subplot.

M. The E gene did not show any significant p-value (Table 1). This might suggest that due to the length of those genes (N – 419 AA, M – 222 AA, E – 75 AA, see Fig. 1), the genes did not have enough codons to go through our statistical tests.

In summary, the analyses could imply that there is an overall selection towards a combination of optimal codons with higher CUB weights in the omicron variant, especially in the early gene ORF1ab.

Table 1. Directional p-values and mean-change p-values for different indices in genes N, M and E.

	ORF1ab		S		N		M		E	
	Directional	Mean-change	Directional	Mean-change	Directional	Mean-change	Directional	Mean-change	Directional	Mean-change
CAI	3.99E-11	0.054	0.003715	0.181	0.001999	0.928	0.034478	0.569	0.189475	0.553
tAI	0.000414	0.015	0.01844	0.187	0.005522	0.519	0.040231	0.625	0.325416	0.516
SDA	0.032989	0.138	0.005503	0.078	0.015799	0.101	0.151491	0.813	0.369355	0.17

3.3   Most Viral Genes Show a Significant Decreasing Trend in Their Codon Usage Adaptability Over Their Evolution

We conducted an analysis on 818 viral genes with documented samples over time. These genes were annotated based on their function in the viral cycle, specifically as early genes (Xr) and late genes (Xs), having a total of 226 early genes and 294 late genes. Our analysis involved calculating the CUB scores, including CAI, tAI, and SDA, for each sequence within each gene throughout the viral evolution. Then, we divided the viral evolution of each gene into two periods: an initial period consisting of sequences from the first half of the observed years, and a progressive period consisting of sequences from the second half. First, we wanted to evaluate what is the stronger trend of the viral proteins over time. Therefore, we calculated the mean CUB scores in each of the viral proteins for the initial and the progressive evolution periods in order to determine whether the trend tends to increase or decrease over time. For each viral protein, we calculated the CAI, tAI and SDA of each of its sequences based on Sect. 2.2. Then, we calculated the mean CUB score for the initial stage and the progressive stage for each viral protein and calculated the ratio between the later stage to the former stage. The distribution of these ratios was able to detect a significant change in CAI scores ratio from normal distribution (Fig. 8), using two-tailed Wilcoxon distribution. This result has shown that the mean ratio is smaller than 1, implying a general decreasing CAI score among viral proteins. The largest decrease could be exhibited in Rhinovirus A, while the biggest increase was shown in TTN like mini virus. However, the overall change was not significant in the other indices, suggesting that it is mainly related to aspects different than mRNA translation that are encoded by codon usage [60].

We aimed to study the changes in CUB score for specific genes. To achieve this, we applied MSA to the sequences of each gene and divided the viral evolution of each gene into initial and progressive period as mentioned before. Next, we calculated the mean CUB score for each codon separately within the MSA in both the initial and progressive periods and evaluated the difference between these scores.

To assess the significance of these differences, we employed the right-tail one-sided Wilcoxon sign-rank test. Additionally, we conducted a mean-change test as described in Sect. 2.4. To examine the results, we analyzed the distribution of p-values obtained from the Wilcoxon test. Viral proteins with less than 10 samples over time were discarded. Then, we analyzed viral proteins which have shown significant trend over time in both the directional and mean-change p-values. Those genes will be presented in the full version of this paper. To assess the False discovery rate (FDR) of these results, we utilized a

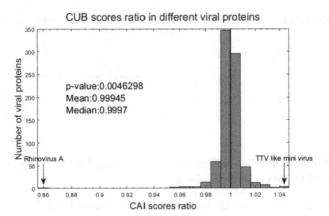

CUB scores ratio in different viral proteins

Fig. 8. CAI scores ratios between initial and progressive stages of viral evolution. The p-value is calculated with a two-sided Wilcoxon signed-rank test.

randomization model in which we performed the analysis as conducted before but with a randomized set, as described in Sect. 2.4.

A set of 56 viral proteins have shown a decreased CAI over time, compared to 19 viral proteins with increased CAI scores over time. After performing FDR randomization model, only 19 viral proteins presented a significant decreasing trend and 15 viral proteins presented a significant decreasing trend, equaling 33.92% and 78.95% FDR. As for the rest of the indices, the FDR for tAI overtime trends was 45.45% for decreased trend and 77.27% for increased trend, and the FDR for SDA overtime trends was 43.75% for decreased trend and 85.71% for increased trend. FDR correction was applied for the rest of the categories as well (more details will appear in the full version of this paper). Overall, there is a weak signal which displays a tendency towards a decreasing trend in their evolution, indicating a decline in the viral adaptability over time. However, this phenomenon might be relevant to only a limited number of viral genes as there are multiple cases with the opposite trend. This could imply different strategies for improving their viral fitness to their host's genome.

Nevertheless, this pattern is not limited to specific viral functions, as both structural and replication-related genes show similar trends in all tables. However, as dsDNA viral proteins being the largest group of viruses, comprising 44.048% of the late genes, 43.805% of the early genes and 47.115% of the non-categorized genes (see Fig. 9), their significance in either trend is proportional to their abundance in most of the categories. This could be exhibited in ssRNA(+) and ssRNA(-) viral proteins as well, comprising 49.048% of the late genes, 53.982% of the early genes and 68.649% of the non-categorized genes. With more significant and consistent trends, it aligns with the current knowledge regarding the high mutation rates of RNA viruses.

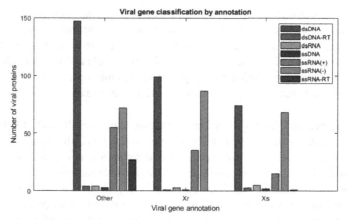

Fig. 9. Baltimore classification of different viral genes annotation: Xr, Xs and Other. Other refers to other viral genes within the dataset which classified with other categories that are not related to their role in their viral cycle or did not appear in VOGdb.

4 Discussion

In summary, we uncover that SARS-CoV-2 has substantial changes within its CUB, particularly evident in specific codons leading to non-synonymous mutations in the omicron variant when compared to its previous variants. This trend is notable when applying directional analysis using column permutations, as we analyzed both the magnitude and the directionality of the change over time among the independent columns. These changes appear to be independently exhibited within distinct codons with improved CUB scores and be particularly pronounced in the early gene ORF1ab compared to the other late genes, suggesting an enhancement in the RdRp mechanism, a crucial component of viral replication. Later, we perform a comprehensive analysis on other human viruses, to determine the directionality and adaptation trends that viral proteins have undergone over time. We observed a relatively higher number of viral proteins displaying a significant decreasing trend in CUB, compared to those exhibiting an increasing trend, suggesting the existence of multiple factors that affect viral evolution in human viruses.

Based on our work, we suggest a few mechanisms that could explain the results and are related to the viral adaptability of SARS-CoV-2 to its host. First, although it is well known that missense mutations are common among SARS-CoV-2 genes, our results have shown that this type of mutations, which is the most abundant among silent, deletion and insertion mutations, could influence CUB of the whole viral population. This was shown by significantly higher CUB scores compared to the rest of the variants in all the relatively long genes.

In addition, as the virus evolves, its genetic variability allows better adaptation to the dynamic environment within the host. This is highly relevant in the case of SARS-CoV-2, since many steps were taken to reduce viral infection such as antiviral drugs, vaccines, isolation etc., as well as the development of host immune responses. Those have caused the virus to deal with a variety of cumulative factors that led it to rapidly evolve in order to keep on spreading and infecting [61, 62]. Therefore, by using non-synonymous mutations with optimized CUB weights, the virus could adapt in more than one factor simultaneously; the virus could evade the host immune response by altering the viral protein's structure and improve translation efficiency towards synthesis of functional viral proteins that are critical to their rapid replication cycle and survival. Our analysis may suggest that such evolution has happened in the omicron variant, as it was shown that the non-synonymous mutations in the omicron comprise an optimized combination of codons.

Furthermore, those non-synonymous mutations can have distinct effects on early genes and late genes due to their different roles in the virus life cycle. In the case of SARS-CoV-2, early genes are involved in viral replication, transcription, and manipulation of host cell processes, while late genes are responsible for producing structural proteins required for the formation of new virions [63, 64]. Given that ORF1ab is expressed shortly after viral entry, the optimized codons within this region are crucial for accurate and efficient translation of NSPs, which are in turn essential for effective viral replication and successful infection. In our analysis, we observed that the omicron variant exhibited significantly different non-synonymous mutations in its early gene, showed with increased CUB scores. This aligns with existing knowledge regarding the increased infectivity of the omicron variant.

However, it is important to consider the fact that the early genes of viruses could modulate the cellular machinery and influence various cellular processes, including the levels and availability of tRNA molecules [65]. It is worth noting that our available data on humans typically represents the host cell environment without viral infection-induced changes. As a result, this data may not accurately reflect the environment in which the late genes are expressed and optimized. In the context of comparing ORF1ab genes to structural genes, it is plausible that the structural genes have adapted to the tRNA levels within the cell after the expression of the ORF1ab gene. This suggests that the late genes may also have undergone evolutionary changes to adapt to the cellular environment following the initial expression of the early genes which cannot be detected without additional experiments.

Finally, our analysis for various viral proteins was able to detect this phenomenon among a few other viral proteins, with more proteins showing a decreasing trend compared to an increasing trend. The opposite trends observed in different viral populations can be attributed to variations in viral environments and the functions of viral genes. This could be exhibited in the orthologous genes for ORF1ab in Human coronavirus NL63 and Human coronavirus OC43, that have demonstrated a significant decreasing trend over time only in their early gene, suggesting a different optimization mechanism despite their genetic resemblance to SARS-CoV-2. However, it is important to note that fitness optimization involves trade-offs between different features, ensuring the long-term survival of the viral population [66, 67]. Therefore, genetic diversity caused by high mutations rates of viruses can emerge in variants which will cause higher replication success, that does not necessarily result in increased CUB [68]. A plausible explanation could be related to immune recognition, as viruses that are easily identified by the host's immune system may face a strong selection leading to reduced fitness. This strategy might reduce the effectiveness of viral infection, but it can still allow the virus to survive and infect [69, 70]. Another reason for these processes is some external interventions such as vaccines and antiviral medications that can enhance viral evolution as well and can potentially cause a bottleneck. Different characteristics of the bottleneck, such as its size and development time between bottlenecks, can determine the directionality of the viral evolution [71, 72].

Nevertheless, it is essential to note that the Covid-19 pandemic has provided a unique advantage in terms of data time resolution for SARS-CoV-2 compared to other documented viruses. This advantage has enabled us to identify additional evolutionary trends in the virus, in contrast to other viral proteins where only a few of such trends were detectable. However, it may introduce some bias, as newer viral genomes might not have necessarily originated from the earlier ones or been collected at the same rate, compared to the well-studied evolution of SARS-CoV-2. This underscores the critical importance of continuous monitoring and frequent sequencing of viruses, as it leads to the discovery of valuable insights that enhance our understanding of viral evolution.

Our results provide a comprehensive analysis that could shed light on the genetic diversity and codon selection of SARS-CoV-2 which are related to the regulatory mechanisms of the virus. Unlike many previous studies that focused solely on the functionality of the spike protein and its structural changes, our analysis examined all genes and their mutations to provide a more accurate understanding of their codon selection and the possible interactions between early and late genes in the viral infection process. Further investigation into the interplay between early genes and late genes could help by identifying novel therapeutic targets for the development of antiviral vaccines and medications. Such research could eventually aid in the development of more effective strategies for combating viral infections.

Acknowledgments. This research was funded by the DFG.

References

1. Whitaker-Dowling, P., Youngner, J.S.: "VIRUS-HOST CELL INTERACTIONS," in Encyclopedia of Virology, pp. 1957–1961. Elsevier (1999)
2. Lucas, M., Karrer, U., Lucas, A., Klenerman, P.: Viral escape mechanisms - escapology taught by viruses. Int. J. Exp. Pathol. **82**(5), 269–286 (2008)
3. Goz, E., Zafrir, Z., Tuller, T.: Universal evolutionary selection for high dimensional silent patterns of information hidden in the redundancy of viral genetic code. Bioinformatics **34**(19), 3241–3248 (2018)
4. Bahir, I., Fromer, M., Prat, Y., Linial, M.: Viral adaptation to host: a proteome-based analysis of codon usage and amino acid preferences. Mol. Syst. Biol. **5**(1), 311 (2009)
5. Crill, W.D., Wichman, H.A., Bull, J.J.: Evolutionary reversals during viral adaptation to alternating hosts. Genetics **154**(1), 27–37 (2000)
6. Sanjuán, R., Domingo-Calap, P.: Mechanisms of viral mutation. Cell. Mol. Life Sci. **73**(23), 4433–4448 (2016)
7. Greenbaum, B.D., Levine, A.J., Bhanot, G., Rabadan, R.: Patterns of evolution and host gene mimicry in influenza and other RNA viruses. PLoS Pathog. **4**(6), e1000079 (2008)
8. Elena, S.F., Sanjuán, R.: Adaptive value of high mutation rates of RNA viruses: separating causes from consequences. J. Virol. **79**(18), 11555–11558 (2005)
9. Stern, A., Andino, R.: Viral Evolution. Viral Pathog. 233–240 (2016)

10. Holmes, E.C., Drummond, A.J.: The evolutionary genetics of viral emergence, pp. 51–66 (2007)

11. Duffy, S., Shackelton, L.A., Holmes, E.C.: Rates of evolutionary change in viruses: patterns and determinants. Nat. Rev. Genet. **9**(4), 267–276 (2008)

12. Wong, E.H., Smith, D.K., Rabadan, R., Peiris, M., Poon, L.L.: Codon usage bias and the evolution of influenza A viruses. Codon usage biases of influenza virus. BMC Evol. Biol. **10**, 253 (2010)

13. Khandia, R., et al.: Analysis of nipah virus codon usage and adaptation to hosts. Front. Microbiol. **10**, 439603 (2019)

14. Cristina, J., Moreno, P., Moratorio, G., Musto, H.: Genome-wide analysis of codon usage bias in ebolavirus. Virus Res. **196**, 87–93 (2015)

15. Biswas, K., et al.: Codon usage bias analysis of citrus tristeza virus: higher codon adaptation to citrus reticulata host. Viruses **11**(4), 331 (2019)

16. Li, G., et al.: Evolutionary and genetic analysis of the VP2 gene of canine parvovirus. BMC Genomics **18**(1), 534 (2017)

17. Cristina, J., Fajardo, A., Soñora, M., Moratorio, G., Musto, H.: A detailed comparative analysis of codon usage bias in Zika virus. Virus Res. **223**, 147–152 (2016)

18. Moratorio, G., Iriarte, A., Moreno, P., Musto, H., Cristina, J.: A detailed comparative analysis on the overall codon usage patterns in West Nile virus. Infect. Genet. Evol. **14**, 396–400 (2013)

19. Jenkins, G.M., Holmes, E.C.: The extent of codon usage bias in human RNA viruses and its evolutionary origin. Virus Res. **92**(1), 1–7 (2003)

20. Belalov, I.S., Lukashev, A.N.: Causes and implications of codon usage bias in RNA viruses. PLoS ONE **8**(2), e56642 (2013)

21. Parvez, M.K., Parveen, S.: Evolution and emergence of pathogenic viruses: past, present, and future. Intervirology **60**(1–2), 1–7 (2017)

22. Pybus, O.G., Tatem, A.J., Lemey, P.: Virus evolution and transmission in an ever more connected world. Proc. R. Soc. B Biol. Sci. **282**(1821), 20142878 (2015)

23. LaTourrette, K., Garcia-Ruiz, H.: Determinants of virus variation, evolution, and host adaptation. Pathogens **11**(9), 1039 (2022)

24. Ojosnegros, S., Beerenwinkel, N.: Models of RNA virus evolution and their roles in vaccine design. Immunome Res. **6**(Suppl 2), S5 (2010)

25. Hie, B., Zhong, E.D., Berger, B., Bryson, B.: Learning the language of viral evolution and escape. Science **371**(6526), 284–288 (2021)

26. Marz, M., et al.: Challenges in RNA virus bioinformatics. Bioinformatics **30**(13), 1793–1799 (2014)

27. Elena, S.F.: "Restrictions to RNA virus adaptation: an experimental approach", *Antonie van Leeuwenhoek*. Int. J. Gen. Mol. Microbiol. **81**(1–4), 135–142 (2002)

28. Hanna, R., Dalvi, S., Sălăgean, T., Pop, I.D., Bordea, I.R., Benedicenti, S.: Understanding COVID-19 pandemic: molecular mechanisms and potential therapeutic strategies. An evidence-based review. J. Inflamm. Res. **14**, 13–56 (2021)

29. De Maio, N., Walker, C.R., Turakhia, Y., Lanfear, R., Corbett-Detig, R., Goldman, N.: Mutation rates and selection on synonymous mutations in SARS-CoV-2. Genome Biol. Evol. **13**(5), evab087 (2021)

30. Magazine, N., Zhang, T., Wu, Y., McGee, M.C., Veggiani, G., Huang, W.: Mutations and evolution of the SARS-CoV-2 spike protein. Viruses **14**(3), 640 (2022)

31. Nambou, K., Anakpa, M.: Deciphering the co-adaptation of codon usage between respiratory coronaviruses and their human host uncovers candidate therapeutics for COVID-19. Infect. Genet. Evol. **85**, 104471 (2020)

32. Tao, K., et al.: The biological and clinical significance of emerging SARS-CoV-2 variants. Nat. Rev. Genet. **22**(12), 757–773 (2021)

33. Planas, D., et al.: Reduced sensitivity of SARS-CoV-2 variant delta to antibody neutralization. Nature **596**(7871), 276–280 (2021)
34. Hu, J., et al.: Increased immune escape of the new SARS-CoV-2 variant of concern omicron. Cell. Mol. Immunol. **19**(2), 293–295 (2022)
35. Chavda, V., Bezbaruah, R., Deka, K., Nongrang, L., Kalita, T.: The delta and omicron variants of SARS-CoV-2: what we know so far. Vaccines **10**(11), 1926 (2022)
36. Kumar, S., Thambiraja, T.S., Karuppanan, K., Subramaniam, G.: Omicron and delta variant of SARS-CoV-2: a comparative computational study of spike protein. J. Med. Virol. **94**(4), 1641–1649 (2022)
37. Davidson, A.M., Wysocki, J., Batlle, D.: Interaction of SARS-CoV-2 and other coronavirus with ACE (Angiotensin-Converting Enzyme)-2 as their main receptor. Hypertension **76**(5), 1339–1349 (2020)
38. Hamming, I., Timens, W., Bulthuis, M., Lely, A., Navis, G., van Goor, H.: Tissue distribution of ACE2 protein, the functional receptor for SARS coronavirus. A first step in understanding SARS pathogenesis. J. Pathol. **203**(2), 631–637 (2004)
39. Zhang, H., Penninger, J.M., Li, Y., Zhong, N., Slutsky, A.S.: Angiotensin-converting enzyme 2 (ACE2) as a SARS-CoV-2 receptor: molecular mechanisms and potential therapeutic target. Intensive Care Med. **46**(4), 586–590 (2020)
40. Lan, J., et al.: Structure of the SARS-CoV-2 spike receptor-binding domain bound to the ACE2 receptor. Nature **581**(7807), 215–220 (2020)
41. Peronace, C., et al.: The first identification in Italy of SARS-CoV-2 omicron BA. 4 harboring KSF141_del: a genomic comparison with omicron sub-variants. Biomedicines **10**(8), 1839 (2022)
42. Chakraborty, C., Bhattacharya, M., Sharma, A.R., Dhama, K., Agoramoorthy, G.: A comprehensive analysis of the mutational landscape of the newly emerging omicron (B.1.1.529) variant and comparison of mutations with VOCs and VOIs. GeroScience **44**(5), 2393–2425 (2022)
43. Plotkin, J.B., Dushoff, J.: Codon bias and frequency-dependent selection on the hemagglutinin epitopes of influenza A virus. Proc. Natl. Acad. Sci. **100**(12), 7152–7157 (2003)
44. Koyama, T., Platt, D., Parida, L.: Variant analysis of SARS-CoV-2 genomes. Bull. World Health Organ. **98**(7), 495–504 (2020)
45. Chatterjee, S., Bhattacharya, M., Nag, S., Dhama, K., Chakraborty, C.: A detailed overview of SARS-CoV-2 omicron: its sub-variants, mutations and pathophysiology, clinical characteristics, immunological landscape, immune escape, and therapies. Viruses **15**(1), 167 (2023)
46. Hatcher, E.L., et al.: Virus variation resource – improved response to emergent viral outbreaks. Nucleic Acids Res. **45**(D1), D482–D490 (2017)
47. Hodcroft, E.B.: CoVariants: SARS-CoV-2 Mutations and Variants of Interest (2021)
48. O'Leary, N.A., et al.: Reference sequence (RefSeq) database at NCBI: current status, taxonomic expansion, and functional annotation. Nucleic Acids Res. **44**(D1), D733–D745 (2016). https://doi.org/10.1007/978-3-319-21602-7_8
49. Alexaki, A., et al.: Codon and codon-pair usage tables (CoCoPUTs): facilitating genetic variation analyses and recombinant gene design. J. Mol. Biol. **431**(13), 2434–2441 (2019)
50. Chan, P.P., Lowe, T.M.: GtRNAdb: a database of transfer RNA genes detected in genomic sequence. Nucleic Acids Res. **37**, D93–D97 (2009)
51. Chan, P.P., Lowe, T.M.: GtRNAdb 2.0: an expanded database of transfer RNA genes identified in complete and draft genomes. Nucleic Acids Res. **44**(D1), D184–D189 (2016)
52. Sabi, R., Tuller, T.: Modelling the efficiency of codon–tRNA interactions based on codon usage bias. DNA Res. **21**(5), 511–526 (2014)
53. Sharp, P.M., Li, W.-H.: The codon adaptation index-a measure of directional synonymous codon usage bias, and its potential applications. Nucleic Acids Res. **15**(3), 1281–1295 (1987)

54. Reis, M.D.: Solving the riddle of codon usage preferences: a test for translational selection. Nucleic Acids Res. **32**(17), 5036–5044 (2004)
55. Hernandez-Alias, X., Benisty, H., Schaefer, M.H., Serrano, L.: Translational efficiency across healthy and tumor tissues is proliferation-related. Mol. Syst. Biol. **16**(3), e9275 (2020)
56. Pechmann, S., Frydman, J.: Evolutionary conservation of codon optimality reveals hidden signatures of cotranslational folding. Nat. Struct. Mol. Biol. **20**(2), 237–243 (2013)
57. Needleman, S.B., Wunsch, C.D.: A general method applicable to the search for similarities in the amino acid sequence of two proteins. J. Mol. Biol. **48**(3), 443–453 (1970)
58. Sievers, F., et al.: Fast, scalable generation of high-quality protein multiple sequence alignments using Clustal Omega. Mol. Syst. Biol. **7**(1), 539 (2011)
59. Davison, A.C., Hinkley, D.V.: Bootstrap Methods and their Application. Cambridge University Press, Cambridge (1997)
60. Bergman, S., Tuller, T.: Widespread non-modular overlapping codes in the coding regions*. Phys. Biol. **17**(3), 031002 (2020)
61. Lauring, A.S., Frydman, J., Andino, R.: The role of mutational robustness in RNA virus evolution. Nat. Rev. Microbiol. **11**(5), 327–336 (2013)
62. Wang, R., Chen, J., Wei, G.-W.: Mechanisms of SARS-CoV-2 evolution revealing vaccine-resistant mutations in Europe and America. J. Phys. Chem. Lett. **12**(49), 11850–11857 (2021)
63. Emam, M., Oweda, M., Antunes, A., El-Hadidi, M.: Positive selection as a key player for SARS-CoV-2 pathogenicity: insights into ORF1ab, S and E genes. Virus Res. **302**, 198472 (2021)
64. V'kovski, P., Kratzel, A., Steiner, S., Stalder, H., Thiel, V.: Coronavirus biology and replication: implications for SARS-CoV-2. Nat. Rev. Microbiol. **19**(3), 155–170 (2021)
65. Mioduser, O., Goz, E., Tuller, T.: Significant differences in terms of codon usage bias between bacteriophage early and late genes: a comparative genomics analysis. BMC Genomics **18**(1), 866 (2017)
66. Manrubia, S., Lazaro, E.: Viral evolution. Phys. Life Rev. **3**(2), 65–92 (2006)
67. Domingo, E., Holland, J.J.: RNA virus mutations and fitness for survival. Annu. Rev. Microbiol. **51**(1), 151–178 (1997)
68. Bull, R.A., et al.: Sequential bottlenecks drive viral evolution in early acute hepatitis C virus infection. PLoS Pathog. **7**(9), e1002243 (2011)
69. Domingo-Calap, P.: Viral evolution and Immune responses. J. Clin. Microbiol. Biochem. Technol. **5**(2), 013–018 (2019)
70. Mordstein, C., et al.: Transcription, mRNA export, and immune evasion shape the codon usage of viruses. Genome Biol. Evol. **13**(9), 1–14 (2021)
71. Nijhuis, M., Deeks, S., Boucher, C.: Implications of antiretroviral resistance on viral fitness. Curr. Opin. Infect. Dis. **14**(1), 23–28 (2001)
72. Domingo, E., Menéndez-Arias, L., Holland, J.J.: RNA virus fitness. Rev. Med. Virol. **7**(2), 87–96 (1997)
73. Gao, Y., et al.: Structure of the RNA-dependent RNA polymerase from COVID-19 virus. Science **368**(6492), 779–782 (2020)
74. Daczkowski, C.M., Dzimianski, J.V., Clasman, J.R., Goodwin, O., Mesecar, A.D., Pegan, S.D.: Structural insights into the interaction of coronavirus papain-like proteases and interferon-stimulated gene product 15 from different species. J. Mol. Biol. **429**(11), 1661–1683 (2017)

Author Index

C. Scornavacca and M. Hernández-Rosales (Eds.): RECOMB-CG 2024, LNBI 14616, pp. 271–272, 2024.
https://doi.org/10.1007/978-3-031-58072-7

Printed in the United States
by Baker & Taylor Publisher Services